# HANDBOOK OF PAPER SCIENCE

# HANDBOOK OF PAPER SCIENCE

The Science and Technology of Papermaking, Paper Properties and Paper Usage

## VOLUME 1

# THE RAW MATERIALS AND PROCESSING OF PAPERMAKING

Edited by

H.F. RANCE

*Rance Consulting, Bridge House, Castle Comb, Wiltshire, England*
*Former Director of Wiggins Teape Ltd., in charge of Research and Development*

ELSEVIER SCIENTIFIC PUBLISHING COMPANY
Amsterdam – Oxford – New York 1980

ELSEVIER SCIENTIFIC PUBLISHING COMPANY
335 Jan van Galenstraat
P.O. Box 211, 1000 AE Amsterdam, The Netherlands

*Distributors for the United States and Canada:*

ELSEVIER/NORTH-HOLLAND INC.
52, Vanderbilt Avenue
New York, N.Y. 10017



ISBN: 0-444-41778-8 (Vol. 1).
ISBN: 0-444-41779-6 (Series)



Printed in The Netherlands

# PREFACE

This work is intended to fill an outstanding gap in the technical literature of paper and papermaking.

The literature already contains many useful works directed towards the technology and craft of the subject, together with a number of valuable monographs dealing in depth with specific aspects. There also lies accumulated within the volumes of scientific, technical and trade journals a mass of valuable material on the fundamentals of paper and papermaking. Especially over the last 25 years, workers in research associations, institutes, universities, and laboratories associated with the paper industry have been prolific in relevant research, much of it fundamental in character. The cream of their work is available in various conference proceedings but no attempt has hitherto been made to present it in ordered fashion, within the framework of a text book or series of text books covering the sequential aspects of the papermaking process and of the products thereof.

The object of this work is to provide such presentation, and the initial intent was that it should be sequentially comprehensive over the chosen field. In the event, that intent has not been fully achieved. Even in this first volume, there are some obvious omissions. The science and technology of pulping of cellulosic trees and plants has been deliberately omitted. Pulping is directly and intimately linked with the papermaking process and the properties of paper, since a major proportion of cellulosic pulp is consumed in papermaking. However, it is a separable subject and one which has been dealt with in the technical literature far more effectively than the science and technology of paper and papermaking as such. Perhaps there is also a need for a new text book on fundamental aspects of pulping, but that has been judged to be outside the scope of this work. Similarly it would have been quite impractical to attempt to cover comprehensively the massive amount of scientific work which has appeared on the non-cellulosic components of plant materials, primarily lignin, which has in any case been adequately covered in earlier text books.

The paper specialist will note some peripheral omissions in this first volume, falling within the field of papermaking as such. There is but scant consideration of non-fibrous solid components of paper, such as loadings, sizing materials and wet strength agents. The hydrodynamics of stock flow through pipes, and the drainage characteristics of two-wire machines are referred to in only cursory fashion. One special aspect of fibre properties of significance as a link between pulp and paper quality, the mechanical characteristics of individual fibres, has been omitted through late withdrawal of a hoped for contribution: this may perhaps be rectified in a later volume.

Despite these omissions, some deliberate, some fortuitous, the editor hopes that this volume will nevertheless be judged by its readers to cover reasonably effectively the bulk of fundamental work on the primary constituents of paper, cellulose fibres and water, and the main components of the papermaking sequence. The properties of paper, and perhaps the usages of paper, will be dealt with in one or more later volumes.

The arrangement of this volume, which comprises texts by several authors, necessarily involves some overlap between successive chapters. To a large degree this overlap is deliberate. The first chapter, by Dr. H. Corte, on Cellulose-Water Interactions, sets the scene for the whole subject, and necessarily contains in summary some consideration of material covered in greater depth by each of the succeeding chapters, and also of material to be covered later. In most cases, however, it considers that material from a viewpoint which differs from that of the other authors, and is thus complementary rather than repetitive.

The reader of this volume will find that one very important subject — the significance of the beating process in papermaking — is discussed at some length by three of the four authors: by Dr. Corte, from the viewpoint of the cellulose-water relationship, by Mr. Emerton, from the viewpoint of fibre properties, and by the writer (as author), from the viewpoint of web consolidation. The writer as editor makes no apologies for this: it is an inevitable consequence of the central and critical importance of beating in the whole field of papermaking and paper properties, and juxtaposition of complementary viewpoints can be highly valuable in this situation. There is little doubt that each of the chapters in Volume 2 (as listed after this Preface) will include further consideration of the significance of the beating process, in the context of each of the sets of properties there dealt with.

Finally, it should be emphasised that this work is not intended merely to be a review of relevant literature. It is certainly that, but it is intended also to be a presentation of critical interpretations and opinions within the chosen field, by authors who have spent the greater part of their professional careers engaged in or in association with original research within the paper industry. The interpretations and opinions expressed are not entirely unanimous even within this first volume, and will certainly not meet with total acceptance by all who read them after publication. This is all to the good. Critical debate on unsolved issues is an essential ingredient for scientific advance. The editor hopes that this publication will provide stimulation, as well as factual knowledge, to readers of all types and viewpoints, within and without the paper industry.

August 1978 H.F. Rance

## CONTRIBUTORS TO THIS VOLUME

Heinz Corte, Dipl. Chem., Dr. rer. nat.
Research Associate, Wiggins Teape Research and Development Ltd., Butler's Court, Beaconsfield, Bucks, England

Hector W. Emerton, M.A.
Bursar, Van Mildert College, University of Durham, Durham, England

Bronislaw Radvan, B.Sc.
Research Associate, Wiggins Teape Research and Development Ltd., Butler's Court, Beaconsfield, Bucks, England

Herbert F. Rance, Ph.D.
Rance Consulting, Bridge House, Castle Combe, Wiltshire, England
Former Director of Wiggins Teape Ltd., in charge of Research and Development

# DRAFT CONTENTS FOR VOLUME 2

ACKNOWLEDGEMENTS

The authors of the five chapters of this volume have cited and in some cases reproduced material from a variety of learned journals, technical association procedings, books and trade journals. In each appropriate case, formal permission for reproduction of material has been obtained from the copyright holders, and the source of the material has been specifically acknowledged, by drawing attention to the designation and origin of relevant articles and conferences in the text and/or the appropriate reference list.

In addition to these formal aknowledgements, however, the editor wishes to express special gratitude to organisations associated with the paper industry, which have been prominent in granting permission to reproduce relevant material from their publications or files, notably:

The Technical Division of the British Paper and Board Industry Federation (B.P.B.I.F.), formerly the Technical Section of the British Paper and Board Makers' Association (London, England).

The Technical Section of the Canadian Pulp and Paper Makers' Association (C.P.P.A. of Montreal, Quebec, Canada).

The Technical Association of the Pulp and Paper Industry (TAPPI of Atlanta, Georgia, U.S.A.)

Verein der Zellstoff- und Papier-Čhemiker und Ingenieure (Zellcheming of Darmstadt, West Germany.)

PIRA, Leatherhead, England, as successors to the British Paper and Board Industry Research Association (BPBIRA).

The Pulp and Paper Research Institute of Canada (Point Claire, Quebec, Canada)

The Finnish Pulp and Paper Research Institute (Helsinki, Finland)

In particular, the authors have in their texts drawn heavily upon material presented at various international conferences mounted by several of the above organisations, notably:

The Fundamental Research Symposia held between 1957 and 1977 at Oxford and Cambridge, England, mounted by the Fundamental Research Committee of the Technical Division of B.P.B.I.F.

The Fundamental Research Conferences held between 1950 and 1975 in the U.S.A, and Canada, mounted jointly by the Paper Physics Committee of TAPPI and the Research Committee of C.P.P.A. Various conferences on specific papermaking themes, mounted by the Technical Division of the B.P.B.I.F., the Technical Section of C.P.P.A., and TAPPI, including the 1964 and 1968 International Symposia on Water Removal at the Presses and Dryers held under the auspices of C.P.P.A. and TAPPI, and the 1975 International Water Removal Symposium held under the auspices of the B.P.B.I.F.

The editor also wishes to express his appreciation to Dr. John Peel for permission to draw on unpublished material presented at a Symposium on Calendering held in 1975 at the Manchester Institute of Science and Technology.

Both editor and authors wish to express their gratitude and appreciation to the directors and to the secretarial and photographic staff of Wiggins Tape Research and Development Ltd., of Butlers Court, Beachsfield, Bucks, England, for the considerable assistance that has been given in preparing the scripts for this publication; and to Mrs. Barbara Jenkins, Mrs. Celia Hamlyn and Mrs. Anne Clark of the W.T. R&D library staff, for their untiring help in literature searches.

Finally, the writer wishes to express his personal appreciation to Miss Judy Taylor for her devoted secretarial assistance over a long period in his activities both as editor and as part author of this work: and his sincere gratitude to his wife for her unceasing encouragement without which the onerous task of editing the work, coupled with authorship of a substantial chapter, could never have been undertaken and brought to fruition.

# CONTENTS

*Chapter 1*

# CELLULOSE-WATER INTERACTIONS

H. Corte

## 1.1 INTRODUCTION

Three hundred kilograms of water per kilogram of paper was the typical figure which the old papermaker had in his mind when comtemplating an increase of his production capacity. It is an impressive figure which, although now largely obsolete, illustrates the most plainly visible practical relationship between cellulose and water in papermaking, namely, the geographical one: Paper mills (and pulp mills) traditionally exist only in the vicinity of an adequate supply of usable water. The first task of water is to act as a dispersing agent for the fibres and to carry them to the paper machine. It is suited for this task because of its abundance and because of the hydrophilic nature of cellulose fibres which have themselves grown in water. At the end of the process, the role of water is more specific. It pulls and, at least partly, holds the fibres together in the web and it can do this because it shares with cellulose the property of possessing hydroxyl groups. In between, the interaction between fibres and water assumes various modes as the suspension turns into a web and the web moves down the machine. Water may thus "be regarded as the amniotic fluid of paper", as Goring recently observed [1]. It continues, during the entire existence of the paper, to exert its influence as it affects or even controls, its behaviour in numerous cases of conversion and usage.

The discussion will begin with the structure of water and the effect of cellulose on it. It will then follow the papermaking sequence and will conclude with cellulose-water interactions that come into play when paper is used and re-used.

## 1.2 CELLULOSE AND THE STRUCTURE OF WATER

Liquid water, the most common liquid on earth, has a number of anomalous properties. These include the high melting and boiling temperatures (0 and 100°C) compared with the hydrides of the two close neighbours of oxygen in the periodic table: ammonia (—77.7 and —33.4°C) and hydrogen sulphide (—85.6 and —60.4°C); the high molar heat capacity (more than twice that of $NH_3$ or $H_2S$ at room temperature); and the biologically important expansion on freezing, with a density maximum at 4° C.

It has long been thought that these anomalies point to a highly-associated structure in which various molecular or structural units co-exist. As early as 1892, Röntgen [2] suspected water to be a mixture of "ice-like and water-like molecules". He put forward the hypothesis that the composition of this mixture varies with the temperature and that this may explain anomalous temperature effects in water. Examples he quoted include the decrease of the compressibility with increasing temperature from 0 to 50° C, the increase of the coefficient of thermal expansion with increasing pressure, and the decrease of the viscosity with increasing pressure *. The density maximum at 4° C finds a simple if qualitative explanation.

Bernal and Fowler [4] attempted to quantify the mixture concept of different structures and postulated three different intermolecular arrangements at different temperatures.

(1) Ice-tridymite-like (four-coordinated) between 0 and 4° C.

(2) Quartz-like (four-coordinated) between 4 and 200° C, approximately.

(3) Close-packed arrangements, as for most liquids, at higher temperatures. The anomalous properties of water were then interpreted as the result of the temperature effect on the composition of the mixture. The theory was largely based on and supported by, X-ray diffraction curves of liquid water. A careful analysis of X-ray diffraction curves was provided by Morgan and Warren [5] who as a result of their improved accuracy denied the existence of quartz-like structures in liquid water. Their results were later checked and confirmed by Brady and Romanon [6] and by Danford and Levy [7].

Eucken [8,9] developed the model of water as a mixture of different aggregates further and assumed, apart from single molecules, aggregates of two, four and eight molecules, each of which occupies a larger volume than the corresponding number of single molecules. By adjusting the mole fractions of the aggregates and their energies of dissociation, he obtained good agreement with experimental values for the temperature dependence of the molar volume from —20 to +70°C at pressures up to 12000 atm, as measured by Bridgman [10] and by Amagat [11], and for the temperature dependence of the molar heat capacity between 0 and +200° C. This model, characterised by a few well-defined (tridymitic) structures of fixed sizes floating in a disordered matrix of single molecules, was designed to "describe all the observations with a minimum number of specific assumptions" [8], and it achieved this remarkably well.

A different approach to the structure of water resulted from investigations of the effects of ions on its properties. These are frequently as anomalous as

* It has since been established that the viscosity-pressure curve exhibits a minimum at about 1000 atm which disappears at higher temperatures [3].

the properties of water themselves. For example, dilute solutions of CsCl have a lower viscosity than water at the same temperature. According to Bernal and Fowler [4], this is caused by the ions somehow breaking the structure of water. Another example is the reduction of the molar heat capacity by 1—1 electrolytes, with the effect increasing from LiCl to KI in the order of increasing ionic radii [12]. Now, when energy in the form of heat is put into water a large proportion of this energy is not used to increase the translational velocity of the molecules, i.e. the temperature, but to "melt" the larger aggregates down to smaller ones, and this was considered to be the reason for the high molar heat capacity of water. Its reduction by electrolytes was therefore again interpreted as structure breaking by ions. Two steps then led to the currently favoured model of the water structure. The first was that the breaking of the structure was no longer considered to proceed from octamers to tetramers, dimers and monomers as suggested by Eucken [8] but to result in a mixture of aggregates of various shapes and sizes. The second was that this form of aggregation, which at first was deduced from the structure-breaking action of ions, was thought to apply to water as such.

The concept of irregular aggregation goes back to Bose [13] who pointed out that in liquids consisting of anisotropic molecules the formation of "swarms" must be expected. Stewart [14] formulated a model of liquid water which consists of relatively large ("polymolecular") aggregates called cybotactic groups forming an irregular array. Between them are single water molecules. There are no sharp boundaries but transition zones between the groups. In addition, there exists a dynamic equilibrium in the sense that regions of higher order can develop into regions of lower order and vice versa.

The various attempts to describe how with increasing temperature the composition of liquid water, consisting of aggregates of different sizes, changes, either in some orderly progression or in a disordered fashion, must be seen in the context of the discussion of the liquid state in general. If one merely speculated about the possible states in which matter can exist, one would probably think of two extremes: one at low temperature where the properties are completely determined by interaction between the atoms or molecules which have no independent movement. This is the state of the ideal crystallized solid; the other at high temperature where there is interaction only on collision and the properties are determined by the independent movement of the atoms or molecules. This is the state of the ideal gas. The existence of a separate liquid state with first-order transitions to the solid and to the vapour states would not easily come to mind.

There are, however, numerous experimental observations which strongly suggest that the liquid state is closer to the solid state than to the gaseous or vapour state. To name but a few, the volume change on melting is much smaller than that on evaporating; both the energy and the entropy of melting are much smaller than those of evaporation; and the heat capacity increases much less on melting than it does on evaporation, indicating that the number of degrees of freedom in the liquid near the melting point is only little larger than in the solid. But it was particularly the similarity of X-ray scattering diagrams of liquids and of micro-crystalline solids that led a number of research workers to regard liquids as disordered solids in which, however, a considerable degree of short-range order between the molecules still exists, notably near the melting point. Solids and liquids are thus different more in a quantitative sense (degree of order) than in a qualitative sense, and this makes the existence of a sharp and well-defined melting point a somewhat difficult concept. According to Frenkel [15] the difficulty is reduced if one considers the reduction of the molecular order as a cooperative process: the larger the number of disordered regions the less energy is required to increase the degree of disorder further. In other words, the disordering process is autocatalytic. Melting is a catastrophic event and occurs when the concentration of disordered localities exceeds a certain value.

Frank and Wen [16,17] proposed the following resonance scheme (Fig. 1.1) for the cooperative forming and breaking of aggregates in water. Molecule (b) is in resonance between structures I and II, representing the partial polarity of the water molecule. This can lead to structure III in which a hydrogen bond is formed by a kind of acid-base reaction. As a result, molecule (b) has become more acidic and can bond with another molecule (d), whereas molecule (a) has become more basic and can bond with another molecule (c). These new bonds will strengthen the (a)-(b) bond and this in turn causes molecule (c) to become more basic and molecule (d) to become more acidic, promoting further bonding of a water molecule at either end, and so forth.

This process of aggregation is not restricted to linear growth. Water molecules obey a rule formulated by Zachariasen [18]: If in a compound of type $AB_2$, $\Sigma e$ is the total number of outer electrons and $n$ is the number of outer electrons of the noble gas following B in the periodic table, then the molecule is linear if $\Sigma e = 2n$; it is angular if $\Sigma e > 2n$ *. For water, $\Sigma e = 8$, $2n = 4$. Its molecules are therefore angular. The angle between the two O—H bonds is 104.5° [19] **. This is close to the tetrahedron angle of across $(-\frac{1}{3})$ =

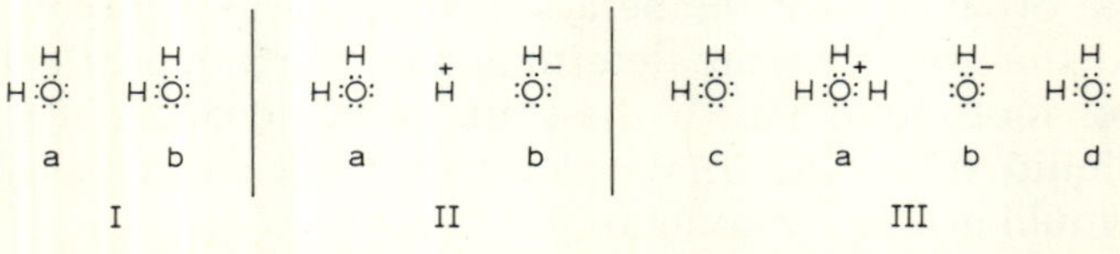


Fig 1.1. Resonance scheme in water (after Frank and Wen [16]).

* Molecules of type $AB_3$ are planar if $\Sigma e = 3n$, but pyramidal if $\Sigma e > 3n$.

** This is the value for water vapour. In liquid water the angle is 105.05° [19a].

109.5° and makes it easy for structures to develop in which each oxygen atom is tetrahedrally bonded to four hydrogen atoms as shown in Fig. 1.2 for ice, which has a structure similar to the tridymite form of $SiO_2$.

According to Frank and Wen both linear and tetrahedral structures are cooperatively formed as pictured in Fig. 1.1. They postulate that when one bond forms many others will form, and when one bond breaks many others will break. This leads to the picture of clusters of various shapes and sizes being continually formed and dissolved. The typical half-life of such a "flickering" cluster is assumed to be $10^{-11}$ to $10^{-10}$ sec, similar to the dielectric relaxation time of water [21]. This is still $10^2$ to $10^3$ times the molecular vibration period and thus long enough for a cluster to have a realistic existence.

This model was adopted by Némethy and Scheraga [22] who subjected it to some extensive speculation and computation with the aim of forcing it to yield thermodynamical data in agreement with experimental observations. Figure 1.3 shows their pictorial representation of the model. The molecules inside the clusters are tetra-coordinated (not shown in this two-dimensional diagram), that is, they take part in four hydrogen bonds. Nearer the surface of the cluster the molecules take part in only two or three hydrogen bonds, and the outermost molecules are bonded to the cluster by a single hydrogen bond. Between the clusters are individual water molecules with all four

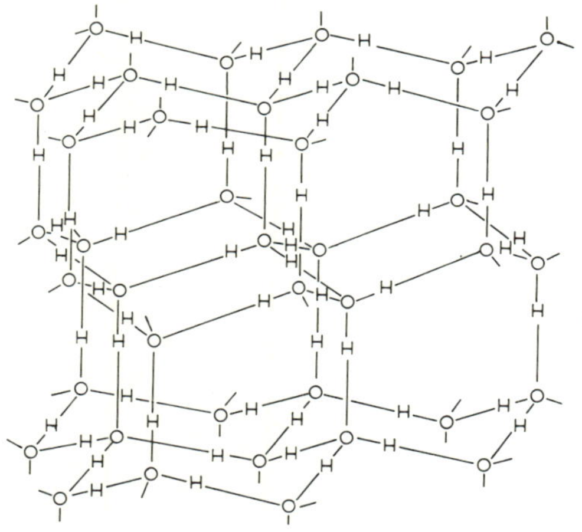

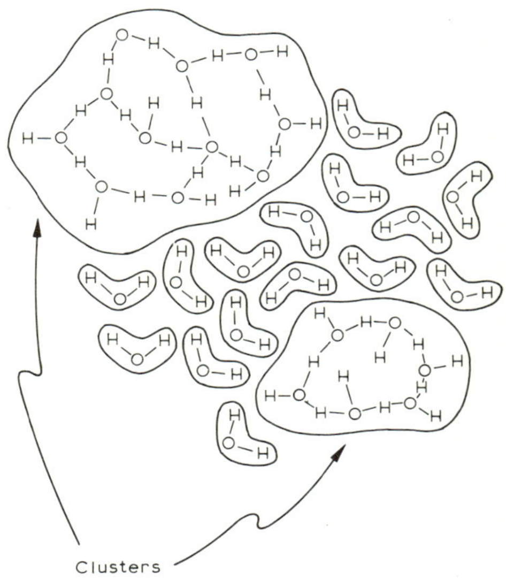


Fig. 1.2. Crystal structure of ice at low pressure (after Némethy and Scheraga [22] from data by Barnes [20]).

Fig. 1.3. Cluster model of Némethy and Scheraga [22].

hydrogen bonds broken. There are therefore five types of water molecules, representing five energy levels, with the tetra-coordinated state serving as the ground level. The authors first counted and averaged the fractions of tetra-, tri-, and di-bonded molecules in compact ("round") model clusters consisting of 12 to 150 molecules. These depend only on the cluster size, i.e. the total number of molecules in the cluster. They then added the mole fractions of singly-bonded and unbonded molecules as two more adjustable variables. The five energy levels were assumed to be equidistant (constant energy difference between consecutive levels) and the distribution of the molecules between the energy levels was assumed to be controlled by Boltzmann factors, exp $(-E_i/kT)$, multiplied by weighting factors related to the degrees of freedom of the molecules in the various bonding states. Thus a three-parameter description of the water structure was achieved and this was used to set up the partition function $Z$ as a function of the three parameters and to evaluate it by differentiating ln $Z$ with respect to the three variables and equating the differential quotients to zero, i.e. maximising $Z$. The thermodynamical functions are then obtained from $Z$ in the usual manner. The calculations of $Z$ and of the thermodynamical functions at various temperatures were done on a computer.

The thermodynamical functions were the free energy, the internal energy, the entropy and the heat capacity. The first three agree with experimental values from 0 to 90° C to better than 3%, the calculated heat capacity, however, falls in this temperature range from 21.5 to 11.7 cal deg$^{-1}$ mole$^{-1}$ whereas the experimental values are 18.1 and 16.2 cal deg$^{-1}$ mole$^{-1}$, respectively. A particularly satisfying result was that the molar volume-temperature curve had a minimum at 4°C. (The three-parameter curve-fitting procedure made it possible to force the curve to pass through the value 18 at 4°C, but did not provide for this value to be the minimum.) The best value for the "energy of the hydrogen bond" (which determines the five energy levels mentioned above) was 1.32 kcal mole$^{-1}$ OH. This is perhaps a little low but falls in the range of values estimated by a number of research workers on the basis of various observations or models. The cluster size varies from 90 molecules at 0° C to 21 at 100° C.

Both the approaches by Eucken [8] and by Némethy and Scheraga [22] are essentially of the curve-fitting type, with the former aiming at a purely phenomenological, and the latter at a more ambitious statistical description. On the whole, there is not much to choose between the two methods, as far as the agreement of calculated thermodynamical functions with observed values is concerned. Even the mole fractions of unbonded OH-groups at 0° C are of a similar order (66 and 47%, respectively) in the two theories. This quantity is interesting because it permits a comparison between different theories and experimental estimates which is not affected by details of the assumed network or aggregate structure and how it breaks down with increasing temperature. Luck [23] compiled a table of these values taken (and sometimes calculated) from 18 different authors using a wide variety of theoretical and experimental methods, which it would take too long to go into in the context of this chapter. They range from 9 to 71%. This is an extraordinarily wide range if one considers the fact that each value is based either on the interpretation of experimentally observed data (such as sound absorption, X-ray scattering, Raman spectrum, energies of melting and evaporation, dielectric constant, etc.) or on theories which were themselves found in good agreement with experimental observations. And it only concerns one reasonably well-defined structural property of water, namely, the fraction of unbonded OH-groups in the liquid at the melting point. Luck concluded that progress in our ideas about the structure of water can only be expected to arise from new experimental evidence.

He then offered for discussion a method to calculate the fraction of unbonded OH-groups from the temperature dependence of the infrared bands, particularly the overtone bands [24]. Figure 1.4 shows the band at 1.155 $\mu$m wavelength which is ascribed to a combination (bending-stretching)

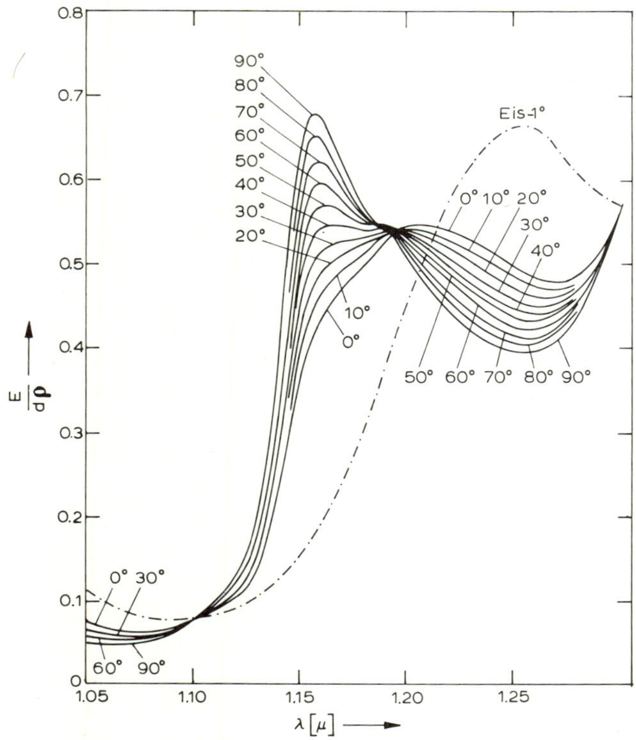


Fig. 1.4. 1.155 $\mu$-band of water at different temperatures (after Luck [24]).

mode. One notices that the extinction curves do not pass through a single isosbestic point, which means that more than two forms of aggregation are in equilibrium at each temperature. (This is even more clearly seen from Fig. 2 of a later publication by the same author [25] where the temperature range was extended to the critical temperature, 374° C.) The curve for microcrystalline ice (dashed in Fig. 1.4) was assumed to belong entirely to tetracoordinated molecules. It is further assumed that the extinctions at the short wavelength end of the band (1.155 $\mu$m) can be ascribed to free OH-groups. The knowledge of the free OH concentration at one temperature would then permit its calculation at any other temperature from the extinction ratio. It was found that a value of 9% free OH at 0° C produced values at temperatures up to 100° C which agreed with values by Haggis et al. [26] who calculated them from the temperature dependence of the dielectric constant. Moreover, if the long wavelength end of the band (near 1.25 $\mu$m) is ascribed to tetracoordinated molecules and if the extinction of ice is taken to indicate 100% tetracoordination, the fraction of such molecules at different temperatures can be calculated from the extinction ratios and they again agree with the ones calculated by Haggis et al. [26] from the dielectric constant.

A more recent value for the concentration of free OH-groups at 0° C was given by Scheraga and his co-workers who continued to develop their cluster model further. In 1974, Lentz et al. [27] published a re-formulated and re-calculated cluster model. The procedure is essentially the same as described earlier, but only two parameters are required this time. The resulting cluster size is much smaller than before, typically six and below, and the concentration of free OH-groups at 0° C is less than 1%, which agrees with estimates by other authors [28—30]. The agreement with thermodynamical data is similar to that before but in addition the calculated intermolecular vibration bands are not very far from observed values. The small cluster size, however, is in sharp contrast to those proposed by most other workers in the field.

This brief survey may have shown what differences of opinion still exist. It would appear to be certain that the concepts of well-defined crystal-like structures have to be abandoned in favour of a dynamic equilibrium of short-lived aggregates or clusters of varying size and shape. It is the distribution of the cluster size and its dependence on the temperature which still awaits a process of convergence and unification.

In view of this uncertainty about the short-range order in liquid water it would be surprising if any statements about its reaction to the presence of a cellulose surface could be made with reasonable confidence that are in agreement with observed facts. Yet this is precisely what was done by Ramiah and Goring [31]. They compared the thermal expansions of dry and water-swollen cellulose and found that both reveal a second-order transition at about 20°C above which the coefficient of thermal expansion increases, e.g. from 5.5 to $7.7 \times 10^{-5}$ cm$^3$ g$^{-1}$ deg$^{-1}$ for dry cellulose but decreases, e.g.

from 18.1 to 12.7 $\times$ $10^{-5}$ $cm^3$ $g^{-1}$ $deg^{-1}$ for wet cellulose. The figures also show that, below the transition temperature, the coefficient of expansion of wet cellulose is about three times that of dry cellulose. These results confirmed very similar ones found earlier by Wahba and Aziz [32]. The increased heat expansion of wet cellulose compared with dry cellulose was interpreted in terms of the cluster theory by Némethy and Scheraga in the 1962 version [22]. The clusters have not only a higher specific volume but also a lower thermal expansion than the monomolecular portion of the water. The cellulose surface was therefore considered to be a "structure breaker": it increases the concentration of unbonded water molecules in its vicinity, and therefore an increased thermal expansion of the system is observed which can be termed an increase of the apparent thermal expansion of the cellulose. This interpretation also supports one of two explanations given for the old observation that the specific volume of cellulose, measured by a buoyancy method, is always smaller with water as the buoyancy medium than with any other liquid. A typical example is a set of values for cotton published by Davidson [33] in 1927. With toluene, helium and water as the buoyancy media, the specific volume was 0.645, 0.638 and 0.621, respectively. The two explanations offered for the low value in water were that: (a) water suffers a compression in the vicinity of the cellulose, or (b) water opens up paths into the cellulose gel that are not even accessible to helium atoms when the cellulose is dry. The conclusions of Ramiah and Goring support the first of the two explanations albeit in the sense, not that water is densified but that its structure is disturbed and more of a species with a higher natural density is generated. The authors extended their investigation to other wood constituents and to carbohydrates including glucose. The results for glucose, whose relevant physical properties are well and accurately known, were then used and manipulated to arrive at very detailed conclusions about the extent of disturbance of the water structure caused by the cellulose surface. For example, at 20° C, 0.045 g unbonded water molecules are generated by 1 g cellulose. This is 0.405 moles $H_2O$ per mole glucose and means, with three OH-groups per glucose unit in the cellulose chain, that every seventh or eight OH-group generates one unbonded water molecule. Recently, Goring [34] applied the revised cluster theory of Lentz et al. in the 1974 version [27] to his observations and recalculated some of the numerical values. The amount of unbonded water generated by 1 g cellulose is now quoted as 0.17 g or 1.53 moles $H_2O$ per mole glucose, i.e. about every second OH-group of the cellulose chain generates one unbonded water molecule. "Unbonded" in this context means of course "broken out of clusters" and, by implication, "linked or bonded to the cellulose surface". Figure 1.5 is Goring's conceptual drawing of the perturbed layer of water near a cellulose surface. He quotes Boesen's [35] value of 0.12 g $H_2O$ per g cellulose in support of his own result of 0.17 g $g^{-1}$. Boesen's value for the "bound water" is, however, nothing but a rough estimate for cotton. He assumes

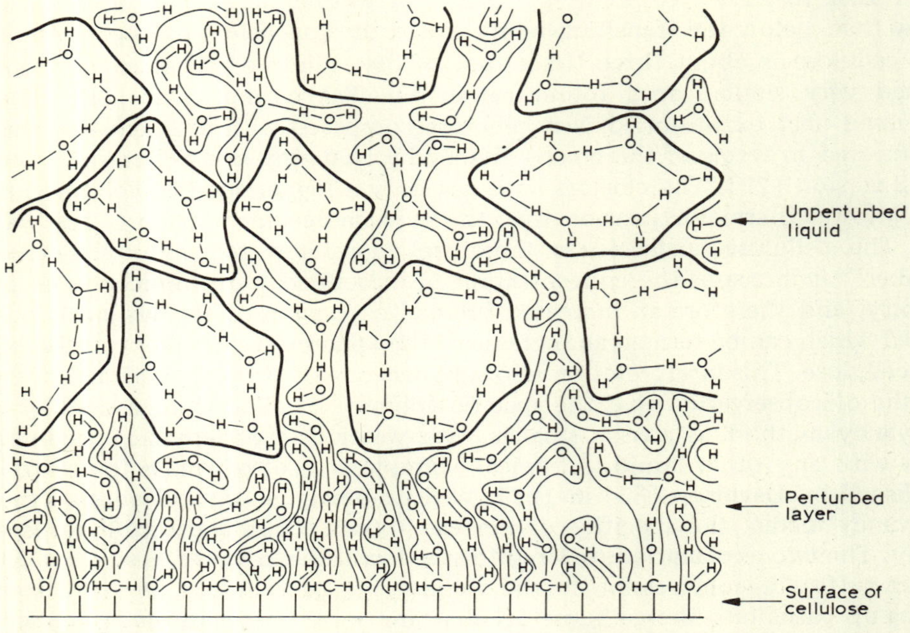


Fig. 1.5. Goring's conceptual drawing of the pertured layer of water adjacent to a cellulose surface [34].

that 35% of it is amorphous, that water can only reach the amorphous part and that each OH-group in this part bonds one water molecule. Thus he arrives at $0.35 \times 3 \times 18/162 = 0.12$ g $g^{-1}$. It is a coincidence that the value is in the same range as Goring's estimate. The concepts behind them are totally different and this example shows perhaps how uncertain the ground still is for detailed speculations. The rather arbitrary assumption of one molecule $H_2O$ per OH-group in the amorphous region of the cellulose is reminiscent of the very first attempts to explain the ca. 30% water adsorption on cellulose at 100% r.h., then generally accepted. It was simply "one molecule $H_2O$ per OH-group" or a ratio of $3 \times 18/162 = \frac{1}{3}$. Apart from the fact that the level of water adsorption at saturation is open to discussion (see later in this chapter) such a notion has long been recognised as totally unrealistic, and similar simplistic approaches will fare no better.

A different approach to the problem of how the presence of cellulose influences the structure of water was made by Caulfield [36]. It is based on the effect of moisture on the tensile modulus of paper. The effect of water on the physical properties of paper will be discussed near the end of this chapter. Only the part of Caulfield's argument concerning the structure of water will be reviewed here. Figure 1.6 shows the logarithm of the tensile

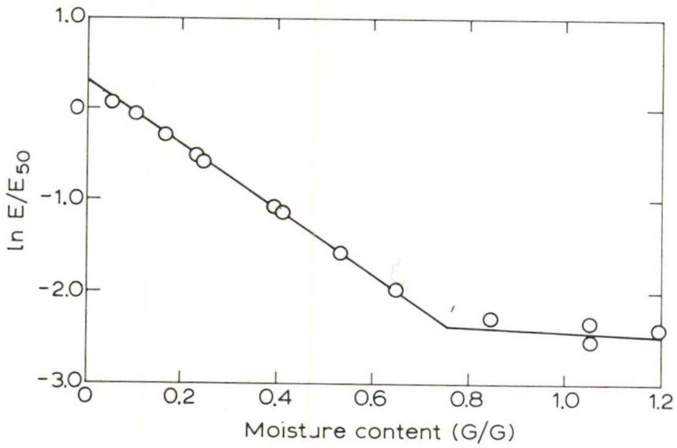


Fig. 1.6. Tensile modulus ratio of wet paper (after Caulfield [36]).

modulus ratio, i.e. of the modulus divided by the value at 50% r.h., plotted against the moisture content for a kraft paper. A straight line is obtained with a break near 80% moisture content. This value was found to be the fibre saturation point as determined by the polymer exclusion method introduced by Stone and Scallan [37]. The constant slope of the curve up to this point means that equal weights of added water reduce the modulus by equal fractions. It will be shown later in this chapter that a monolayer of adsorbed water is completed at around 5% moisture and further water is adsorbed, not on cellulose, but on already adsorbed water molecules. This means that the bonding forces which resist the separation of fibres during straining are transmitted through chains of water molecules in which the bonds between adjacent molecules become weaker as the distance from the fibre surface increases. Fig. 1.7 shows Caulfield's pictorial representation of this situation.

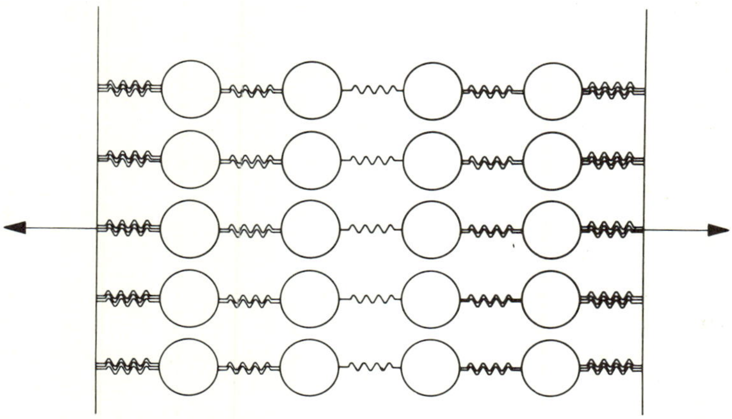

Fig. 1.7. Chains of water molecules between adjacent fibre surfaces (after Caulfield [36]).

The comparison of Figs. 1.5 and 1.7 shows that, in the former, cellulose appears to be a structure breaker and in the latter a structure maker. Not surprisingly, perhaps, the two views were said to be diametrically opposite. In the author's opinion, this does not necessarily follow. Goring and Caulfield consider two different situations. In the former situation the cellulose surface is bounded by bulk water extending, theoretically, to infinity. The structures assumed to be broken down are clusters. The breakdown produces single water molecules or, maybe, doublets which may well adhere to the cellulose surface in some orderly fashion. In the latter situation the water is confined in spaces between fibres. If the fibres are to form a coherent network which resists attempts to alter its volume and shape by the development of restoring forces, a considerable proportion of the fibre surfaces cannot be separated from their neighbours by distances more than a few molecular diameters. There is simply not enough room for bulky, tetracoordinated clusters to form, and some degree of order within the water film between two fibre surfaces is perfectly understandable as a result of the polarity of the water molecules. Stamm and Loughborough calculated in 1935 [38] that the reduction of the free energy during the first stages of the adsorption of water vapour on dry cellulose is smaller than the energy released in the form of heat. This means a reduction of the entropy which, for the first portions of adsorbed water, amounted to 0.27 Clausius $g^{-1}$. If this is multiplied by $T = 273$ the result is $TS = 74$ cal $g^{-1}$, which is close to the heat of melting of ice, 80 cal $g^{-1}$ (the volume change is less than $-0.1$ $cm^3$ $g^{-1}$, which makes the change of the free energy negligible). The first adsorbed water molecules are therefore in a more or less ordered state.

To repeat it, this is not in contrast to any concept of cellulose (or other hydrophilic surfaces) as a structure breaker. We must clearly distinguish between water molecules in flight hitting and settling on, a surface full of hydroxyl groups like outstretched arms, and a cellulose molecule diving into a sea of clusters, large or small, which may well lose some of their members to the intruder. Water on its own tends to structure itself because of the polarity of its molecules. Cellulose tends to impose some structure on the water in the vicinity of its surface because of its hydroxyl groups. When both are together the two structuring principles co-exist. The views of Goring and of Caulfield are therefore not contradictory but complementary.

## 1.3 CELLULOSE-WATER INTERACTIONS IN FIBRE SUSPENSIONS

Typically, the papermaker's fibre suspension originates in the pulper (or "slusher") and ends at the dry-line (or set-line) on the wire of the paper machine. The most important cellulose-water interaction at this stage of the papermaking process is that which enables us to make a fibre suspension at all, as everybody will appreciate who tries to suspend fibres in chloroform, for example. This interaction reveals one of the two faces of the hydroxyl groups in the cellulose chain. It is the one that is turned "outward" to the surrounding water molecules and associates readily with their hydroxyl groups. The other face is turned "inward" to the hydroxyl groups of adjacent cellulose chains of the same fibre or of another fibre that is close enough. Given the right conditions, such as close proximity for a sufficiently long time in an atmosphere not saturated with water vapour, they can bond adjacent fibres together as will be discussed in the next section.

When the cellulose chains were synthesised during the growth of the cell wall there was probably enough time for them to position themselves with respect to their neighbours, at least over part of the chain length, so that the potential energy was minimised locally. Thus, well-ordered crystallised regions were formed. They are responsible for the X-ray diffraction pattern of cellulose, first analysed by Meyer and Misch [39] who found the unit cell of the crystal lattice to be as shown in Fig. 1.8. Kiessig [40] found that the crystal lattice is not altered by water. Water, therefore, cannot penetrate into the crystallised regions of cellulose and this is the reason why cellulose fibres are insoluble in water. But one only has to methylate cellulose very slightly to make it soluble, as Jayme and Froundjian found [41]. The reason is that the bulkier methyl groups wedge the cellulose chains apart so that more hydrophilic OH-groups are set free than are blocked by hydrophobic methyl groups. With increasing methylation the methyl groups gradually win the upper hand and the solubility decreases.

The interest of the papermaker in fibre suspensions is chiefly directed towards their hydrodynamic properties. Some of them are reviewed in Chapter 4. There the presence of fibres is, in many cases, almost incidental; in others their presence *is* of importance but not their cellulosic character. There are only a few instances where the behaviour of fibre suspensions is

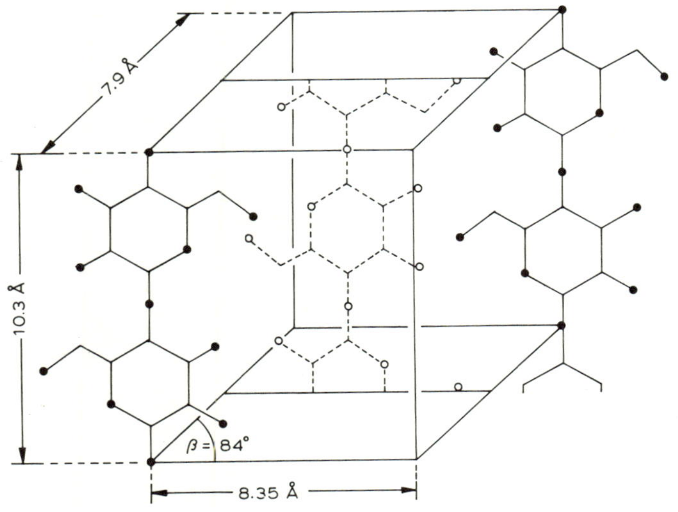


Fig. 1.8. Unit cell of crystalline cellulose (after Meyer and Misch [39]).

affected by cellulose-water interactions, and these will now be reviewed in the order: beating, flocculation and drainage.

### *1.3.1 Beating*

This subject forms part of Chapter 3. The present discussion will therefore be short and confined to the special cellulose-water relationship which makes the beating (or refining) of pulp fibres the characteristic papermaking operation it is. Historically, beating machines were used in Europe to disintegrate rags. The important invention was the "hollander", named after the country of the unknown inventor. Its design and action remained essentially the same long after rags had been replaced by woodpulp as the main raw material for papermaking. New designs appeared with the advent of conical and disc refiners but, from a mechanical engineering point of view, the basic function of all of these machines remained comminution. This is indeed the only visible effect when pulp is beaten in a solvent such as carbon tetrachloride, or dry, which can be done for example, in a ball mill or in a Jokro mill. Figure 1.9 shows it. When the process is carried out in water, the result could

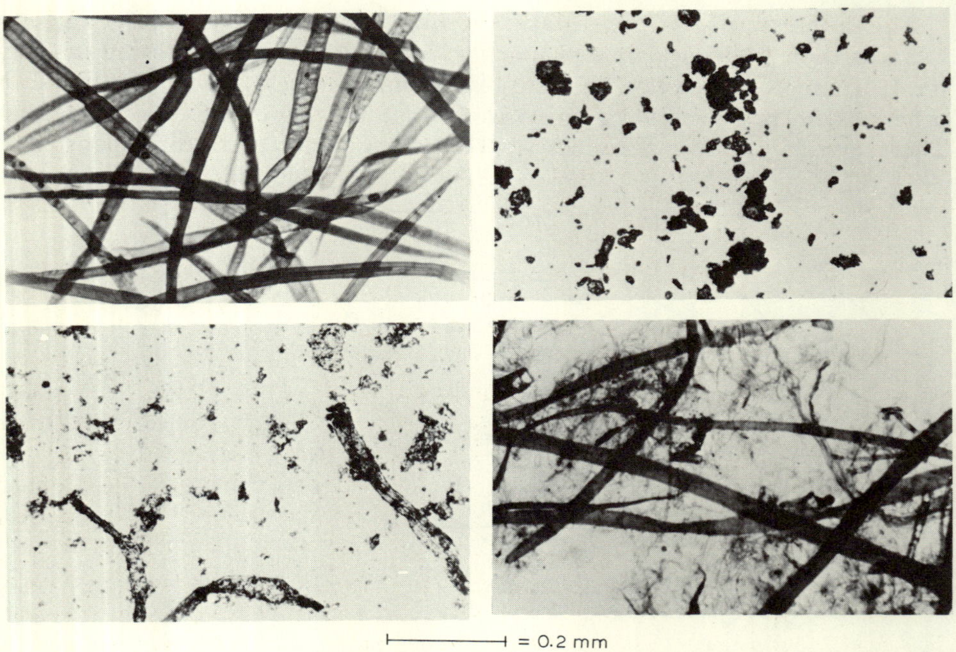


Fig. 1.9. A bleached softwood sulphite pulp before (top left) and after mechanical treatments: ballmilled dry (top right) and in carbon tetrachloride (bottom left) for several hours, and beaten in the PFI mill to 40°SR in water (bottom right).

still be called comminution, if the meaning of the term is not restricted to the generation of separate particles of smaller size from the original charge but extended to include the reduction of fibres to smaller entities which, however, still hang together. Kress and Bialkowsky, in their classical and often cited experiments [42], demonstrated conclusively that the latter type of beating effect is produced only in liquids that react with, i.e. penetrate and swell, the pulp fibres. This ability is the essence of the cellulose-water relationship in the beating process.

In the thirties and forties, the literature abounded with articles on what was often called the "theory of beating". A survey was given by Clark in 1954 [43]. As Steenberg observed [44] they were, however, as a rule not concerned at all with an analysis of the beating process but with attempts to interpret effects of beating that are of interest to the papermaker. The two most obvious ones are the drainage resistance of the pulp and the tensile strength of paper made from it, both of which increase with the beating time. The first step was usually a description of the impression the beaten pulp made on the senses, particularly the tactile and the visual senses, with the latter considerably sharpened by the use of the microscope and later the electron microscope. There were two schools of thought for further speculation: the chemical and the physical one. The former used the word "hydration" as the key-term to describe the state of beaten fibres and the slippery feel of beaten pulp suspensions; it was considered to be the cause of fibre bonding in the sheet-making process. The latter used the term "fibrillation" which was held responsible for the increasing drainage resistance of the pulp suspension. Both terms were essentially descriptive and sufficiently vague to cause much debate, even controversy. Yet it was (and still is) generally felt that together they describe the type of comminution typical of the beating process, and the role which water plays in it. It is no exaggeration to say that most research work on beating since then was concerned with the elucidation of a more precise meaning of these two terms.

An important step in this direction was made by Robertson and Mason in 1949 [45] when they developed a method to determine the specific surface and the specific volume of pulp fibres from the flow rate of water through pulp pads of different concentrations, and applied the method to pulps beaten for different lengths of time. The result was that the specific surface increases slowly at first and then much faster as beating progresses, whereas the specific volume increases rapidly at first and more slowly later. It was an interesting result because the difference in the shape of the curves was similar to that between the development of the wetness, as measured with the Schopper-Riegler tester, and of the burst or tensile strength. It was confirmed by Corte [46] and Fig. 1.10 shows an example of his results. The development of those strength properties which depend primarily on fibre-to-fibre bonding seems to be related to the swelling of the fibres ("hydration") and the increase of the drainage resistance to the increase of the

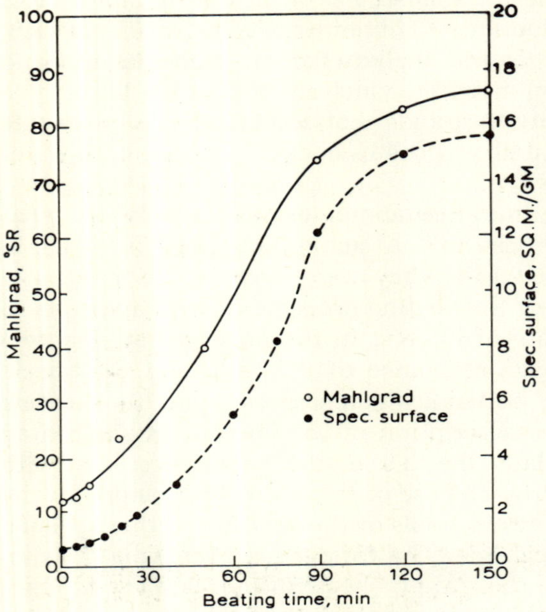


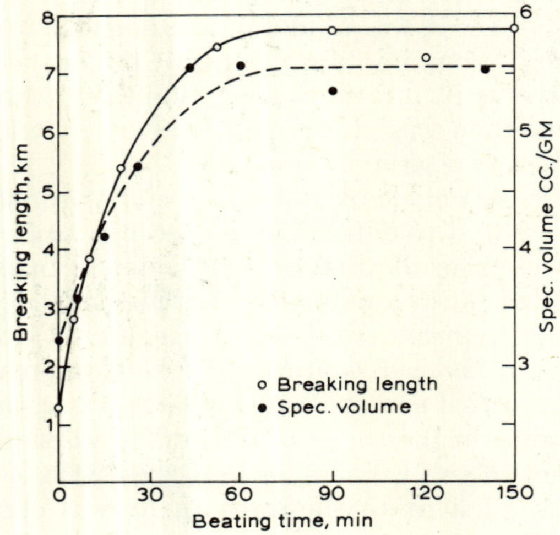


Fig. 1.10. Effect of the beating time in a Jokro mill on the Schopper Riegler wetness/ specific surface (top) and on the breaking length/specific volume (bottom) of a bleached spruce sulphite pulp [46].

specific surface ("fibrillation"). Fibre suspensions and their reaction to mechanical treatments can vary so much, however, that this relationship can only be of a general but not of a quantitative nature. This was discussed by Robertson at the Fundamental Research Symposium, Cambridge, 1957 [47].

Further research into the nature of swelling and fibrillation was greatly stimulated when indirect "surface — and — volume" experiments as described in the last paragraph were supplemented (some may have said supplanted) by direct microscopic observation of the fibre morphology and its changes during beating. Emerton must be given the credit for initiating this phase with his book "Fundamentals of the Beating Process" [48]. One result of the work was that the term "fibrillation", which traditionally denoted the partial separation of the primary wall and of the fibrils of the outer secondary wall, was used to include "internal" fibrillation also. Similarly, the term "swelling", which traditionally denoted the increase of the outer dimensions, was extended to mean the uptake of water, generally, "of which the volume changes of swelling are but an extrinsic feature" (Emer-

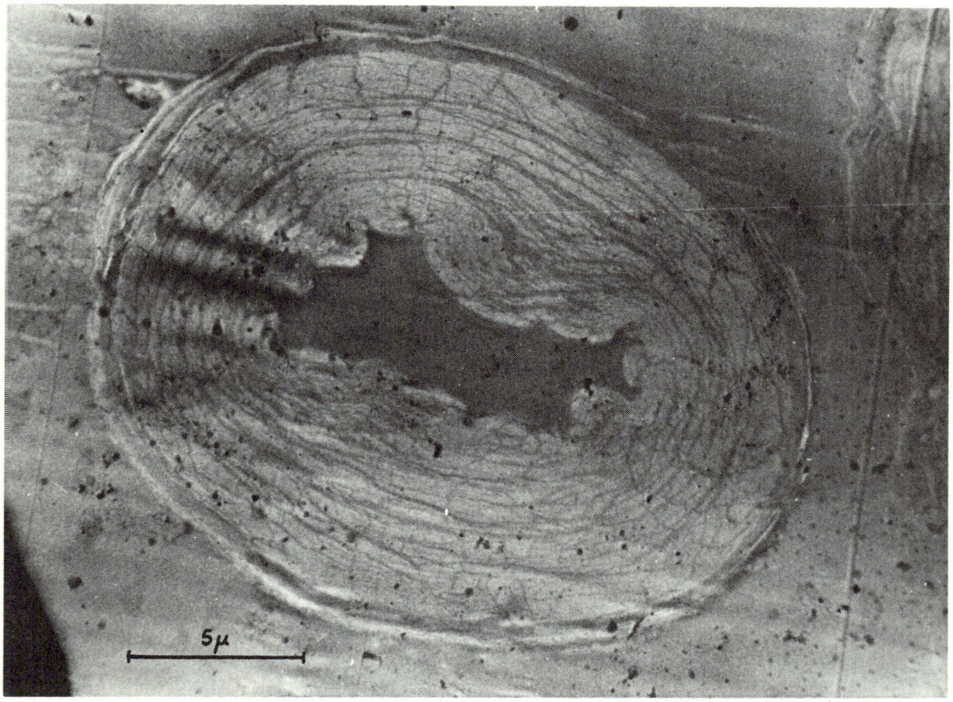


Fig. 1.11. Electron micrograph of thin section of a beaten sulphite fibre (after Page and De Grâce [49]).

ton). The mechanism was thought to be delamination of the fibre wall, that is, its separation into concentric lamellae caused by shear forces during beating and accompanied by the penetration of water into the annular gaps. As a result, the flexibility of the fibres was increased, which in turn increased fibre bonding, i.e. strength. Some of the views were backed by photographic evidence but some were more speculative. For example, it took ten years before the delamination of the fibre wall was actually observed by Page and De Grâce [49]. Figure 1.11 is taken from their article. Two years earlier, Stone and Scallan [50] had reached the same conclusion from measurements of the nitrogen adsorption on solvent-exchanged fibres. Their work will be returned to later.

The morphological approach and the efficient use of the microscope and electron microscope have greatly increased our knowledge of what happens to the fibres during beating. So far it has been purely descriptive. Attempts to subject it to some form of mathematical analysis have not been reported. This is in contrast to the comminution effect proper of the beating process, which is the change of the fibre length distribution. Equations for this stochastic process were published by Corte [51], Kane [52], Nordman and Laininen [53], and Meyer et al. [54]. The first of these contains an example of how a detailed mathematical analysis of the variation of the fibre length distribution with beating time can yield considerable details of the fibre length reduction process, in this case the preferential fracture of softwood sulphite fibres half-way between ray-cell passages, an effect that is absent from sulphate pulp of the same wood species *. This comminution effect of beating has not yet been related to cellulose-water interaction. It is mentioned here because it is the one effect of the beating process that has been quantified. The reason for this to be possible was simplification: The objects used for the statistical analysis were not fibres with a complex morphological structure but straight lines with different lengths, and the length was the only variable. Structural differences could then be deduced and looked for as a result of the statistical analysis. It would be interesting to see whether the structure of fibres can be simplified and generalised to permit a statistical analysis of delamination, external fibrillation and swelling with the aim of deducing structural differences from such an analysis. This would assist and complement microscopic investigations which are by nature time-consuming and limited to extremely small samples.

---

* Corte [51] developed a method to calculate the shortening probability of a fibre from fibre length distributions after different beating times. The occurrence of the weak spots in softwood sulphite fibres was deduced from the observation that the shortening probability of such fibres during the initial stages of a cutting beating action was independent of the fibre length. This was thought to result from increased hydrolysis as the cooking liquor moves along the fibre axes away from the ray-cells, loses its selectivity and increases the concentration of lignin sulphonic acid. The suspicion was confirmed by pulling apart bundles of undercooked fibres under the microscope (unpublished work).

### *1.3.2 Flocculation*

The term flocculation has an ambiguity which it shares with the paper property, formation, often if incorrectly associated with it. On the one hand, it stands for the flocculated state of the fibre suspension under a given set of conditions; on the other hand, it stands for the floc forming process.

It is the latter where cellulose-water interactions may be expected to play a part, if we assume for the moment that flocs are actually formed from individual fibres.

The control of flocculation is as old as papermaking itself and numerous observations made under production conditions were published in the early papermaking literature. They were reviewed by Woolwage in 1939 [55] who initiated the study of flocculation in the laboratory under controlled experimental conditions. Well-dispersed pulp suspensions of about 0.01% concentration were made to flow with different speeds but in a laminar fashion through a glass tube and the distance travelled in the tube at which visible flocs began to form was used to describe the flocculation tendency. The experiments confirmed earlier observations that flocculation is a result not only of mechanical, i.e. hydrodynamic, but also of a number of non-mechanical factors. Again, it is the latter where cellulose-water interactions may enter the picture. Woolwage found the flocculation tendency to increase with the temperature, with beating and with the addition of bentonite; and to decrease with the addition of alum under certain conditions and of deacetylated Karaya gum. A year later, the results were confirmed by Erspamer [56] on pulp suspension of 0.05% concentration. He additionally observed the deflocculating effect of locust bean gum, methylcellulose and tetrasodium pyrophosphate in the presence of calcium ions.

In 1948, Mason [57] described a more sophisticated instrument to study flocculation in suspensions of 0.05% concentration sheared in the annular gap between two vertical glass cylinders of which the outer cylinder rotated (a Couette-type viscometer). In the next ten years he and his co-workers published a number of articles on the flocculation of fibre suspensions and, more fundamentally, on the motion of particles in sheared suspension. The forming of flocs was treated exclusively as a mechanical process consisting of the rotation, collision and adhesion of fibres, which is outside the scope of this chapter. Some of the work, particularly on the motion of fibres, is of great interest in its own right but its relevance to the mechanism of floc formation in papermaking is difficult to visualise because at no stage, from the tree to the slice of the paper machine, are the fibres in an environment in which they are able to rotate, collide with and adhere to, others. In other words, the suspension leaving the slice contains hardly any single fibres to perform the complex motions so carefully observed and described by these authors. Robertson and Mason later related flocculation to the flow behaviour of fibre suspensions [58,59] which were then recognised as forming connected networks with measurable elastic and strength properties. The network concept was further developed by Wahren and his co-workers in several publications. A review was published by Wahren in 1964 [60] and another presented at the Cambridge symposium 1965 [61]. The approach is almost entirely mechanistic and mainly directed towards flow properties. It is obvious that in such network-like "suspensions" flocs are not formed from individual fibres but by breaking up the network. Flocculation is thus no longer a process of aggregation but one of disruption, controlled by mechanical shear forces.

There is virtually no room for non-mechanical forces in the approaches of both Mason et al. and Wahren et al. Such forces, in particular electrokinetic or, more generally, electrochemical, forces were claimed by a number of research workers to play an important role in the behaviour of suspended fibres. The deflocculating effect of certain chemicals and plant mucilages was mentioned earlier. A "phenomenal" effect of the root extract of *hibiscus manihot*, L. (rose mallow) on the dispersion of long-fibred mitsumata pulp was reported by Broadbent and Harrison [62] who stressed the point that "no amount of agitation" dispersed the highly flocculated and entangled suspension with which they started. Such plant extracts had been used for centuries by Japanese papermakers. Their effect was by several researchers vaguely associated with some kind of electrokinetic phenomenon. This was repeatedly rejected by Mason [63] and by Robertson and Mason [64,65] who considered the action of these substances to be simply mechanical lubrication. That it is not quite so simple was shown by Huber and Weigl [66] who found that the extract from *hibiscus manihot* contains mainly calcium salts of polygalacturonic acids with molecular weights of up to 250,000 which have a network structure caused by a kind of internal cross-linking via calcium ions. They form a fairly rigid skin around the fibres, which not only keeps them apart but also, as the authors demonstrated, straightens them, an important effect when using very long fibres. Electrokinetics is brought into the picture as being responsible for the initial adsorption of the deflocculant on the fibres and thereby indirectly responsible for the dispersing effect. The shear forces in these experiments were small, however, and not comparable to those at which Robertson and Mason deny any but mechanical forces to be of consequence for flocculation and dispersion.

Electrokinetic properties of cellulose fibres in water were studied for several decades without arousing much interest. According to Goring and Mason "anomalous results were reported and the general reproducibility was poor". They attributed this partly to faulty experimental techniques and, in a notable three-part contribution [67], the last co-authored by Biefer, put the measurement of electrokinetic phenomena involving cellulose fibres, in particular the streaming potential method, on a sound experimental basis. A very considerable literature has developed in the past 25 years. Melzer published a comprehensive survey in 1972 [68] with 102 references which he updated in 1977, adding another 313 references (as yet unpublished). Most cf the literature is concerned with such papermaking problems as sizing, dyeing, pitch control, filler and fines retention, and effluent treatment. In the majority of cases, the method of investigation was the measurement of the electrophoretic mobility of particles in water and in aqueous solutions of electrolytes. This method was in a number of cases applied to colloidal particles obtained from the filtrates of cellulose fibres and it is also applicable to microcrystalline cellulose. Because of their dimensions, fibres are not well suited for this method. Their electrokinetic properties are therefore determined from streaming potential or, occasionally, electroosmotic measurements. The results are often expressed as the so-called $\zeta$-potential, a mathematically derived quantity which is often proportional to the electric charge of the fibres.

Here is not the place to review the voluminous literature associated with the concepts of the electric double layer and the $\zeta$-potential, since they apply to colloidal systems in general and not specifically to the cellulose-water system. A brief reference, relevant for paper, will be made later. The one fundamental result of all the work that has been reported on cellulose fibres or particles is that they assume a negative charge in water. The origin of the charge is still unknown. A rule found by Coehn in 1898 [69] says that, of two substances in contact with each other, the one with the higher dielectric constant assumes a positive charge. Therefore, because of the higher dielectric constant of water, the negative charge of cellulose is not surprising, which does not answer the question what causes it, but puts it into a wider context. It is quite likely that both of the other two effects which can influence the charge of suspended particles contribute to the charge of cellulose fibres, namely, the dissociation of hydroxyl or possibly carboxyl groups, and the adsorption of ions.

The effect of anions on the electrophoretic mobility (or $\zeta$-potential) was investigated, e.g. by Chripunowa et al. [70] who found it to decrease in the order $I^- > Br^- > Cl^- > F^- > CNS^- > CH_3COO^- > HCOO^-$ and $Fe(CN)_6^{4-} > Fe(CN)_6^{3-} > SO_4^{2-}$. The effect of cations was investigated by many workers. Cations generally reduce the negative charge and hence the mobility of pulp fibres. Multivalent cations are particularly effective and reduce the charge to zero, a point often but incorrectly called the "isoelectric" point *, where the mobility is zero. The charge can even be reversed with increasing cation concentration and the mobility increases again, this time towards the cathode. What is not so well known is the fact that the mobility reaches a maximum and with increasing concentration, for example of $Th^{4+}$-ions, is reduced and

---

* This term was introduced by Michaelis [71] to denote that hydrogen ion concentration at which the cationic dissociation of amphoteric electrolytes, such as glycin, equals the anionic dissociation ("Zwitterions"). For suspended particles, one should speak of the electroneutrality point.

approaches zero again. The top half of Fig. 1.12 from an article by Chang and Robertson shows the effect [72], expressed as the ζ-potential, calculated from measurements of the streaming current generated by pulp pads that had been compressed to different concentrations [73]. Multivalent anions, such as $Fe(CN)_6^{3-}$ and $Fe(CN)_6^{4-}$, sometimes increase the mobility slightly at low concentrations but always reduce it to zero at higher concentrations. The reduction of the mobility or of the ζ-potential with increasing ion concentration can in these cases not be caused by a reduction of the surface charge. Conventional electrokinetic theories explain the effect as a compression of the double layer by ions but the interpretations vary in detail, see, for example, Kruyt [74].

The electrophoretic mobility of paper discs, 5 mm in diameter and 0.1 mm thick, was used by Corte [75] to calculate the charge density of paper surfaces in water by equating the driving force in the electric field and the frictional force according to Stokes's law. The value was of the order $10^8$ elementary charges per $cm^2$. It means an average separation of the charges by about 3000 atomic diameters which he considered not compatible with the concept of a coherent electric double layer. This was contradicted by Overbeek at the Oxford symposium 1977 [76] who stated that the retardation of the movement in the electric field is $10^4$ to $10^5$ times the value yielded by Stokes's law, due to the formation of an ion cloud around the paper discs. The controversy is unresolved.

An interesting effect, observed by Goring et al. [67], is the ca. 25% increase of the a.c. conductivity of pulp pads in water with increasing frequency from 0 to 1000 Hz, with most of it occurring in the range 0 to 100 Hz. This is in sharp contrast to all theories of the surface conductance based on the concept of the electric double layer, which predict a decrease of the

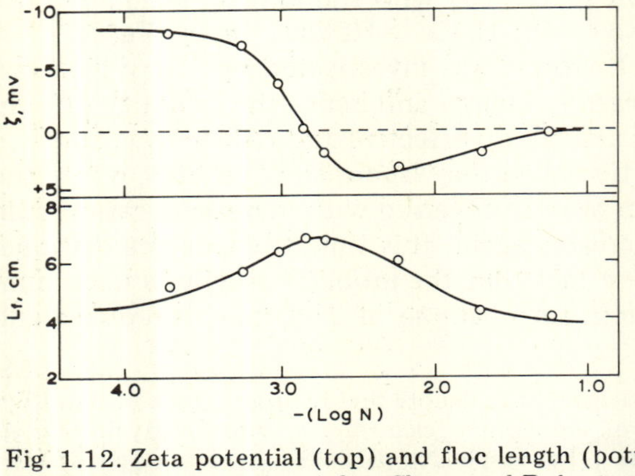


Fig. 1.12. Zeta potential (top) and floc length (bottom) of a kraft pulp as a function of the $ThCl_4$-concentration (after Chang and Robertson [72]).

conductivity with increasing frequency, if a dispersion is observed at all. It is consistent, however, with Urquhart's concept of mobile molecule ends on the surface of swollen fibres [77] which are surrounded by water molecules and would float away, were they not anchored in more ordered or crystallised regions. Campbell called this state "a sort of "partial solubility"" of the cellulose chains [78] at the surface which may well be able to move in an a.c. field.

Returning to flocculation, the bottom half of Fig. 1.12 shows the "floc lengths" of the suspensions used for the streaming current measurements. This quantity is observed and measured when a well-dispersed suspension of 0.7% consistency is extruded vertically down from a tube at a rate of 1 mm $sec^{-1}$ into a cylinder of water. The method is described in [59b]. The floc length is a measure of the fibre network strength in water. It is seen in the figure to reach a maximum when the $\zeta$-potential becomes zero for the first time as a result of added $Th^{4+}$-ions. At this point, the charge of the fibres and thus the electrostatic repulsion forces between them are also zero. The same effect was observed with a cationic linear polymer (poly-1,2-dimethyl-5-vinyl-pyridinium methyl sulphate) as additive. Significantly, the network strength continues to decrease at higher $Th^{4+}$-ion concentrations when the $\zeta$-potential approaches zero again since the charge density of the fibres remains high. The variation of the floc length as shown in Fig. 1.12 is interpreted in terms of the lubrication concept mentioned earlier, lubrication (or reduced friction) being effected by the neutralisation of the negative electric charge of the fibres. Although the shear forces are small, the floc strength is believed to be related to the hydrodynamic behaviour of fibre suspensions under papermaking conditions. Poppel and Bicu [79] made similar observations with $Th^{4+}$-ions and with a cationic wet strength resin. They found in addition that the dry and wet strength of the paper reached maximum values at addition levels which reduced the charge to zero.

The study of flocculation received a new stimulus when the range of chemical papermaking aids was augmented by the introduction of high-molecular polyelectrolytes, such as polyacrylamides and polyethyleneimines, which are used both as drainage and as retention aids. In either application the mechanism of their action is associated with the way in which they influence the flocculation tendency of fibre suspensions, with or without filler additions, under papermaking conditions. Research was carried out not only in the laboratories of the paper industry but also in the laboratories of the chemical industry where these substances were developed. Hennig [80] emphasised the qualitative similarity of the flocculation in fibre suspensions (typical dimension $10^{-1}$ cm) to that in colloidal systems (typical dimension $10^{-7}$—$10^{-6}$ cm). He made the point that zero charge alone is not sufficient to cause flocculation and that adsorbed polar groups are necessary to form bonds of some kind between the particles (fibres). The experiments were of the sedimentation type and the shear forces were small. The bond formation on addition of polymer flocculants is now usually called "bridging" and is envisaged as a physical linkage between particles provided by the linear polymers which, at molecular weights in the million range, can be several microns long. According to Horn and Melzer [81], bridging is the prevailing mechanism of the action of polyacrylamides as flocculants whereas polyethyleneimines act chiefly through charge neutralisation.

The retention of fibre fines by adsorption or flocculation on the fibres was recently investigated by several authors. King and Williams [82] observed that fines, once they are adsorbed on the fibres by the action of polymers, remain attached to them even at polymer concentrations at which the fibres are redispersed again by stirring. The resistance of the fines flocculation on fibres by polyelectrolytes to shear forces (different rates of stirring) was also demonstrated by Britt et al. [83]. Lindström and Söremark [84] drew attention to kinetic effects which control the fibre-fines coflocculation by polyacrylamides. The polymer adsorption on the fines is much faster than that on the fibres, and it is reversible. When the fibres begin to adsorb polymer from the solution the fines must desorb some of it. The proposed coflocculation mechanism is the adsorption of positively-charged fines onto negatively-charged fibres during short contact times even at a stirring speed of 500 rpm in a 2-l vessel. Such detailed formulations need perhaps some more experimental support but there can be no doubt that flocculation phenomena in fibre suspensions may be affected by electrochemical forces which can compete with considerable shear forces. The quantification of these forces, both in controlled laboratory experiments and (at least for the order of magnitude) in typical production situations, would seem to be an important next step forward.

### *1.3.3 Drainage*

Similar to flocculation and the general behaviour of fibre suspensions, drainage has been studied mainly from a hydrodynamical or mechanical point of view. Some of these studies were very fundamental and have led to or were associated with, the introduction of new technologies, such as twin wire forming and the replacement of table rolls by hydrofoils. Part of the literature is reviewed in Chapter 4. Previous literature reviews were published by Han [85] and by Parker [86]. On the whole, the cellulosic nature of papermaking fibres and their interaction with water do not have a very significant effect on the mathematical form of the drainage process (although they very much affect the drainage rate and hence the machine speed). Indeed, experimental laboratory investigations on drainage are frequently carried out first with glass or synthetic model fibres to verify empirical or theoretically derived drainage equations. Deviations observed with pulp fibres are then interpreted in terms of their morphology and interaction with water.

The fibre properties which can be of some importance are their shape and their mechanical properties, both of which are influenced by the interaction with water. A recent study of the effect of the fibre shape on drainage was carried out by Arnevik and Helle [87]. They found a marked influence particularly of the cross-sectional shape. Model rayon fibres with cylindrical cross-sections allowed the water to drain much faster than flat fibres with a rectangular cross-section. Similar results were obtained when using unbeaten woodpulp fibres with a wide variety of cross-sectional areas and wall thickness/lumen ratios. The drainage rates varied over a range of almost 3 : 1. However, beating was found to level the differences and to reduce this range to 2 : 1.

One effect of water on the shape of fibres is their swelling. It is appreciable only when the fibres have been beaten. A typical result of this is demonstrated in Fig. 1.13 which shows the temperature dependence of the gravity

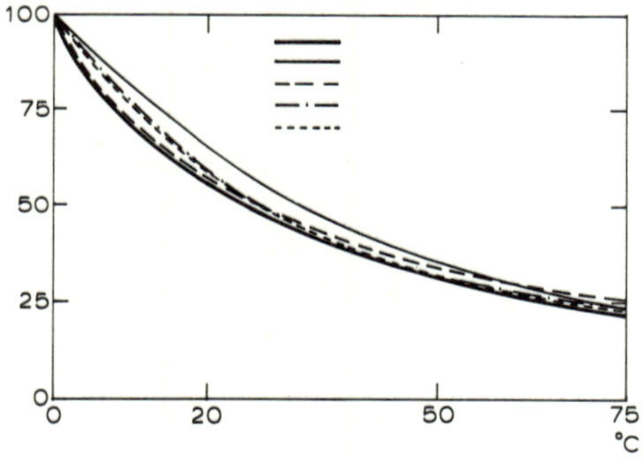


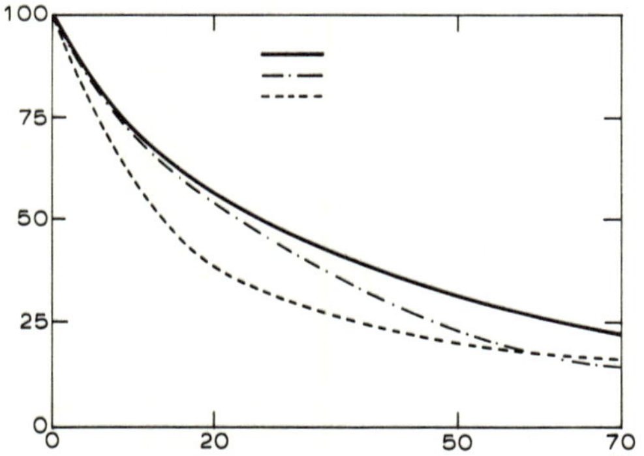


Fig. 1.13. Drainage resistance of pulp suspensions and viscosity of water as a function of the temperature [89].
Top: unbeaten pulps (curve markings from top to bottom: water, poplar, spruce, beech, birch).
Bottom: Pulps beaten to 48°SR (curve markings from top to bottom: water, spruce, poplar).

drainage of various pulps through a stationary wire in an instrument similar to the Williams freeness tester [88]. Assuming a simple filtration process in which a mat of constant concentration is formed and the supernatant suspension retains its original concentration, Corte derived the following drainage equation

$$t = b\ (x - \ln x - 1). \tag{1.1}$$

Here, $t$ is the time, $x$ is that fraction of the water which is still above the wire at time $t$, and $b$ is a constant [89]. In the figure the values of $b$ are plotted against the temperature in per cent of their values at 0°C. For comparison the viscosity of water is plotted in per cent of the value at 0° C. In the upper part of the figure (unbeaten pulps) the curves nearly coincide, but in the lower part those for the pulps, beaten to 48° SR, are well below the curve for water, indicating the deswelling effect of increasing temperature. It would be interesting to see whether this deswelling effect reduces the strength of paper made from heated fibre suspensions or wheter it is overshadowed by the heating of the web in the drying section before the major part of bonding occurs. Another effect of water on the shape of fibres was discussed earlier in connection with "fibrillation" during beating, which only occurs in water (or other liquids that interact with cellulose). The parallel development of the Schopper-Riegler wetness and the specific surface during beating suggests that it is this change of the fibre shape which is mainly responsible for the increased drainage resistance of beaten pulps. The well-known reduction of the Schopper-Riegler wetness with increasing temperature (beyond that due to the reduced viscosity of the water) could then be interpreted as a reduction of the state of fibrillation at elevated temperatures. On the other hand, the effect of alum on the drainage resistance of pulp suspensions is often held to be the result of some form of dehydration (or deswelling) of the pulp fibres by $Al^{3+}$-ions. As was said before, these relationships are probably of a more general rather than of a specific nature. It is very likely that both swelling and fibrillation influence the drainage characteristics of a pulp suspension and that both are affected by the temperature and, possibly, by multivalent ions, in the same direction.

The effect of water on the mechanical properties of fibres suspended in it is closely related to swelling and fibrillation and results in their increased flexibility and softness. As a consequence, they can pack more closely under the influence of a hydrostatic pressure or of the viscous drag of water draining through the mat. The compaction of fibre mats in laboratory drainage of filtration experiments is the most notable perturbation of the older drainage or filtration equations, which were usually derived under the assumption that an incompressible bed is formed. Extensive work in this area was done at the Institute of Paper Chemistry, Appleton, WI, U.S.A. Ingmanson developed an elaborate theory of filtration through compressible mats [90], supplemented by Nelson [91] and by Han [92] and later followed by articles by Han and Ingmanson [93] and by Chang [94]. Common to all of these investigations is that they are concerned with constant pressure (or suction) filtration and that the compression is thought to be in equilibrium with the hydrostatic pressure although Chang [94] also considered the difference between static and dynamic compressibility. Experimental checks were carried out in the form of static laboratory filtration tests — static in the sense that the forming mat was not moving. If one were to contemplate the application of these filtration theories to the table roll drainage on a Fourdrinier paper machine, one would describe this process as a type of filtration in which a constant suction is applied, not continuously but in a number of discrete steps equal to the number of table rolls, with roughly equal time intervals of zero drainage between the steps. However, when this was done by Mauranen et al. [95] for ten different paper machines they found the theory "unsuitable for the prediction of drainage by table rolls". Figure 1.14 shows one set of results for a paper machine with 32 plain table rolls, making manilla board of 163 g $m^{-2}$. They are plotted according to eqn (1.1). The points scatter, at least partly because in the absence of tabulated data the values were read off the graphical plot (Fig. 4 of the article quoted above), but a straight line approximation appears to be justified. Another example is even more illustrative. Probably the first full-scale drainage experiment on a production machine was carried out in 1951 by Bennet and his co-workers from the British Paper and Board Industry Research Association, Kenley, Surrey [96]. One-foot long trays were installed in the centre under each of the 21 table rolls of a newsprint machine and the drainage into these trays was measured once or twice a day for 23 days, the entire life of the wire. In

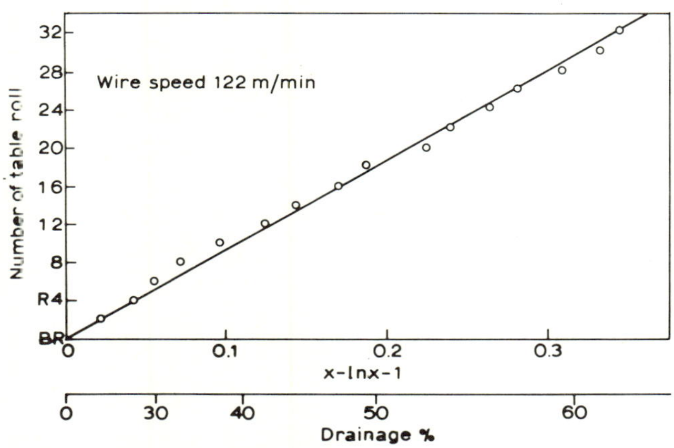

Fig. 1.14. Table roll drainage on a paper machine making manilla board plotted according to eqn (1). Data from Mauranen et al. [95].

Fig. 1.15 the results obtained on the seventh day are plotted as circles according to eqn (1.1). The data are taken from tables published in 1953 [97]. The points begin to deviate from the straight line near the end of the wire after almost 75% drainage and the deviation is in a direction opposite to that which would be expected to result from a compression of the web. The results published [96] are those obtained on the penultimate day of the wire life. They are plotted in Fig. 1.15 as crosses. The slope of the straight line, that is, the drainage resistance, has increased by about 10% and the points begin to deviate (again towards higher drainage rates) after about 60% drainage. Both effects are consistent with Corte's findings [89]: The fibres deposited during the latter stage of the drainage process come down relatively gently and in a more flocculated form so that the specific drainage resistance of the later deposits decreases; and anything that retards the drainage process as a whole will enhance the effect. There is no indication that the web, which is loosened up and sometimes even lifted off the wire on approaching a table roll, suffers a compression anywhere. It should be mentioned that the table rolls were not equidistant, which means that the ordinate in Fig. 1.15 is not the time. In other words, for stepwise drainage the time in eqn (1.1) is replaced by the sequence steps.

Table roll drainage and the drainage experiment described by eqn (1.1) are two entirely different processes, however, and it must be considered coincidental that the equation describes table roll drainage so well. In particular, the pressure in the laboratory drainage experiment decreases whereas the suction in the nips of the table rolls is assumed to be constant, at least as long as the wire passing over the table roll is saturated with water. One could speculate that the nip length and hence the dewatering effect of the table

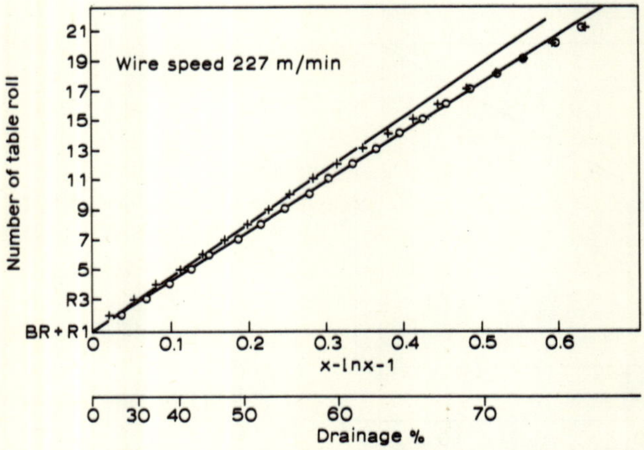


Fig. 1.15. Table roll drainage on a paper machine making newsprint, plotted according to eqn (1). Data from Gill [97] for the seventh day of the wire life (circles) and from Bennett [96] for the 22nd day of the wire life (crosses).

rolls decreases down the wire. Only a detailed analysis, which is still outstanding, can decide whether such an effect would indeed lead to a relationship of the form of eqn (1.1). With the steady advance of two-way drainage the interest in table roll drainage is certainly and inevitably waning, however, and it is quite possible that this form of dewatering will be obsolete before it is fully understood.

## 1.4 CELLULOSE-WATER INTERACTIONS DURING CONSOLIDATION

For the purpose of this discussion the consolidation phase of the paper-making operation begins after the fibre suspension has been turned into a continuous web and enters the first wet press, and it ends at the reel-up. Although a variety of specialised compacting devices have been designed for inclusion in this part of the paper machine, such as wet and dry smoothing presses or MG cylinders, the consolidation is essentially achieved in two steps: pressing and drying. The two steps will be discussed separately, followed by a discussion of the mechanism of bonding in the consolidating web.

### *1.4.1 Pressing*

The literature on wet pressing is extensively reviewed in Chapter 5. The review shows that very little has been published on specific cellulose-water interactions during this stage of consolidation. The attention is almost entirely focused on the mechanism by which water is expressed from the wet web by the application of a mechanical load, on the role of the felt, and on engineering features designed to speed up the removal of expressed water from the nip. In fact, one of the first attempts to apply physical reasoning to an understanding of wet pressing was made by Nissan [98] when he speculated on the function of the felt in water removal. The fibre-water interaction enters the picture somewhat obliquely when the responses of the wet web and of the felt to a compressive load are compared. According to Nissan [98] the water leaving the press on the ingoing side comes mostly from the felt as it is compressed. The web is considered to be saturated with water and therefore much less compressible than the felt. On leaving the press, the felt expands and is assumed to suck water from the web which is in turn removed from the felt on re-entry into the press. A specific conclusion from this speculation was that the felt should compress easily (i.e. at a pressure too low to crush the paper web) and that it should recover quickly and fully.

In 1964, Wrist published a compilation of compressibility results on various wet fibre webs and woven felts [99]. A comparison between them is possible because both types of materials were found to react to an applied pressure in the same way: the concentration of the mat or felt is proportional to the $n$-th power of the pressure. The value of $n$ is then a measure of the compressibility. For woollen felts it was found to lie in the range 0.194—0.22, if the pressure is expressed in c.g.s. units; for webs of pulp fibres the range was 0.292—0.457, and the values were independent of the degree of beating. Wrist concludes that "if these values are generally typical . . . and if the same relationship is also true under the conditions existing in the press nip, one of Nissan's assumptions is invalid".

These compressibility results were obtained under static experimental conditions, however, and the question how the compression occurs under the dynamic conditions in a press nip has not been answered yet. Turner et al. [100] observed the time necessary to reach compression equilibrium. They found that the equilibrium web concentration-pressure relationship is independent of the degree of refining but that the time to reach equilibrium with similar pulps increases with progressive refining " in the same manner as their respective drainage times vary". This must be a very important factor in wet pressing where the total pressing time in the nip even at moderate machine speeds is of the order of only a few milliseconds. Only very few high-speed dynamic compression experiments have been reported, the first probably by Asklöf et al. in 1964 [101]. They found that, although the exponent of the pressure-concentration relationship was not affected by the compression speed, the fibre/water ratio reached at a given dynamic pressure fell by a factor of about 3.5 when the compression speed was increased by a factor of 50, from 0.1 to 5 cm $sec^{-1}$. Increasing the degree of refining from 14 to 30° SR reduced the fibre/water ratio at a given compression speed by about the same factor. Wahlstrom [102] confirmed these two effects qualitatively in machine experiments. The solids content of the web after wet pressing fell from 35 to 30% when the degree of beating was increased from 19 to 30°SR. Another set of his results shows that the solids content of the web increases for different bleached pulps in the order spruce sulphite — pine sulphate — birch — eucalyptus, the latter two being viscose-grade pulps. The solids content is reduced by 1—2 absolute percent for every 10 m $min^{-1}$ increase of the machine speed. This rather large effect, however, was found at low machine speeds, from 40 to 120 m $min^{-1}$. It is much smaller at high machine speeds, typically 0.24—0.35 absolute percent when the speed of a newsprint machine was increased from 400 to 900 m $min^{-1}$ [102].

So far, it does not seem to have been possible to separate the effects of machine speed and Schopper-Riegler wetness on the consolidation in wet presses quantitatively into purely hydrodynamic factors and specific contributions made by the cellulose-water interaction. The same applies to another phenomenon in wet pressing that has caused a considerable controversy, namely, rewetting of the web by the felt on leaving the press nip. This is not a case of hydrostatic equilibrium not being reached because of insufficient time, but of a supposed dynamic capillary flow process. As was said before, Nissan [98] speculated that on leaving the press nip, the expanding felt picks up water from the paper web. Prima facie this is not very likely because the pores in the felt (within the yarn as well as between the strands) are larger than the pores in the web. It was Wahlström's contention [103] that the paper web gains the upper hand over the felt in the competition for the water leaving the press nip with the two and that the web is rewetted. Although such a mechanism is more likely than the opposite one no convincing experimental evidence of it existed at the time. Wahlström based his statement on work by Sweet, which was published a year later [104]. He passed rectangular sandwiches of water-saturated felt and paper through a press nip and determined the moisture content of the paper as a function of the basis weight. The paper was wet-strengthened tissue of 17.5 g $m^{-2}$. The basis weight of the paper web was varied by placing several sheets on top of each other. The result was that, typically, the moisture content of the web made up of two sheets was about 175% and decreased to below 90% when the web was made up of 30 sheets. This was interpreted as showing a transfer of water from the felt to the paper. It is doubtful for two reasons, however, that these experiments are typical of what happens in the press nip of a paper machine. Firstly, a stack of rewetted wet-strengthened tissue is not comparable to a wet web of the same basis weight as far as the movement of water from it or into it is concerned. This is because the interaction between fibres and water is considerably changed when the fibres are dried and rewetted, as will be discussed at the end of the chapter. Secondly, the speed in these experiments was 15 m $min^{-1}$ (and presumably web and felt were separated by hand after emerging from the nip). Therefore, the time element which must be important for any movement of water in the expanding web-felt sandwich is rather different from that on a paper machine even at moderate speeds. One observation of rewetting in a suction press running at about 250 m $min^{-1}$ was reported by Morrison [105]. "By separating sheet from felt immediately, instead of allowing the sheet to stay with the felt for approximately 6 ft., a 1—2% increase in sheet dryness was realized."

The fluctuating fortunes of the rewetting concept took another turn when Heller and Tewksbury in 1972 published results of experiments on a laboratory wet press running at a speed of 1.8 m $min^{-1}$ [106]. Their experiments were of the same type as those by Sweet [104]. A typical result was that sheets of about 100 g $m^{-2}$ entering the press with a solids content of 20% emerged from it with one of 70% at a line pressure of about 90 kg $cm^{-1}$. They concluded that rewetting could not have occurred since such a time-dependent effect would have reached, at only 1.8 m $min^{-1}$, an order of magnitude which would have made it impossible to reach 70% solids. These results and the conclusion drawn from them were discussed by Schiel [107]. He starts by calculating what nip length would be required if the dewatering effect observed by Heller and Tewksbury (H&T) were to be achieved on a production machine at 500 m $min^{-1}$. The nip length in the H&T experiments was estimated from an empirical formula * to have been 21.2 mm (at 1.8 m min$^{-1}$). At 500 m min$^{-1}$, it would have to be increased to 21.2 × 500/1.8 = 5900 nm, which would require press rolls of 11.6 km diameter! Schiel then proceeds to reason why no rewetting will be observed in a laboratory press running at 1.8 m min$^{-1}$ but must be expected on production machines. His argument is based on experiments on a fabric press run with a felt but without paper at different speeds, line pressures and felt moistures. The water transferred to the microroc roll (on the paper side of the felt) was doctored off with a well-fitting blade and its volume was measured. The result was that above a certain solids content of the felt, typically 55—60%, the transfer of water to the roll (considered to represent rewetting) becomes negligible. The higher the machine speed the lower is the solids content of the felt at which this happens. The line pressure has little influence on the relationship between felt solids content and "rewetting". The water transfer to the solid roll is taken to indicate that no competitive capillary suction of felt and paper web is required to explain rewetting. In other words, "rewetting" is simply splitting of the water film between felt and paper, a mechanism already suggested by Wrist [99], and independent of any particular fibre-water interaction. The argument then is that, in an actual production situation, large quantities of water flow from the web into the felt so that the felt solids content is never increased to a level at which rewetting (of the paper this time) becomes negligible — in contrast to the conditions during the H&T experiment where "the felt was so much compressed that no rewetting water could leave the press".

At the Water Removal Conference, London, 1975, Heller et al. [108] summed up the situation as follows: "Until some means is devised to measure the actual moisture content in the paper web in the mid-nip region, the existence of a rewetting phase in only conjecture". In spite of very considerable engineering progress (to mention only suction, fabric and grooved presses) a similar state of uncertainty characterises our knowledge of the precise role of fibre-water interactions in this phase of the consolidation process.

### *1.4.2 Drying*

The overall characteristics of the paper (its "type" or "grade") are determined by the choice of the furnish and the stock preparation; but the consolidation process on the paper machine offers the papermaker a considerable scope for "fine adjustments" which can influence the properties and performance of the product to such a degree as to make the difference

* Nip length = $\sqrt{8hr_e}$, where $h$ is the felt compression in the centre of the nip and $r_e$ is the "effective radius" given by $r_1r_2/(r_1 + r_2)$. $r_1$ and $r_2$ are the radii of the two press rolls.

between acceptance and rejection by the customer. The drying of the web is that phase of the consolidation process which is particularly sensitive to cellulose-water interactions.

Table 1.1 demonstrates the commercial side of the cellulose-water relationship. It shows the water removal sequence on a fairly large paper machine making a duplicator-type paper at 400 m $min^{-1}$. 99.2% of the total water removal are achieved by mechanical means and only 0.8% require heat, but the latter costs over ten times as much as the former. The figures will vary in detail between machines and will depend on fuel prices, but the overall picture is typical and illustrates the strong bond between water and cellulose at low moisture and how expensive it is to drive the water off at this stage.

Both the high cost of drying and its potential as a control device for the papermaker have attracted — and still do — considerable research. Much if not most of the investigational work on drying was conceived, at least in the earlier days, as an exercise in chemical engineering: The web entering the drying section was just another wet material from which the water was to be removed by heat. It was important to establish whether the conventional phenomenological description of drying curves could be applied to paper. An early example is the work by Sherwood et al. [109] who compared laboratory air drying of a sheet of pulp with machine drying of a 75 g $m^{-2}$ bond paper on 28 drying cylinders at a machine speed of 120 m $min^{-1}$. They found in both cases a constant rate drying period followed by a falling rate drying period, so the behaviour of paper appeared normal. It is typical that they comment on the moisture content (60% on dry cellulose) at the critical point, which separates the constant rate from the falling rate drying period, as being "considerably higher than found in drying of clay and granular materials, for which the critical moisture is usually found to be from 5 to 30%, dry basis". An early attempt to quantify the drying process for design purposes was made by Cowan and Cowan in 1937 [110] and it is again typical of these earlier approaches when they mention the "strength and cohesion of the fibre mass . . . and the hygroscopic characteristics of the sheet, governed

TABLE 1.1

Water removal on a paper machine

| Position | Solids (%) | kg water per kg solids | % total water removed | Cost per ton dry paper |
|---|---|---|---|---|
| Breastbox | 0.5 | 199 | 0 | |
| End of foil section | 5.0 | 19 | 90.5 | |
| Couch | 20 | 4 | 98.0 | £1.17 |
| End of press section | 40 | 1.5 | 99.2 | |
| Reel-up | 94 | 0.064 | 100 | £12.45 |

in turn by the freeness, the percentage of sulphite pulp *, and the formation of the sheet" as controlling the maximum rate at which steam can escape (which controls the drying rate), and then say that "this is largely a paper-making, not an engineering problem" but that "ignorance of these factors does not affect the validity of the equations given in the paper".

Equations are then given. They continued to be given in increasing numbers and complexity by subsequent authors. Nissan presented a very elaborate speculation in 1956 [111], and another set of equations was published by Han in 1964 [112] who reviewed the subject in 1970 [113]. These rather ambitious attempts were based on the belief that by writing down the classical equations of heat and mass transfer, conservation and continuity, the drying process of paper can be described quantitatively by adjusting the numerous constants and coefficients, such as the specific heat, the diffusivity, the porosity and pore-shape factor, the tortuosity factor, the heat conductivity, etc. It is the method traditionally favoured by the chemical engineer who is more interested in extracting practical design data from observations on models or real systems than in detailed understanding. It can be quite successful in cases where the interaction between water and the material to be dried is not very strong, as in the drying of plastics, or where the heat and mass transfer is largely controlled by aerodynamical factors, as in impingement or spray drying, or even in through-drying as will be briefly discussed below. In systems with a strong solids-water interaction the number of variables increases considerably, with the result that observations on one particular system cannot so easily be generalised, and that their predictive value is much reduced.

The observation of a constant and a falling rate drying period in paper and the critical moisture content is an example. According to Sherwood et al. the two drying periods are clearly separated and the critical moisture content is 60% on dry cellulose, as was mentioned before. Attwood and Smith [114] and Smith and Attwood [115] carried out very carefully designed experiments on a laboratory simulator of cylinder drying and came to the conclusion that there is no constant rate drying period, "a falling rate being apparent from the outset". Nissan [111] *assumes* the critical moisture content $W_c$ to be 40% on dry cellulose and the drying rate in the falling rate period to be $W/W_c$ times the constant drying rate, where $W$ is the moisture content. Dreshfield [116] concludes from laboratory experiments that the critical moisture content depends on the moisture content at the hot surface and its gradient through the thickness of the sheet, and can vary between 25 and 95% on dry cellulose. Gavelin [117] describes a typical drying curve for water to consist of a constant rate period down to about 80% moisture content *on wet basis*, followed by three distinct falling rate periods with increasing

* These authors consider newsprint machines.

slopes. Some of these differences of opinion are probably due to different and not comparable experimental or observational conditions, but there can be no doubt that the fibre-water interaction introduces a variability into the process of hot surface drying which makes any generalisations extremely difficult.

This is even more evident when the role of the felt in paper drying is considered. Sherwood et al. [109] conclude that 83% of the water are removed by the felt and only 17% evaporate in the draws between the felts. Cowan and Cowan [110] by contrast claim that the cylinder merely heats the web and that most of the water evaporates between cylinders. Smith and Attwood [115] found that "evaporation between cylinders makes only a minor contribution to the drying rate". Nissan's speculations are so general that they can easily accommodate anything that may happen. In 1956 [111], the felt actually reduced the drying rate on a newsprint machine and nearly all the drying occurred in the draws. In 1957, however [118], the felt was all-important on a machine making tracing paper, so important in fact, that a liquid-phase transfer mechanism was assumed to explain the high drying rate achieved by the felt. This was in turn rejected by Kirk et al. in 1963 [119] who claimed that only a vapour-phase mechanism of moisture transfer through the felt was operative.

Now it must be emphasised that the contradicting pronouncements quoted in the last two paragraphs were all based on observations and reasoned argument. Apart from an occasional optimism in adopting simplifications, there are no grounds for the assumption that these authors did anything but conscientiously interpret and report their findings. The conclusion must be that cellulose-water or fibre-water interactions introduce a new variability into a process which is complicated enough without such interactions. The variability is probably greater than is generally appreciated and one result is that the drying of the paper web on a Fourdrinier paper machine has so far stubbornly refused to be boiled down to a mathematical problem of solving a set of simultaneous differential equations. Careful experimentation is perhaps a more useful and realistic method to improve our understanding and through it, our mastery, of the process.

It was mentioned earlier that through-drying of paper is less beset by fibre-water interactions and therefore more amenable to a numerical description. One of the first analyses of systematic laboratory experiments was published by Ju Chin Chu and Weng Ling Kuo [120]. They relate the constant rate period and the subsequent two falling rate periods to the porous structure of the paper. In the constant rate period, the water is removed from the surface of the fibres and the larger pores between the fibres by a two-phase laminar flow. During the first falling rate period, the water is removed from the smaller pores between fibres and the larger pores (lumina) in the fibres, controlled by capillary forces; and, during the second falling rate period, it is removed from the pores of the cell wall, controlled by diffusion. The three ranges of pore sizes were determined by the mercury injection method. The fractional void volumes of the three ranges determine the critical moistures at the transitions between the constant rate and the falling rate periods. The authors achieved a remarkably good description of the observed rates at air flow-rates varying from 0.07 to 0.3 kg $m^{-2}$ $sec^{-1}$ and at 65 and 82°C.

The second falling rate drying period is of little practical importance as Martin and Corte pointed out [121], since the moisture in this phase is very low and paper is usually dried to somewhere near the end of the first falling rate period. Schlünder therefore only considers two types of pores: those between and those within the fibres [122]. He characterises his chemical engineering approach to the process of through-drying as follows: "In the past, many attempts have been made to predict drying rate diagrams from some fundamental physical properties of the product. Today these attempts have been given up since the experimental determination of a whole drying rate diagram is much faster and cheaper than the determination of the physical properties, not mentioning the enormous difficulties arising with the set-up of a respective theory for the prediction of drying rate diagrams based on these physical properties". He then describes laboratory equipment with which precise measurements of the drying rate can be made even for very fast drying processes (1 second and less). Typical is the dimensionless and normalised presentation of drying rate diagrams of which Fig. 1.16 shows an

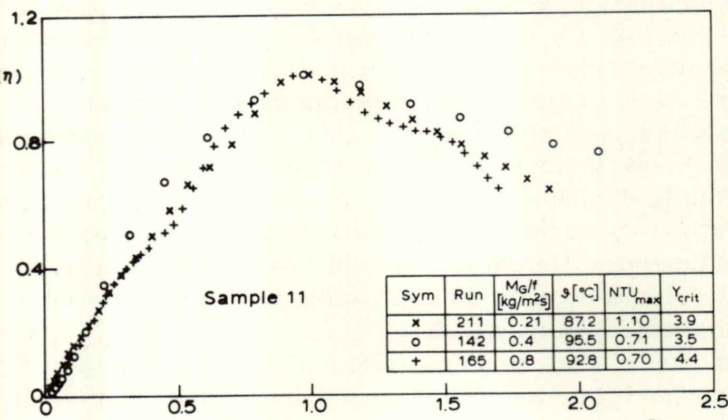


| Sym | Run | $M_G/f$ [kg/m²s] | $\vartheta$[°C] | $NTU_{max}$ | $Y_{crit}$ |
|---|---|---|---|---|---|
| × | 211 | 0.21 | 87.2 | 1.10 | 3.9 |
| o | 142 | 0.4 | 95.5 | 0.71 | 3.5 |
| + | 165 | 0.8 | 92.8 | 0.70 | 4.4 |

Fig. 1.16. Normalised drying rate as a function of the normalised moisture content for tissue paper at various flow rates and 90°C air temperature (after Schlünder [122]).

$$\dot{\nu}(\eta) = \frac{\ln[(X^* - X_{in})/(X^* - X_{out})]}{\ln[(X^* - X_m)/(X^* - X_{out,\,max})]}$$

(normalised drying rate), where $X$ = air humidity and $X^*$ = wet bulb humidity.
$\eta = (Y - Y_{hygr})/(Y_{crit} - Y_{hygr})$ (normalised product moisture content), where $Y$ is the product moisture content, "hygr" refers to the equilibrium value and "crit" to the value at the maximum drying rate.

example for tissue paper at 90°C and at air flow rates of 0.21, 0.4 and 0.8 kg $m^{-2}$ $sec^{-1}$. The normalised drying rate — moisture content function $\dot{\nu}(\eta)$ is seen to be (for a given temperature) almost independent of the air flow and is therefore also independent of the relative humidity of the air. "This is important . . . for the purpose of extrapolation and for the application in dryer design and scale-up" (Schlünder).

The two main reasons why such a degree of generality can be achieved in through-drying are, firstly, the fact that heat and mass transfer coincide in space and time whereas in cylinder drying they are more or less separated between the cylinder mantles and the draws. Secondly, the structure of the material is so open that the problem can largely be reduced to one in geometry and aerodynamics. Special cellulose-water interactions, apart from the moisture — relative humidity relationship, have hardly any influence on the shape of the normalised drying rate diagrams. In other words, the through-drying process is little "material-specific". According to Schlünder [122] the diagram in Fig. 1.16 is very similar to one for a woollen textile of similar porous structure. Conversely, fibre-water interactions become more decisive in compact web structures drying under less well-defined conditions in the drying section of a Fourdrinier machine.

One feature of the fibre-water relationship which contributes probably more than any other to the complicated nature of the drying process is the way in which water is actually accommodated in the web, that is, between and within the fibres. The latter will be considered first. How the water resides inside the fibres of a wet web is obviously related to their morphological structure. This fact alone makes for a considerable variety of situations and must reduce the hope of easy generalisations. It attracted the attention of morphologists and cellulose chemists long before papermakers took much notice of it. At first the question of how liquid water is held by a fibre was not at all clearly formulated. The feeling was that cellulose has a specific affinity for water and a number of attempts were made to characterise this affinity numerically. For example, Champetier [123], adopting a method developed by Schreinemakers for the determination of the composition of double salts [124], soaked cellulose fibres in fairly concentrated (typically 20—40%) aqueous solutions of sodium thiosulphate. The solution becomes more concentrated because the cellulose has a greater affinity to the water than to the thiosulphate *. It is squeezed out of the cellulose by an increasing pressure, and analysed. The mole ratio thiosulphate/cellulose is plotted against the mole ratio water/cellulose. Figure 1.17 shows one of Champetier's plots for thiosulphate solutions of three different concentrations. The straight lines meet at a point on the abscissa which is taken to indicate that

* The preferred uptake of water from electrolyte solutions by cellulose fibres was first observed by König in 1882 [125].

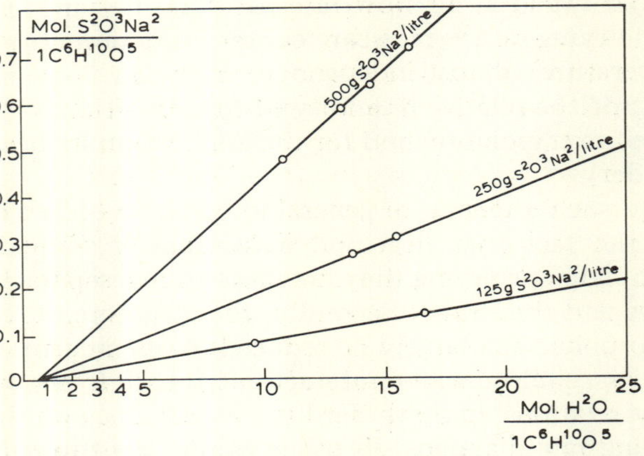


Fig. 1.17. Thiosulphate/cellulose ratio of solutions expressed from soaked cellulose pads at increasing pressures as a function of the water/cellulose ratio in the pad for different thiosulphate concentrations (after Champetier [123]).

0.5 moles of water per mole of glucose are not available as solvent for $Na_2S_2O_3$. The method was further developed by Tankard [126] who obtained similar results and conjectured that about 0.5 moles of water per mole of glucose (i.e. 5.5%) are "entering into chemical combination". Hermans is more specific and postulates the existence of two stoichiometric cellulose hydrates I and II with 3.7% ($\frac{1}{3}$ mole) and 11.1% ($1\frac{1}{3}$ mole) of water, respectively. These are formed in the "amorphous" portion of the cellulose which for native fibres amounts to 40%. Thus a total of 0.4 (3.7 + 11.1) = 5.9% of water is bonded as "hydrate water" [127]. This will be returned to later in connection with the adsorption of water *vapour* on cellulose. In the present discussion, we are concerned with *liquid* water.

Jayme [128] applied a centrifuge method used in the textile industry to paper and called the amount of water held by a given weight of cellulose under standardised conditions of centrifugation the "water retention value" (WRV from <u>W</u>asser<u>r</u>ückhalte<u>v</u>ermögen). Another possibility is to mix water-swollen cellulose fibres with a solvent miscible with water, such as acetone or dioxane, and to determine with the aid of the dielectric constant how much water the cellulose has lost to the solvent [41]. Matthes [129] compressed pads of swollen cellulose fibres in a hydraulic press in order to determine the amount of water left in the pad after extrapolation to infinite pressure; and finally, LeCompte [130] suggested to compare the Schopper-Riegler wetness values of pulps before and after treatment with boiling water. All these methods, and others whose enumeration would go too far, are aimed at determining a certain amount or portion of the water in contact with cellulose fibres that is somehow distinguished from the rest. But what this distinction is cannot be deduced from any of the methods, which did not prevent some of them, particularly the determination of the water retention value, to gain a certain measure of acceptance for comparison purposes [131].

Gradually, the investigational methods became more sophisticated and specific in the sense that the question they were designed to answer was not so much "what fraction of the water in contact with cellulose fibres is distinguished from the rest?", but rather "where does the water go?". One approach was to make the hydroxyl groups of the cellulose react with some added chemical. The extent of the reaction was then taken to indicate the "accessibility" of the sample. The method by Assaf et al. [132] is based on the reaction with thallous ethylate in organic solvents. It was further developed by Minhas and Robertson [133] who treated cotton fibres in a number of liquids which were then solvent exchanged to benzene where they were made to react with thallous ethylate. The thallium cellulosate was reacted with methyl iodide and the methoxyl content of the cellulose was determined by the TAPPI Method T209 M45. From this, the percentage of substituted hydroxyl groups was calculated. It varies with the liquid in which the cellulose was treated, typically from 0.84% for cyclohexane to 16% for ethylenediamine. For water, the value is 13.6%. The authors discuss at some length the obvious question whether the rather complicated procedure, particularly the solvent exchange, leaves the structure of the fibres unchanged, and they come to the conclusion that it does and that in particular the swelling of cellulose in benzene is negligible. This, however, is difficult to reconcile with the finding in the same article that the surface area calculated from the accessibility to thallous ethylate in benzene is 16.5 $m^2$ $g^{-1}$ compared with only 1.3 $m^2$ $g^{-1}$ obtained from the nitrogen adsorption on the fibres after immersion in benzene. Frilette et al. [134] used heavy water as the reactant and observed a very rapid reaction followed by a much slower one. The extent of the rapid reaction is taken as a measure of the accessibility. For cotton the value is 21% and for woodpulp it is 46%. The exchange reaction with tritiated liquid water was used by Sumi et al. [135] to determine the accessibility of holocellulose from spruce and birch. The values for samples that had been dried after delignification with peracetic acid (yield 83%) were 64.6% (spruce) and 53.8% (birch). Holocellulose is of course not the same as woodpulp, but the interesting point is that the values for birch and spruce are significantly different, and this was consistently so for other preparations from the two wood species, each pair having been subjected to identical treatments.

An original method to determine accessibility was published by Neal et al. [136]. It is based on comparing the apparent heat expansion of cellulose with that of glucose in water, assuming that glucose hydroxyl groups interact with water in the same manner as cellulose hydroxyl groups which are fully accessible to water. Typical values for the accessibility were: Spruce holocellulose — 38%; mixed hardwood sulphite pulp — 34%; spruce sulphite pulp —

22%; cotton — 15%. In contrast to the results quoted above, the accessibility for spruce pulp is lower than for hardwood pulp (it is not stated whether the latter contained birch). The authors reason that "the presence of a substantial quantity of hydrophilic hemicellulose in the spruce holocellulose undoubtedly accounts for its accessibility being higher than the other pulps". Sumi et al. in the previously quoted article [135] reason that "one possible way of accounting for the greater stability and lower accessibility of the birch carbohydrates is to assume that there is a difference in the accessibility of the respective hemicellulose fractions".

The methods outlined above by no means exhaust the subject of accessibility studies. A review of the literature on the subject up to 1950 was presented by Tarkow [137] who discusses particularly the confusion created by juxtaposing the pair accessible—inaccessible with the pair amorphous—crystalline. A more recent contribution to this particular controversy was published by Rånby in 1964 [138], and the latest review of the subject is by Rowland [139].

One method that must be mentioned in the present context is the "solute exclusion technique" introduced by Stone and Scallan [140,141]. It is a modification and considerable extension of the "non-solvent water" method mentioned earlier [123] except that carbohydrates of different molecular weights are used as solutes instead of an electrolyte. Morehead had found in 1947 [142] that dextran molecules of sufficiently high molecular weight cannot enter the pores in the cell wall. Stone and Scallan used narrow dextran fractions with molecular weights ranging from 1400 to $24 \times 10^6$, corresponding to molecular diameters from 20 to 1600 Å *. At the low molecular end, the dextrans were supplemented by a mono-, a di-, a tri- and a tetrasaccharide. Unlike Champetier's procedure, the solutions were not used to soak dry pulp samples. Instead, they were added to pulp suspensions of about 20% consistency. The concentration of the supernatant liquid was compared with that which would have been obtained if all the water had been available for dilution. The difference increases with the molecular weight of the solute and represents the inaccessible water or the cumulative pore volume. Figure 1.18 is an example for spruce wood and kraft pulps with different yields made from it. The curves show that dextran molecules larger than 140 Å in diameter (molecular weight ca. 100,000) are excluded from the fibre wall. The inaccessible water at the plateaus is the total amount of water (per gram of dry cellulose) which the fibre takes up when immersed in liquid water. It is said to be the "fibre saturation point". The fibre saturation point increases with increasing removal of wood components and is seen to be much higher than the value of about 0.3 g $H_2O$/g cellulose that had previ-

* The molecules were assumed to behave in solution like hydrodynamic spheres and their diameters were calculated from their diffusion coefficients.

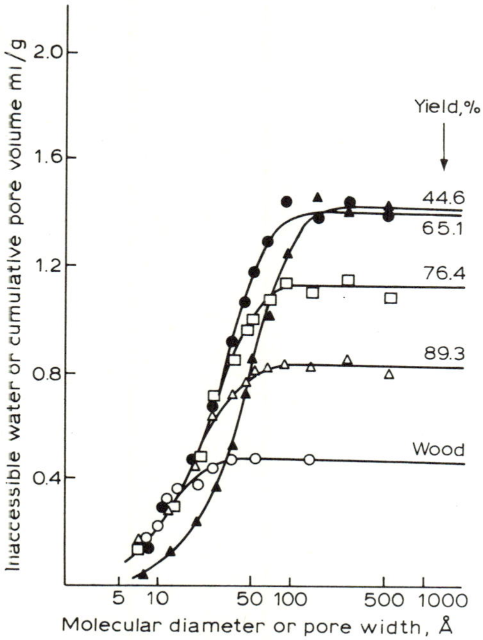


Fig. 1.18. Accessibility curves of spruce wood and kraft pulps of different yields (after Stone and Scallan [141]).

ously been estimated from adsorption isotherms of water vapour (see last section). Stone and Scallan confirmed this high value by an independent method. This is the so-called porous plate method, introduced by Barkas and Hallan [143] and extensively used by Robertson [144]. It permits the removal of water from a wet fibre mat by an increasing vacuum and thus the determination of the moisture in the pad in equilibrium with very high relative humidities in the 99.00—99.99% range. Figure 1.19 is taken from the article by Stone and Scallan [140]. The curves above the one for wood (spruce) belong to kraft pulps with decreasing yield from 92.4% (K20) to 48.7% (K125). The fibre saturation points are the water retention values at the near-horizontal parts of the curves which indicate the phase of the desorption process between the emptying of the smallest lumina and the emptying of the pores in the cell wall. They agree within a few percent with the values obtained from the solute exclusion method.

The authors investigated in a number of articles how the water is actually embedded in the structure of the cell wall and came to the conclusion that the cell wall is delaminated into several hundred lamellae concentric with the fibre axis and about 100 Å thick, and that the water fills the annular gaps

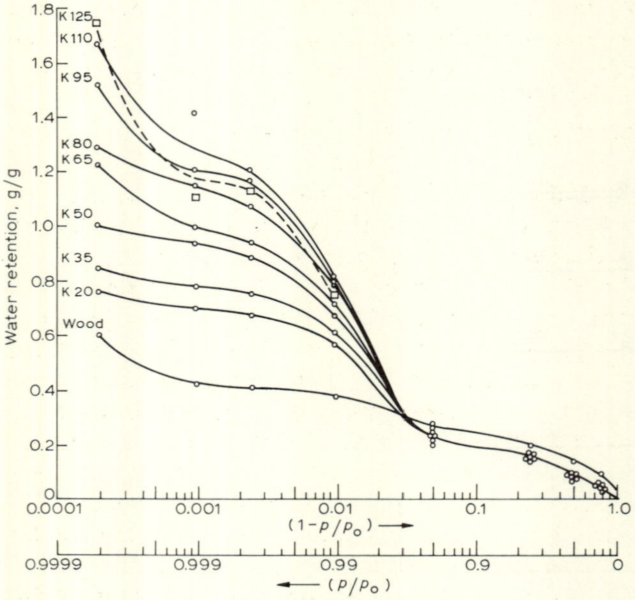


Fig. 1.19. Water desorption isotherms of spruce wood and kraft cooked microtome cross-sections using the porous plate technique (after Stone and Scallan [140]). 100 $p/p_0$ = relative humidity.

between them. The average (median) spacing is given as 35 Å. When the values are plotted on logarithmic probability paper, the pore size distribution by volume is seen to be very nearly lognormal — a rather surprising result. When the fibre dries from water, the annular gaps close, the cell wall becomes virtually non-porous and the specific surface of typically 1 $m^2$ $g^{-1}$ as measured by the BET nitrogen adsorption method is almost the same as that calculated from the outer fibre dimensions. The method used to remove the water without closing the gaps was to replace it by methanol and the methanol subsequently by pentane which was finally evaporated in a stream of dry nitrogen. The BET specific surface values were then typically 180 $m^2$ $g^{-1}$ [50]. The pore sizes, i.e. the average separation between the lamellae, were calculated from the nitrogen adsorption isotherms. The argument was supported by electron micrographs of very thin cross-sections of fibres. Although they did not show exactly the pore sizes predicted from the adsorption isotherm (according to the authors this is due to distortions caused by the embedding technique), they did show the lamellar structure of the swollen cell wall and the existence of slit-like pores with a certain wall separation. The picture was later slightly modified and refined in the sense that the lamellae are not completely separated to form concentric annular gaps but are still bonded to each other along the periphery at more or less regular intervals, so that lenticular openings are formed. A recent review was presented by Scallan at the Oxford Fundamental Research Symposium 1977 [145].

These remarkable contributions to our knowledge of the interaction between pulp fibres and liquid water may not answer all the questions one may wish to ask but they certainly throw new light on a number of observed phenomena and questions arising from them. For example, why do the various methods outlined above to determine the accessibility of, or some distinguished portion of the water in contact with, cellulose fibres produce so different and non-comparable values? The answer seems to be that each method employing an extraneous chemical as a tracer, be it sodium thiosulphate or thallous ethylate or heavy water, sets up a different ternary system. The information it yields is about that particular system whose behaviour is controlled by the interaction between all three partners, and not about the binary cellulose/water system alone. Methods which use physical criteria instead of chemical indicators, such as heat expansion or centrifugation, rest each on their own built-in assumptions which can hardly be expected to refer to comparable situations. It is true that the solute exclusion method, too, employs a ternary system to explore the properties of the binary fibre/water system, and one very important condition for this method to yield realistic results is the absence of interaction between cellulose and dextran, meaning adsorption of dextran molecules on cellulose surfaces. This, on the face of it, seems rather unlikely, but Stone and Scallan [141] state clearly that the results are not influenced by the concentration of the dextran solution, indicating the absence of adsorption effects. If this is accepted the method must be considered to provide the most realistic description at present available of the geometry of swollen fibres suspended in water. The question "what water is distinguished from the rest?" asked in any of the other methods is reduced to "what range of pore sizes is the method able to "see"?".

Another result of this investigational technique was that it confirmed conclusively Emerton's concepts of internal fibrillation and swelling of the cell wall [48], referred to on pp. 17/18 nearly ten years after they had been published. In particular, the parallelism between fibre swelling during beating and the developemnt of the tensile strength, which had previously been established from measurements of the specific volume of beaten fibres (see Fig. 1.10), was found to be reflected in the effect of beating on the fibre saturation point as determined by the solute exclusion technique [145,146]. An added effect of the delamination of the cell wall is the fact that the lamellae can move against each other with relative ease. The fibres become more flexible and "conformable" with progressive beating and this, together with effects to be discussed in the following sub-section contributes to the similarity between the developments of the tensile strength and of fibre swelling, internal or external, during beating. It is here where the papermaker's interest in the porous structure of the cell wall lies.

Finally, the distribution of the water between the interlamellar spaces in the cell wall has an immediate bearing on the subject of this sub-section, drying. Figure 1.19 shows that, with the drying process beginning at a water retention value of about 1.75 (K125), which corresponds to 36% solids content of the mat, the water begins to leave the cell wall at a relative humidity of 99.9% when the mat has a solids content of 45%. Although this is under equilibrium conditions which are certainly not realised in machine-drying, it is very likely that the water begins to leave the cell walls as well as the interfibre spaces at an early stage during the drying process. If, as Stone and Scallan suggest, the lamellae themselves are virtually non-porous the water must leave the interlamellar gaps in the axial direction. This means diffusion paths which are much longer than the thickness of the webs and the water-filled interstices in it. They must therefore have an overriding influence on the drying rate of the web. Moreover, the length (and width) of these diffusion paths through which most of the water removed in the drying section has to escape are determined, for a given degree of chemical and mechanical treatment, solely by the morphology of the fibres. They vary not only from fibre to fibre but, more important, from species to species. They vary with the degree of chemical treatment as Figs. 1.18 and 1.19 show; and they vary with the degree of mechanical treatment as was shown by Stone et al. in 1968 [146]. This variability must be the main reason why global generalisation attempts of the drying process in the paper machine by some mathematical formalism have so far proved to be rather divorced from reality and therefore of so little predictive value.

### *1.4.3 Bonding*

Undoubtedly the most important change suffered by the fibre web in the drying section is not that it loses water but that it bonds together to form paper. For this to happen the water must leave the spaces between the fibres. The necessary external condition for bonding to occur is indeed the existence of very many very small interstices between the fibres which are filled with water and exposed to an atmosphere that is not saturated with water vapour. Under these conditions a pressure develops in the water which is approximately given by

$$p = \frac{RT}{vM} \ln(P/P_s) \,, \tag{1.2}$$

where $R$ is the gas constant, $T$ the absolute temperature, $v$ the specific volume of water, $M$ its molecular weight, $P$ the vapour pressure and $P_s$ the saturation pressure. As $P < P_s$ the pressure $p$ is negative, that is, the water is under tension: it evaporates. If the water forms a meniscus between two fibres this tension tries to pull the fibres together to a distance $x$ given by the equilibrium between the tension $p$ and the capillary pressure $2\gamma \cos \theta / x$, where $\gamma$ is the surface tension of the water and $\theta$ the contact angle between water and the fibres. Therefore in equilibrium,

$$-x = \frac{2vM\gamma \cos \theta}{RT \ln(P/P_s)} \,. \qquad (1.3)$$

Equation (1.2) is not quite exact since $v$ is, strictly speaking, that increase in volume which a meniscus spread over a very large area suffers upon addition of 1 g water at constant $p$. Since the meniscus is under tension ($p$ is negative) the value is greater than the specific volume of liquid water. Also, $p$ is the tension for constant $v$ and therefore varies when the mensicus becomes thinner. The equation which goes back to Thomson [147] was extensively discussed by Porter [148] and by Barkas [149] for aqueous solutions and by Barkas [149] in addition for aqueous gels. For water, the deviations from eqn (1.2) are negligible.

Table 1.2 gives values for $p$ and $x$ for two temperatures at different relative humidities, assuming zero contact angle ($\cos \theta = 1$). At low relative humidities the tension is seen to be quite high and the equilibrium distances so small that eqns (1.2) and (1.3) for calculating forces and distances begin to lose their applicability and molecular forces and dimensions take over, so to speak. Campbell introduced this fundamental fact to the papermaking world

TABLE 1.2

Capillary tensions and equilibrium distances between two flat surfaces at two temperatures and various relative vapour pressures

| $P/P_s$ | 20°C | | 100°C | |
|---|---|---|---|---|
| | Capillary tension (kg cm$^{-2}$) | Distance (cm × 10$^6$) | Capillary tension (kg cm$^{-2}$) | Distance (cm × 10$^6$) |
| 0.999 | 1.27 | 116 | 1.62 | 91.7 |
| 0.99 | 13.99 | 10.6 | 17.81 | 8.33 |
| 0.95 | 70.90 | 2.09 | 90.26 | 1.64 |
| 0.90 | 145.6 | 1.02 | 185.4 | 0.801 |
| 0.80 | 308.1 | 0.482 | 392.2 | 0.378 |
| 0.70 | 492.5 | 0.308 | 627.0 | 0.237 |
| 0.60 | 705.2 | 0.210 | 897.8 | 0.165 |
| 0.50 | 755.8 | 0.155 | 1217 | 0.122 |
| 0.40 | 1265 | 0.117 | 1611 | 0.0922 |
| 0.30 | 1663 | 0.0892 | 2117 | 0.0701 |
| 0.20 | 2223 | 0.0668 | 2829 | 0.0525 |
| 0.10 | 3180 | 0.0467 | 4048 | 0.0367 |

in one of the classics of the paper physics literature 45 years ago [78]. If the role of the surface tension in promoting bond formation needed further evidence it was provided by Van den Akker in 1952 [150]. He dried wet fibre mats by sublimation of the water after it had been frozen in a cold box and found strength values that were only 40% of those of sheets dried at room temperature. The equilibrium distance is of course reached only if the fibres are capable of following the pull of the water, that is, if they are soft and pliable. If they are rigid like matchsticks they will stay where they are and the water just evaporates, leaving gaps between the fibres. The all-important fact is that in a drying paper web most fibres will indeed approach their neighbours to distances at which they can bond. The external condition necessary for this to happen, that is a close enough proximity between adjacent fibres, is thus provided by the capillary forces in the drying web. The force is of course present inside the fibres as well as between them. There is therefore a competition between intrafibre closures of gaps between lamellae and interfibre closures of gaps between the outsides of adjacent fibres. The outcome of this competition depends on the proximity of the partners. Inside the fibres the average distance between potentially bonding surfaces is of the order 35 Å as was quoted before. Unbeaten or moderately beaten softwood fibres may in the drying web form ribbons with a thickness of typically 5 microns, which is over a thousand times that distance. They are huge and rigid by comparison and, with undamaged or only moderately damaged surfaces, the odds are very much in favour of intrafibre bonding at the expense of the mechanical coherence, that is, strength of the resulting dry web, such as tissue paper, blotting paper or filter paper, for example. It is the disruption and part-removal of the primary and outer secondary walls of the fibres and the exposure of their inner secondary walls during the beating or refining process, that makes closer approaches between adjacent fibres or parts of them possible and thus tips the balance in favour of interfibre bonding. The increasing ease with which the fibre lamellae can slide over each other, that is, the increasing flexibility of the fibres, works in the same direction. These two occurrences, the loosening of the outside as well as of the inside of the cell walls, are of course the salient features of Emerton's concepts of the beating process [48], as was said before. They were supplemented by a brief statement about the papermaking significance of the outer secondary wall at the Fundamental Research Symposium, Cambridge, 1957 [151]. At the same symposium Giertz [152] presented a number of photomicrographs which clearly show the part-removal of the outer secondary wall and the loosening of the inner secondary wall. More pictorial evidence of the loosening of the fibre structure in the form of a series of electron micrographs was presented by Jayme and Hunger at the Fundamental Research Symposium, Oxford, 1961 [153]. Another impressive series of pictures (phase contrast micrographs of very thin cross-sections of very light-weight papers) was published by Asunmaa and Steenberg [154]. The external and internal loosening of the fibres cause the slippery feel of beaten fibre suspensions, they presumably cause what was once called "hydration" (in the non-stoichiometric sense), and they promote the formation of bonds between the fibres.

The close proximity of adjacent fibres is a necessary but not a sufficient condition for the formation of such bonds. Another condition is that cellulose fibres are chemically suited to bond to each other. In this context it is perhaps worth remembering that at the time when Ts'ai Lun made the first paper in A.D. 105 [155] * silk was used as a writing material. It is quite likely that people first tried to use waste silk for papermaking and did not succeed. It is the presence of hydroxyl groups in cellulose that makes the formation of hydrogen bonds between pulp fibres possible. Hydrogen bonds are now generally accepted as being the agent which holds a sheet of paper together. This has not always been so. An early speculation was that the coherence and strength of paper was simply caused by mechanical entanglement of the fibres. It can easily be shown (for example by solvent-exchanging the water in a wet web to carbon tetrachloride) that this is not so. Some people suspected a glue-like substance of the pectin type in the pulp suspension to be responsible, but found nothing. A review of the earlier literature was given by Clark in 1954 in the article already mentioned [43]; another one, listing various possible bonding mechanisms, was published by Swanson in 1956 [157].

In 1955 Corte and Schaschek [158] published what can be regarded as the experimental proof of hydrogen bonding in paper. They deuterated cellulose samples with water vapour containing six to seven percent $D_2O$ in small successive steps, each time waiting for the equilibrium to be established. Free hydroxyl groups react with a lower equilibrium constant $K = [D]/[H]$ than bonded hydroxyl groups. $K$ therefore increases with progressive deuteration and reaches a plateau value at about 0.05. Figure 1.20 shows a set of their curves. The measured value of $K$ permits the calculation of the degree of deuteration; the degree of deuteration at which the plateau is reached is the percentage of free hydroxyl groups in the sample; and the reduction of this value from 3.80 to 2.05 as fibres are made into sheets with increasing strengths is considered to be the direct proof of hydrogen bonding. The energy of the bond was calculated quantum-mechanically from infrared-spectroscopic data and found to be typically 4.5 kcal mole$^{-1}$ OH. This value was used to calculate the rupture energies of a number of papers. They were found to be of the same order of magnitude as those determined experimentally by extrapolating rupture energies from load/elongation curves, measured at different sample lengths, to zero sample length (fracture zone), but lower than the latter by an amount that increased with increasing rupture energy.

---

* More recent research dates the very first attempts further back [156].

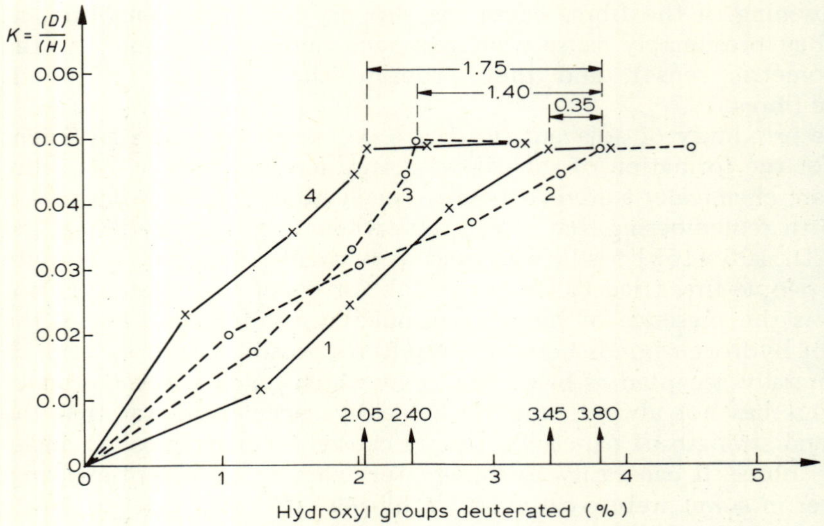


Fig. 1.20. Deuteratation curves of a spruce sulphite pulp sheet (1), single fibres of the pulp beaten to 35°SR (2), paper sheet made from them (3), and paper sheet made from the pulp beaten to 45°SR (4) [166].

At about the same time a number of authors suspected hydrogen bonding in paper from a variety of indirect experimental observations, for example Broughton and Wang (from rupture energies of paper immersed in various liquids) [159], Nordman et al. (light scattering in strained paper) [160], Nissan (molecular approach to viscoelasticity) [161]. Although water plays an essential role in bringing about hydrogen bonding it has so far not been established whether some water molecules are actually part of the bond (not to be confused with bridges of water molecules as depicted in Fig. 1.7). This is at least a possibility and may explain the uncertainty about the amount of water remaining in paper after prolonged conventional drying. It will be returned to in the last section.

At the Fundamental Research Symposium, Oxford, 1961, Page and Tydeman reported a new and important observation [162]. It was the forming of "microcompressions", a kind of small-scale creping, on the contacting surfaces of pairs of crossing fibres in the drying process. Figure 1.21 gives an impression. When a fibre in a wet web dries it shrinks because the water leaving the inter-lamellar spaces pulls the lamellae together by the surface tension mechanism discussed earlier. A fibre on its own shrinks only laterally, typically by 20—30% as observed by Tydeman et al. [163]. The axial shrinkage is negligible. A fibre not on its own but in contact with another fibre which crosses it, say, at a right angle, will begin to be bonded to that fibre by the

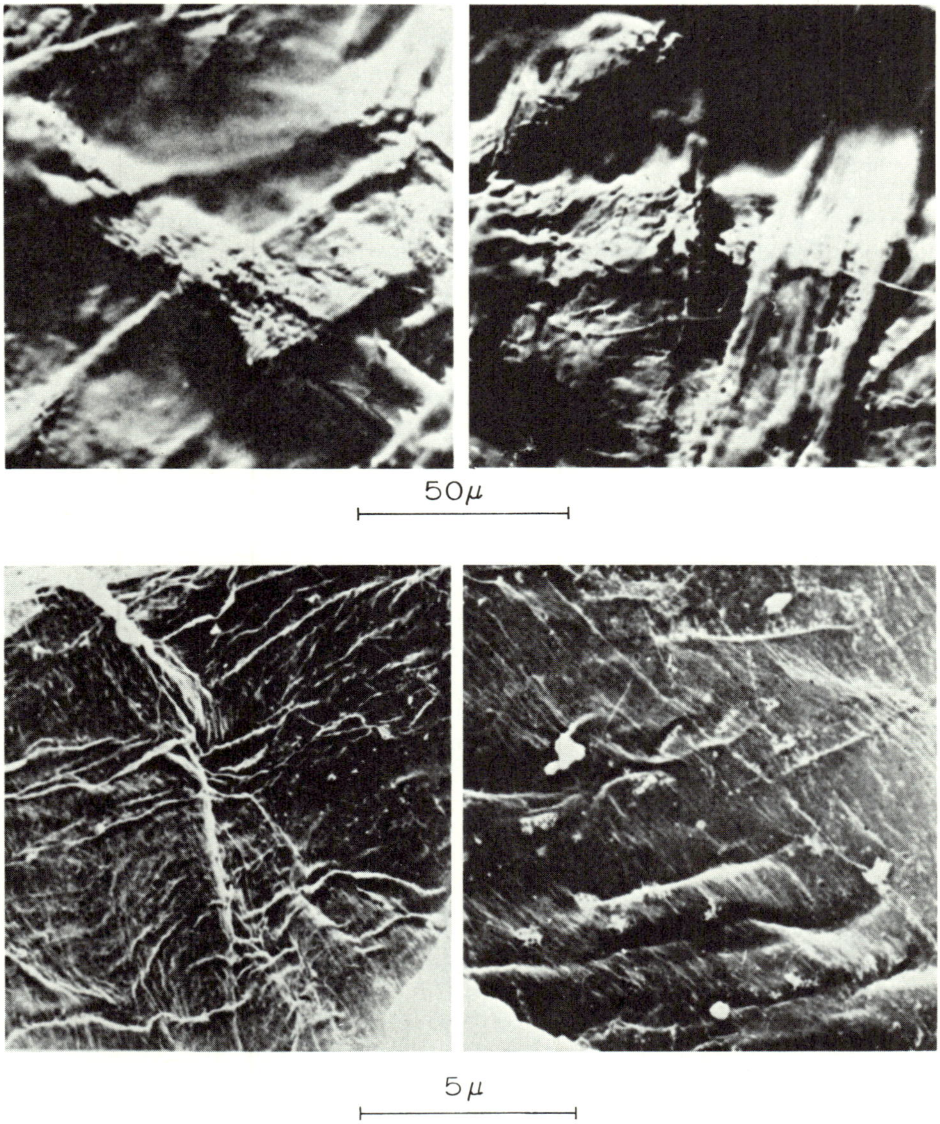


Fig. 1.21. Top: Microcompressions at bonded fibre crossings in a freely dried handsheet. Bottom: Similar features as seen it in the electron microscope (after Page and Tydeman [162]).

same forces which cause it to shrink. It therefore compresses the crossing fibre axially and is itself compressed axially. This is an illustration of the competition between intrafibre and interfibre closures which was mentioned on p. 46. Any one fibre crosses and therefore contributes to the compression of, several other fibres and, looked at the other way, is being compressed by several other crossing fibres. This results in a measurable shortening of the fibres by as much as 12% [162]. One immediate result of this mechanism is that the wet web as a whole shrinks during drying. If allowed to shrink freely the longitudinal shortening of the fibres was found to be the same as the shrinkage of the sheet. If the web is restricted in one direction, as it is in the drying section of a Fourdrinier machine, many more microcompressions running in the machine direction are formed by cross-directional shrinkage of the web than microcompressions running in the cross-direction. Add to this the effect of fibre orientation: More fibres lying in the machine direction cause microcompressions on fibres lying in the cross-direction than vice versa. The result is a pronounced anisotropy of the orientation of microcompressions, and this explains quite satisfactorily the well-known anisotropy of the mechanical properties of machine-made paper the explanation of which had eluded many researchers for such a long time — a fine success of the skilful use of the microscope (and the eyes).

Another result of microcreping at the bond sites is the fact that load-elongation curves of paper for the first few tenths of a percent elongation are not straight but convex with respect to the strain axis, more so for well bonded papers, such as glassine, than for less well bonded papers, such as filter; and always more so in the cross-direction than in the machine direction. This results from the relative ease with which the microcompressions can be "ironed out" and causes mechanical instability in some cases of usage, such as on continuous stationery printing machines, where the external forces and the deformations are of this order. It probably also contributes to the ease with which paper responds to internal swelling forces caused by the adsorption of water vapour: moisture expansion. This will be returned to in the last section.

Corte and Herdman [164] observed that the extent of shrinkage of wet webs depends on their basis weight. Heavier webs shrink more than light webs. They were able to describe the relationship quantitatively with the aid of statistical geometry. This permits the prediction that in a drying web which is clamped in a frame so that the external dimensions remain unchanged, small areas heavier than the average basis weight will shrink and in doing so expand areas with basis weights below the average. The heavier areas become heavier still and the lighter areas lighter. The result is that the variance of the distribution of the mass density in small regions increases. "Frame-dried" webs finish up with a less uniform areal mass distribution than "plate-dried" sheets where the effect of differential areal shrinkage is absent. The effect was confirmed experimentally.

Closing this section, an interesting observation by Broens [165,166] must be discussed that in a web drying under unaxial restraint as in a Fourdrinier machine, bonds are not only formed but also broken (or prevented from being formed). He clamped wet webs in a recording tensile tester with a fixed gauge length and dried them by placing an electric heater some distance in front of them. Figure 1.22 shows the drying tension (top) which in this case develops at a fairly high solids content (bottom). The tension destroys some of the bonds (or prevents bonds from being formed), the more so the longer they act. The area under the ascending part of the tension curve is a measure of the severity and duration of this action. It was varied by placing the heater at various distances from the web. When the rupture energy, i.e. the area under the load-elongation curve, was plotted against the area under the drying tension-time curve a linear negative relationship was

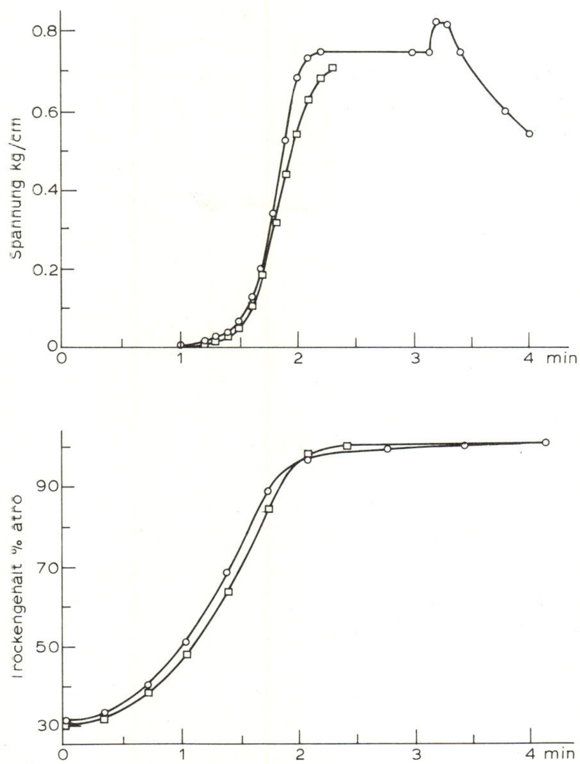


Fig. 1.22. Drying tension (top) and solids content (bottom) as functions of the time during shrinkage-free drying of unbleached spruce sulphite pulp, 68°SR, dry basis weight 55 g $m^{-2}$ (after Broens [165]).

found, that is, the rupture energy decreases linearly with the drying tension-time integral. Broens quantified this relationship by treating the fracture of hydrogen bonds as a monomolecular chemical reaction, which introduces the time into the bond-breaking process. In its simplest form the making and breaking of paper can thus be described by the equation

$$2\,OH \rightleftharpoons (OH)_2.$$

One consequence of this is that a stronger paper (higher rupture energy) is produced on a faster machine than on a slower machine with the same drying tensions. Another is that bonds which are strained but not broken exist in the finished paper in a state of high potential energy. They are very "unhappy" and the slightest encouragement causes them to part, or their dissociation in only a matter of time with the energy of activation of the dissociation reaction much reduced. This is accompanied by small movements of the paper. In other words, paper is dimensionally unstable. After some time, more and more of these bonds disappear and the paper becomes more dimensionally stable. This is called curing. Anything that encourages strained bonds to part will accelerate the process, and such encouragement is often given in the form of heat and high humidity. Part of the internal "relaxation" is undone when the paper is cooled and dried again. Repeated curing cycles improve the result but even so perfect mechanical stability can probably never be achieved.

## 1.5 CELLULOSE-WATER INTERACTIONS IN THE USE OF PAPER

Our paper is now dry and we can use it. There are uses which involve water directly either in vapour or in liquid form, such as packaging and aqueous coating. In other cases of usage and conversion, water plays more of an indirect role in that it influences paper properties which in their turn influence the utility of the product, such as strength or porosity, or in the case of conversion, the runnability of the paper, as on printing machines. There are finally uses to which paper can only be put if the cellulose-water relationship is more or less drastically altered or eliminated. These three classes will be discussed in turn and the section will conclude with a brief discussion of recycling.

### *1.5.1 Usage properties involving water*

Impermeability to water vapour is a frequent requirement for packaging materials, either to keep the moisture inside, such as in tobacco, or to keep it out, such as for table salt. Although paper on its own is never considered as a water-vapour-proof packaging material, the transport of water vapour through paper has been the subject of a number of investigations, partly because of its intrinsic interest and partly because of the role it plays in the final stage of the drying process. Two observations will show that the transport of water vapour through paper is a phenomenon quite different from the transport of inert gases, such as air, through paper. The first is the fact that paper made from well beaten pulp has an air permeability which is lower than that of paper made from unbeaten pulp by four or five powers of ten. The water vapour permeability (WVP) by contrast decreases only by 50% or so. The second observation is provided by the aluminium dish method often used for the determination of the WVP. The dish is sealed with the paper sample and contains either a desiccant or water. In Europe the test is carried out at 65% r.h. This means that in the first case there is a partial pressure gradient corresponding to 65% r.h. driving the water vapour into the dish and in the second case a gradient corresponding to 35% r.h. driving the water vapour out of the dish. If water vapour behaved like an inert gas with the transport rate proportional to the pressure gradient, one would expect the dish of the first type to gain weight at about twice the rate as the dish of the second type loses it. But upon comparison it is found that the flow-rate in the second case is about four times that in the first case, and not half.

The problem was studied in the author's laboratory by Vollmer who measured flow rates of gases and vapours through paper caused by very small pressure differences (typically 1 mm Hg) at different absolute pressures [167]. The results, fully supported by experiments, may be summarised as follows.

(1) The flow of inert gases through paper at low pressures (up to about 20 mm Hg) is of the Knudsen type [168]. The mean free path of the molecules is of the order of magnitude of the pore diameter. The molecules therefore do not so much collide with each other as in diffusion and flow processes at higher pressures, but bounce from wall to wall, moving in the direction of decreasing pressure. Their progress, i.e. the flow rate, is thus independent of the presence of other molecules, i.e. the pressure.

(2) The vapour pressure of water at 20°C is limited to the range of Knudsen molecular flow. One would therefore expect the flow rate of water vapour to be of the Knudsen type, $\sqrt{29/18}$ times that of air, for example *, and independent of the pressure, i.e. relative humidity. At very low pressures the flow rate of water vapour is indeed $\sqrt{29/18} = 1.27$ times that of air but soon increases very steeply with increasing vapour pressure, i.e. humidity, to such high values that the contribution made by the Knudsen flow becomes negligible.

(3) The mechanism of this additional flow is surface diffusion, that is, the water molecules do not leave the surface of the pore walls but jump from

* Knudsen flow is inversely proportional to the square root of the molecular weight.

OH-group to OH-group. Their progress is therefore not affected by the size of the pore as long as it is larger than a water molecule, and this explains the first of the two observations mentioned before.

(4) The surface diffusion is controlled by the adsorption of water on cellulose and its humidity dependence by the shape of the adsorption isotherm. Figure 1.23 is taken from Vollmer's article [167]. The shape of the isotherm is typical. The flow-controlling gradient is not that of the vapour pressure (abscissa) but that of the surface concentration (ordinate) and the steep rise of the isotherm at high humidity provides the explanation of the second observation mentioned before. Vollmer found that the surface diffusion coefficient is proportional to the surface concentration, which enhances the effect of the humidity on the WVP. He also found that the surface diffusion mechanism does not start below a relative humidity of 25% (at 20° C), at which a monolayer of water has been formed on the cellulose, amounting to 4.36% by weight. After this the contribution of the surface diffusion to the WVP is described by

$$\dot{n}/\Delta P = \text{const.} \cdot Dc \cdot \mathrm{d}c/\mathrm{d}P. \tag{1.4}$$

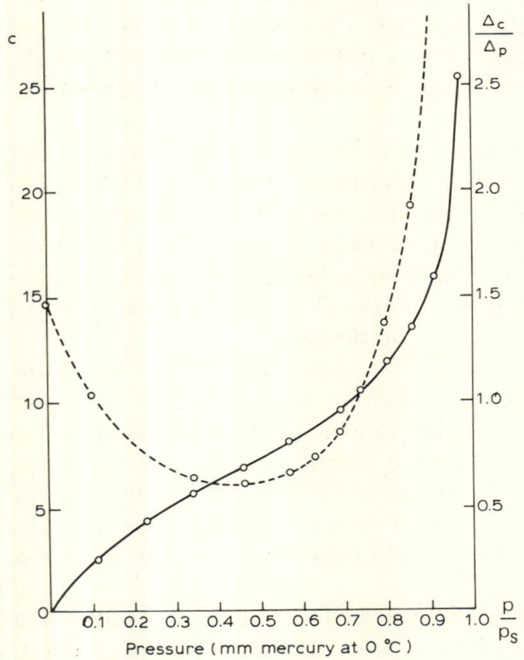


Fig. 1.23. Adsorption isotherm (solid curve, $c$ = % moisture) and its slope (dashed curve, $\Delta p$ in mm Hg) of water on cellulose (after Vollmer [167]). The figure is reproduced from [170].

Here, $\dot{n}/\Delta P$ is the molar flow rate per unit pressure difference, $D$ the surface diffusion coefficient, $c$ the surface concentration and $P$ the vapour pressure. $dc/dP$ is the slope of the adsorption isotherm, shown as the dashed line in Fig. 1.23.

The work was continued and extended to higher temperatures by Schaschek [169]. Increasing temperature will increase the surface diffusion coefficient in eqn (1.4) but reduce the slope $dc/dP$ for a given value of $c$. In order to decide the issue, Schaschek determined flow rates at different temperatures and plotted the results according to eqn (4) with $c \cdot dc/dP$ taken from adsorption isotherms measured at the same temperatures. This procedure eliminates the effect of the temperature on the adsorption and leaves only that on the surface diffusion coefficient. Figure 1.24 shows a set of his results taken from Corte's account of the work at the Fundamental Research Symposium, Cambridge, 1957 [170]. For each temperature the two straight lines intersect at a point on the abscissa which indicates the onset of surface diffusion. The slopes, which are proportional to the surface diffusion coefficient, increase with increasing temperature and permit the calculation of the energy of activation from Arrhenius's equation, $D = \text{const.} \cdot \exp(-E_a/RT)$, yielding a value of $E_a = 8\text{—}8.5$ kcal mole$^{-1}$. The temperature dependence of the slope of the adsorption isotherm is controlled by the energy of adsorption $E_{ads}$, and the temperature dependence of the vapour pressure by the latent energy of evaporation, $E_l$. The WVP therefore varies with the temperature in proportion to $\exp(E_{ads} - E_l - E_a)/RT$. Schaschek showed that the

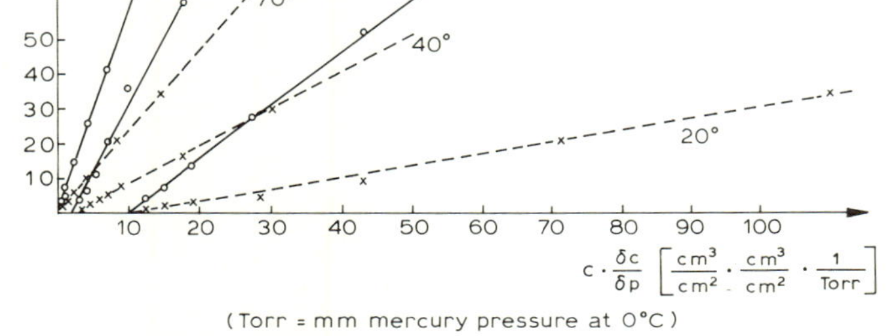


Fig. 1.24. Surface diffusion of water vapour through two papers from pulp beaten to different °SR-values for three temperatures (after Schaschek [169]). The figure is reproduced from [170]).

WVP of paper for different temperatures and relative humidities can be predicted from the adsorption isotherm of water vapour and the air permeability (to account for the Knudsen flow contribution) [169]. Kammermayer [171] analysed Vollmer's approach further and extended the surface diffusion mechanism to other materials. Another detailed analysis of the simultaneous space and surface diffusion of water vapour through paper was published by Rounsley in 1964 [172]. Ahlen [173] found that the surface diffusion between 30 and 50°C is independent of the temperature, a result that is difficult to reconcile with the fact that paper dries faster at higher than at lower temperatures: $E_a$ in the exponential expression quoted above would be zero and the exponent would become positive. Bristow [174] found that the WVP of paper and board decreases with increasing thickness of the sample but with a power of the thickness smaller than unity, typically 0.7 for a gradient of the relative humidity from 75 to 0%; and that the power was nearly the same, whether the thickness was varied by making the sample up from an increasing number of sheets of equal basis weights or by testing single sheets with increasing basis weights. This effect is difficult to explain. It must be remembered, however, that the factor $dc/dP$ in eqn (4) is in the diffusion equation proper $dc/dL$, where $L$ is the thickness coordinate of the sample. Vollmer [167] considered papers of typically 80 g m$^{-2}$ basis weight and because of the low value of $L$, replaced $dc/dL$ by $dc/dP \times \Delta P/L$, after which $\Delta P$ appears on the left-hand side of the equation and $L$ (the thickness of the sample) is incorporated in the constant. In other words, he assumes a linear gradient of the vapour pressure in the sample. Bristow's results could mean that this is not justified for thicker papers.

An interesting observation is that the diffusion of water vapour through cellulose film is considerably faster than that of alcohol vapour despite its higher vapour pressure. This was the basis of a French patent to extract water from wine in the vapour phase without altering the temperature [175]. Unpublished experiments by the author have shown that the process also works with dense paper such as glassine.

Summarising it can be said that, in spite of some differences of opinion in detail, it is generally agreed that the adsorption of water vapour on cellulose is the central mechanism that controls the movement of water vapour through paper. Moreover, all reactions of paper and its properties to changes of the humidity in its environment must be controlled by how much moisture it picks up in the first place, and sometimes also by how fast it does this. The literature on the water vapour adsorption on cellulose and paper is so voluminous that it is plainly impossible to do it justice on the few pages available in the context of this chapter. Most of the early work was done by research workers in the textile field with cotton and viscose rayon as their chief materials. It was extensively reviewed and discussed by Hermans [127]. One has to distinguish between observed facts and conjectures. Even on observed facts, there is some disagreement between researchers. Not surprisingly, there is even more variety and disagreement when it comes to conjectures which are mostly about the properties, the distribution and generally the state, of the adsorbed water.

The general shape of the adsorption isotherm as shown in Fig. 1.23 is agreed upon by everybody. Uncertainties still exist about the beginning and the end of the curve. It appears that so far nobody has claimed to have produced absolutely dry cellulose. A residual moisture content of about 0.5% is assumed for cellulose dried in air at 105°C to constant weight [127]. But even if dried in vacuum over phosphorus pentoxide or in a stream of dry nitrogen or hydrogen at 110°C the Fischer titration method, for example, still gives a water content of about 0.5%. As was said before, it is still not clear whether or not some water is part of the hydrogen-bonded cellulose or paper structure, or whether it is just very strongly bonded or, perhaps, trapped — a borderline between physical and chemical bonding.

The other end of the adsorption isotherm, i.e. the amount adsorbed at 100% r.h., is similarly uncertain, largely because of practical difficulties when experimenting with water vapour near the saturation pressure. Values reported in the literature for native cellulose lie between 25 and 35% on dry basis [77,78, 176—180]. For regenerated cellulose, the values are markedly higher, at 40% and above [180,181], which is generally thought to be related to the lower degree of crystalline order compared with that of native cellulose. All of these are much lower than those for the fibre saturation point mentioned on pp. 40/41 with which they must, however, not be confused. The exposure of cellulose fibres to water vapour at saturation pressure and the immersion of them in liquid water are, energetically speaking, two entirely different situations which will cause different amounts of water to associate with the fibres. If they were not, the recycling of paper would not be the problem it is — more about this later.

The adsorption isotherm shown in Fig. 1.23 was measured in the absence of air in an atmosphere of water vapour at increasing pressure. The *desorption* isotherm which is obtained when the vapour pressure is reduced from near-saturation, lies above the adsorption isotherm. Typically, dry paper will take up about 6—7% water when the relative humidity is increased to 65% at 20°C, but retain 8—9% water when it is reduced from saturation to 65%. This adsorption hysteresis has given rise to a considerable amount of literature. Some authors claim that it is not real, caused by experimental error and disappears in the absence of air or other non-condensable gases [182,183], and this view was recently repeated [184]. It was refuted by Urquhart in 1929 [185] who explained the hysteresis in terms of his "partial solubility" concept referred to on p. 23. A similar view was expressed by Matthes [129] who explained the narrowing of the hysteresis loop with repeated cycling as the result of progressive "hornification", see below. This still leaves a reproducible adsorption hysteresis. Perhaps the most plausible thermodynamic formalism of the adsorption hysteresis was presented by Barkas

[186—188]. His analysis is based on Porter's treatment of the osmotic pressure of compressible solutions [189] which he applied to the swelling of wood under stress. One of his remarkable results was a quantitative relationship between the three moduli of elasticity of wood (in the axial, radial and tangential directions) and the adsorption isotherms of water vapour obtained when the sample was permitted to swell only in one of these directions. A result of interest here is that the adsorption isotherm of a sample that, as a result of external constraints, cannot swell as much as it would like to is flatter than that of a sample which is allowed to swell without constraint. In other words, the adsorption of water at a given relative humidity is reduced by external pressure. Barkas used wood for his experiments but the results were verified by Treloar for regenerated cellulose fibres [190] and by Kubát and Nyborg for paper [191]. Here the argument is reversed. A strip of paper or a rayon filament tensioned in an atmosphere of constant relative humidity adsorbs more water as the tension is increased. The higher the modulus, the greater is the effect. If the deformation is perfectly elastic the unload-contraction curve coincides with the load-elongation curve and the paper loses as much moisture upon unloading as it had gained upon loading. At zero load the sample would assume its original length and neither the load-elongation curve nor the adsorption isotherm would exhibit any hysteresis. However, paper (or rayon) upon unloading does not generally return to its original length. The unloading curve is steeper than the loading curve. The modulus during unloading is higher and so is the moisture kept in the sample at a given load. The adsorption hysteresis therefore reflects irreversible processes during the deformation and would disappear if the deformation were reversible.

Although Barkas's explanation of the hysteresis is plausible and elegant, it must be emphasised that it is a thermodynamic formalism which describes relationships between changes of state in continua. It tells us what to expect if the extension of paper is reversible and what to expect if it is partly irreversible, but it cannot tell us whether in fact it is reversible or not, or to what degree; and it does not provide insight into molecular mechanisms.

How the water taken up from the vapour phase is actually accommodated in paper (or in the fibres) is still a matter of conjecture. Much of it was originally based on the shape of the adsorption isotherm and on density measurements. The shape of the isotherm has been discussed and speculated on by many people. Briefly, the three main accommodation mechanisms favoured at one time or another were capillary condensation {see, e.g. [129,192,193]}, solution {[77,127,129]} and multilayer adsorption {[167,194]}. None of them on its own or in combination has proved satisfactory so far and the reason for this is that, similar to the situation discussed in the sub-section on drying, the fibre-water interaction introduces a variability into the process which is not catered for by simplifying models. Hermans's notion of stoichiometric hydrates being formed while water is "dissolved" in cellulose [127] seems to have been abandoned and the likening of the adsorption isotherm with the vapour pressure curve obtained when water is added to sulphuric acid [38] is an extreme example of the chemical-formalistic view. But cellulose fibres are not just a chemical but morphological entities and batches from different species, unlike batches of sulphuric acid, can be very different. Besides, the definition of a solution is that every solute molecule (water) is much more likely to be totally surrounded by solvent (cellulose) molecules as its nearest neighbours than by molecules of its own kind — a notion really too bizarre to be adoptable for the adsorption of water vapour on cellulose fibres.

The capillary condensation concept on the other hand may have some justification for rigid porous bodies but, again, paper is not a rigid body in which water vapour enters pre-existent pores of fixed sizes on adsorption which it leaves unchanged on desorption. Instead, water vapour seems to open spaces which did not geometrically exist before, and to close them on leaving. The same fact interferes with the application of the multilayer adsorption theory as originally developed by Brunauer et al. [195]. It was tried with varying degrees of success {[129,167,194]} and in the author's opinion provides probably the most promising basis for a realistic description of the adsorption process. A strong point in its favour is the sharp onset of surface diffusion at a moisture content of about 4% which corresponds to that at which a monolayer has been formed, see Fig. 1.24. If this layer of immobile water molecules is accepted as a physical reality it is not difficult to visualise the further adsorption of water to proceed by the statistical formation of more layers, wedging or pushing open spaces where this is energetically possible and forming structures within and between fibres as schematically presented in Fig. 1.7, but not necessarily so well ordered. At very high relative humidities, capillary condensation may well occur * or, on the return, "liquid" water may remain in very small spaces when during desorption the vapour pressure has fallen below saturation, "capillary retention" as Christensen and Giertz aptly called it [196].

The increase of the apparent density of water associated with cellulose was mentioned on p. 9 as one of the reasons given for the high density of cellulose when measured in water as the buoyancy medium. Results with water vapour were published by Filby and Maass as early as 1932 [197] and thirty years later by Alinče [198] who found the density to be 1.4 for the first portions of adsorbed water and to fall to its normal value at over 10% moisture content. Filby and Maass [197] quoted a value as high as 2.6 at about 3% moisture, falling to unity at 9% moisture, whereas Hermans [199]

* Some of Schaschek's experiments [169] strongly indicate the onset of capillary condensation in dense materials, such as cellulose film and vegetable parchment, at high relative humidities.

finds, for regenerated cellulose, a value of 1.6 at 4% moisture which does not fall to its normal value until the moisture is 20—25%. Stamm and Seborg [200] used Davidson's [33] pycnometric method to measure with benzene as the buoyancy medium the density of cellulose on which water had been adsorbed. For a sulphite pulp they found typically a density of 2 for the first adsorbed water, falling to 1.5 at 1.5—2%, to 1.1 at 6% and to 1 at 24% moisture (dry basis). How much of these differences is attributable to experimental techniques and how much to differences of material properties is not known. It must also be remembered that, similar to what was said on p. 9 for the thermal expansion, the anomalous behaviour of the cellulose-water system in these instances is always interpreted as an anomalous property of one of its constituents, here water — hence the term "apparent" density. The fact remains, however, that a volume contraction occurs when water vapour is adsorbed on cellulose. The concomitant entropy reduction was mentioned on p. 12 [38]. Together with the fact that the specific heat of the first portions of adsorbed water (0.8 cal $g^{-1}$ $deg^{-1}$) is between those of liquid water and of ice points to an ordering mechanism during the initial stages of adsorption, compatible with the formation of an immobile layer as mentioned before.

Mills [201] quotes a value of 1.25 for the specific heat of the first 10% of the adsorbed water and a value of 1.13 for the next 10% *. Referring to work by Takeda [203] and experiments of his own, he claims that the dielectric constant of adsorbed water is very low (about 4.5 at 3% moisture) which he interprets as indicating a discontinuous state of the water at this moisture level.

The heat of adsorption can be determined either from the temperature-dependence of the adsorption isotherm or calorimetrically. The first method makes use of the Clausius-Clapeyron equation for the desorption pressure as a function of the temperature. Whereas the vapour pressure $P$ of liquid water is independent of the amount of water present, the desorption pressure of adsorbed water depends on the amount adsorbed. Plots of $\log P$ against $1/T$ for given surface concentrations are ideally straight lines, called isosteres, and the heat of adsorption is calculated from their slopes. The method requires very accurate adsorption measurements but produces well-defined results, particularly at low levels of adsorption where secondary effects, such as swelling, are negligible. It was used by a number of authors, among them Houtz and McLean [204], Neale and Stringfellow [205], Wahba and Nashed [206] and Wahba et al. [207]. There is general agreement that the initial differential heat of adsorption is between 15 and 16 kcal $mole^{-1}$ water. If

* Shipley et al. [202] found values below 1 for up to 8% adsorbed water only if they were determined by heating the sample from —35 or —20°C to +25°C. If heated from —5 or +5°C to +25°C the values were above 1, a result not fully understood.

the heat of evaporation (10.5 kcal) is subtracted, 4.5—5.5 kcal mole$^{-1}$ water remain as the bonding energy of the first adsorbed water molecules, which is similar to the average hydrogen bond energy in paper quoted on p. 47. As the coverage with water vapour increases, the differential heat of adsorption of further water decreases gradually to the heat of evaporation. Theoretically, this should occur at the onset of bulk condensation but, experimentally, the value is usually reached anywhere between 15 and 25% moisture. Reported results vary much more than they do for the initial heat of adsorption. This may well be due to material differences but probably also to secondary effects such as wetting and swelling. Such effects come certainly into play when the heat of adsorption is determined calorimetrically. This is done by immersing cellulose samples in water after they had adsorbed varying amounts of water vapour, and measuring the amount of heat evolved. The method was used, among others, by Argue and Maass [208], Guthrie [209], Wahba [210] and Iyer and Baddi [211]. It, too, requires great experimental care but the results, because of the secondary effects mentioned above, are not quite as well-defined as those obtained from adsorption isotherms. There is, however, overall agreement between the two methods. Argue and Maass found [208] that the heat of adsorption for a given moisture content is higher when the moisture content was reached by desorption than when it was reached by adsorption, a kind of hysteresis effect. The difference between the heat of adsorption and the heat of desorption was discussed by Wahba and Nashed [206] who also compared the isotherm and the calorimetric methods. Iyer and Baddi [211] determined differential heats of adsorption by numerical differentiation of integral heats of adsorption curves and found a break in the curve for cotton at about 4% moisture which they associate with the completion of an immobile monolayer, in good agreement with Vollmer's results mentioned earlier. The specific surface calculated from this water content is 122 g m$^{-2}$, demonstrating that water opens spaces as it interacts with cellulose. The authors conclude that their results "indicate that cellulose does present a true surface for the adsorption of water vapour".

The heat evolved when cellulose is immersed in water is known to textile research workers as the heat of wetting. To the papermaker, the term wetting usually refers to the surface of a sheet of paper. Wetting of the paper surface by water is the condition for and in a sense the first step of, the penetration of water into the sheet and therefore of great practical interest. In spite of this, very little work on the wetting phenomenon of the cellulose-water system has been reported. One reason for it might be the fact that paper surfaces are never really pure cellulose surfaces and that wetting processes are extremely sensitive to contaminations. Even a monomolecular layer of grease, a few Å thick, can significantly increase the contact angle [212]. The wetting of paper surfaces by water is therefore not really a case of cellulose-water interaction but of interaction between water and some extraneous substances. The wetting of pure cellulose surfaces by water and other liquids was described in a remarkable four-part article by Borgin [213]. He used laboratory-made cellulose films and model filaments prepared by Hermans [127]. His chief method of investigation was the accurate measurement of contact angles. When measured in dry air, a typical value was 34° after one minute, falling to 20° after 10 min. At high relative humidities the values were 18° and 10°, respectively. The difference is, of course, caused by the increasing moisture content of the cellulose with increasing relative humidity. In another set of measurements the contact angle for dry cellulose was found to be 26°. It was constant up to 10% moisture content and then fell sharply to 20° at 14% and to 13° at 26% moisture. Increasing temperature reduces the contact angle, typically from 44° at 1°C to 22° at 75°C when measured after one minute in dry air, or from 28° at 1°C to 10° at 75°C when measured after 10 min at high humidity. The author is at pains to explain why, in spite of the greatest care taken in preparing the cellulose films, the contact angles for a given set of conditions varied by typically 10° or more, and blames unavoidable microscopic contaminations for it. Similar experiments were carried out by Luner and Sandell [214] with similar results. They found in addition that the contact angle depended on the method of preparing the film and on any physical treatment suffered by it, for example, stretching.

What we can learn from all this is, firstly, that untreated paper surfaces are so ill-defined chemically, that contact angle measurements are relatively meaningless. This was discussed at some length by Knight [215] who produced "wettability" patterns be spreading a thin film of a dyestuff solution on paper surfaces and demonstrated that they can vary considerably for any given grade of paper. Secondly, the frequently-made assumption that complete wetting occurs inside untreated paper is not justified. Not only is the contact angle not zero but it probably varies at least as much as it does for very carefully prepared cellulose films.

An example from the author's experience may illustrate this. It concerns the manufacture of dissolving pulp in sheet form and the property of interest is the buoyancy of the sheets when steeped in 18% sodium hydroxide solution, see, for example, the Scandinavian standard SCAN — C20 : 65. The floating tendency was unacceptably high for the customer. It is caused by air entrapped as a result of an uneven capillary rise of the liquid over the thickness of the sheet. This in turn is caused, at least partly, by non-uniform wettability. The cure was a very fine mist of mineral oil sprayed on the web over the suction boxes. The amount of oil in the sheet was well below the limit of detection by chemical analysis but sufficient to reduce the wettability very slightly and, most important, to make it more uniform than it had been before.

In coating and impregnating operations wetting is followed by penetration. It is very rare that water is forced into or through, paper by an external pressure. The mechanism by which liquid water penetrates paper is therefore nearly always capillary suction, and to this day experimental observations are usually presented in terms of Lucas's classical equation for the height of a liquid, $h$, rising in a capillary tube of radius $r$, as a function of the time, $t$,

$$h^2 = r\gamma \cos\theta \; t/2\eta \tag{1.5}$$

where $\gamma$ and $\eta$ are the surface tension and the viscosity of the liquid and $\theta$ is the contact angle between the liquid and the capillary tube [216]. Very aptly, Lucas first tested his equation with laboratory filter papers but he was careful to "exclude liquids which cause a change of the pore widths by swelling of the cellulose fibres", the closest to water being methanol, ethanol, and glycerol, apart from aniline, benzene, chloroform and others. His results were confirmed by Peek and McLean, using again filter paper and a similar range of liquids [216a]. Values for $r$ (or, more accurately, $r \cos\theta$) were found to be of the order $10^{-4}$—$10^{-3}$ cm, which is a reasonable range for filter paper.

Lucas was fully aware of the fact that his formal relationship (1.5) was only valid for the porous structure of paper, if it was not changed by the liquid. It is surprising that people should have expected it to be valid for the penetration of water also and it is even more surprising that several of them found that it was. Polčin [217] confirmed it for pulp sheets of 500 g $m^{-2}$ pressed to various densities and dried at different temperatures and so did Kuniak [218] for sheets of 600 g $m^{-2}$, varying the density and the disintegration time (which increased the wetness up to 31°SR). Chatterjee [219] experimented with standard handsheets from 12 different pulps and explained the validity of Lucas's equation by the fact that the swelling of the (presumably unbeaten) pulps was slight and fast so that the water following the advancing meniscus was moving in a stable albeit changed porous structure. All these experiments were of the Klemm type [220] with a vertical strip touching the water surface, the whole being enclosed to prevent evaporation. Bristow [221] used the Cobb test [222] and found the linear relationship between the water uptake (in g $m^{-2}$) and $\sqrt{t}$ to exhibit an intercept of around 5 g $m^{-2}$ uptake at zero time. Since 5 g $m^{-2}$ is about half the basis weight of a single fibre, it is just possible that this seemingly rapid initial period of penetration is largely the filling of the surface pores.

Similar curves, showing a fast initial uptake followed by a flatter branch linear in $\sqrt{t}$, were found by Clark et al. [223] and by Windle et al. [224] for capillary and for pressure-forced migration of water from a clay-coating mix into sized coating base paper. The opposite curvature (slower initial uptake followed by a steeper branch linear in $\sqrt{t}$) was found by Österberg et al. for water penetrating into hardboard [225] and by Bristow for alkaline solutions penetrating into rosin-sized liner board [226]. Experimenting with sized papers, Wink and Van den Akker [227,228] find that the viscosity of the liquid (water compared with CMC solutions) did not affect the initial rate of water uptake, and the same was found by Verhoeff et al. [229]. All these authors quote in their articles the Lucas equation although the materials and/or the experimental techniques were quite, sometimes totally, different from those for which it was derived. It is not surprising that their observations do not agree with it and the various speculations about the reasons for the deviations show that the initial stage of penetration is often not a cellulose-water interaction problem at all but related to the hydrophobic nature and the roughness of the surface and that later some interaction can complicate the picture as it did in examples discussed previously. A more comprehensive review of the subject was published 1976 by Hoyland and Field [230].

The most obvious interaction is swelling and this was investigated by Bristow for corrugated medium and for liner and fibre building board [231, 232]. Hoyland [233] measured simultaneously the penetration of water and aqueous solutions into untreated cellulose sheets from pulp beaten to 31° SR and the swelling which it causes. He showed clearly that the Lucas equation does not describe the penetration process because of swelling and suggests that the latter is a diffusion process in the sense that "molecules of the fibre move into and occupy space previously occupied by water molecules and . . . molecules of water diffuse into the fibre and occupy positions previously occupied by fibre molecules". This must be considered highly speculative for two reasons. Firstly, the plot of the uptake of water expressed as the fraction of the total uptake, against $\sqrt{t}$, is not a straight line (which would support the argument) but has an initial curvature (Hoyland's Fig. 9). Secondly, such a plot is based on substituting the variable $x/\sqrt{t}$ ($x$ = space coordinate) into Fick's diffusion equation. This, however, is a similarity transformation and is permissible only if both types of molecules can fill the entire (theoretically, infinite) space, i.e. if the cellulose dissolves *. A better fit of the data is achieved if swelling is assumed to approach the final (equilibrium) value in an exponential fashion, that is (in Hoyland's notation) $\Delta Z/\Delta Z_{\max} = 1 - e^{-kt}$, where $k$ is a rate constant. This would make the process similar to the increase of the specific volume during beating. The curves for the breaking length and for the specific volume in Fig. 1.10 were in fact calculated on this basis [46].

Finally, one feature of the Lucas equation or its modifications is the assumption of laminar flow. It is known that a liquid penetrating a non-swelling porous system does not always move smoothly with a well-defined liquid front but can move in jerks when "bottlenecks" are passed and voids behind them are filled. This was discussed by Schubert at the Oxford symposium, 1977 [234] for the penetration of liquids into powders. It was shown to be controlled by the *dynamic* capillary pressure, always smaller

* Hoyland's article [233] is marred by a printing error. The right-hand side of his eqn (4) should read: $1 - \mathrm{erf}(x/2\sqrt{Dt})$.

than the static capillary pressure and dependent on the non-equilibrium advancing contact angle. The distribution of the liquid in the porous body can be such that isolated air bubbles are left in it. It is quite possible that such a mechanism sometimes operates when water penetrates into paper or board of considerable thickness and produces buoyancy effects as described on p. 62.

*1.5.2 Other usage properties*

Mechanical properties are the best studied group of physical properties of paper and this includes the effect of water (moisture) on them. Figure 1.25

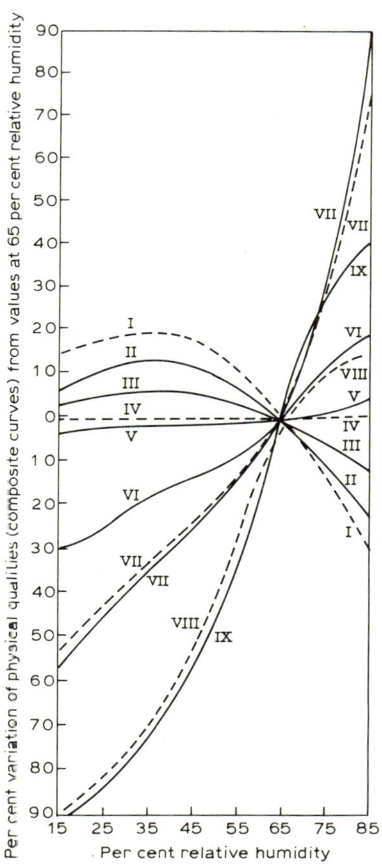


Fig. 1.25. Effect of the relative humidity on some paper properties (after Scribner and Carson [235]). I/II, tensile strength; III, burst strength; IV/V, thickness; VI, tear strength; VII, elongation at break; VIII/IX, folding endurance.

is taken from the third volume of "Pulp and Paper Manufacture" [235] *. The shapes of the curves are fairly typical for pure cellulose papers such as handsheets. The tensile strenght and the burst strength are reduced by increasing moisture — a result of the competition between water molecules and the hydroxyl groups of cellulose for hydrogen bonds with other hydroxyl groups. The flat maximum at 35% r.h. (corresponding to about 5% moisture) was reported by Houston et al. [236] but has not been observed by anybody else [237,238]. It is therefore doubtful. The tear strength and the folding endurance increase considerably between 3 and 12% moisture. These strength properties are indicative of the delaminating resistance which increases with increasing flexibility and, in the case of the tear strength, with increasing width of the peeling zone, both of which increase initially with moisture. Both strength properties will eventually decrease as the relative humidity approaches saturation, and for weakly-bonded papers, such as blotting paper, the folding endurance actually falls steadily in the humidity range covered in Fig. 1.25 [238].

The initial slope of the load-elongation curve obtained with a recording tensile tester (under conventional test conditions) is considered by many to be the modulus of elasticity (Young's modulus) of the paper. Andersson and Berkyto [239] measured such curves at relative humidities from 0 to 95% for kraft paper and newsprint and found that the modulus falls in that range by a ratio of about 3 : 1 for both types of paper and both in the machine and in the cross-directions. The shape of the curves is typical: A slow decline at first, becoming much steeper above 40% r.h. and apparently flattening off at very high humidities. A relationship of very similar shape was found by Higgins and Balodis [240] on handsheets from eucalyptus kraft pulp beaten to 40°SR, only that the modulus fell by a ratio of 5 : 1. A relationship of the opposite shape (steep decrease of the modulus at low and at high moistures and a flatter portion between) was found by Craver and Taylor [241] for kraft wrapping paper using a sonic technique, although the authors elected to interpret their data as a linear relationship between the logarithm of the modulus and the moisture content. The total change of the modulus was only by a ratio 1 : 0.85 from 0 to 22% moisture content. Riemen and Kurath [242] used the vibrating reed technique (60 cps) to measure the dynamic modulus of handsheets made from a softwood sulphite pulp and found it to drop linearly by about 3 : 1 from 0 to 14% moisture and a linear relationship of the same extent was found by Benson for the machine and cross-directions of softwood kraft liner boards [243]. Nissan [244] used some of these experimental data to support his own speculations about paper behaving mechanically like a three-dimensional continuum of hydrogen bonds. These were published in several articles, for example, an early

* A similar figure is included in the second edition, Vol. 2, p. 137, of the work, 1969.

quantitative outline in 1955 [245] and later additions and summaries in 1964 [246] and in 1967 [247], and they will be reviewed in Volume 2. The effect of water on the modulus is interpreted as a monomolecular reaction of a water molecule with the ring oxygen of one cellulose chain bonded to the primary OH-group of another chain. Above a critical moisture content, which corresponds to the completion of a monolayer of adsorbed water molecules, the reaction is complicated by a cooperative mechanism similar to Frank and Wen's flickering cluster model referred to on p. 5 [16,17]. This in turn is characterised by a Cooperative Index which is: (a) the average number of water molecules in a breaking-up cluster; and (b) the slope of the straight line when the logarithm of the modulus is plotted against the moisture content. When a semilogarithmic regression was fitted to several of the experimental data quoted above, the average slope was found to be 6.71 which Nissan claims agrees with values that can be extracted from the articles by Némethy and Scheraga [22] and by Starkweather [248]. As none of the experimental results produce unmistakably exponential curves, their supportive value is perhaps not very great, but Caulfield's results (Fig. 1.6) show the same trend and strong support was provided by Salmén and Back at the Oxford symposium, 1977 [249] who found that the modulus decreases exponentially from the critical moisture content (ca. 5%) onward, and that it does this in a temperature range from −10 to +46° C. It is intriguing that Nissan's very stimulating concept of a hydrogen-bonded continuum representing paper is not only not contradicted by experimental evidence but positively supported by it. Yet, there remains the fact that paper does consist of fibres and that it is this fibrous structure and its visible unevenness which so much distinguishes it from other sheet- or foil-like materials and determines so many of its usage properties. There is more to paper than load-elongation behaviour and whilst hydrogen bonding is undoubtedly the cause of the mechanical coherence of paper, no comprehensive theory to describe the reaction of paper to change in its physical environment, including humidity, can claim to be realistic without taking full account of its fibrous structure. Moreover, as was said on p. 50, load-elongation curves for any but very weak papers start off as straight lines only on instruments with conventional inertia, but as curves convex to the strain axis on more sensitive instruments — a direct consequence of the fibrous structure. This is perhaps not very important when comparing papers for stiffness in a practical sense but means that the term "Young's modulus" has to be carefully defined before it is used to support rather detailed speculations, particularly on the effect of water on it.

De Ruvo et al. [250], using a torsional pendulum method, found the rigidity of handsheets to decrease with increasing humidity to the same extent as that of the constituent individual fibres and found the effect to be greater for bleached sulphite fibres than for unbleached high-yield sulphite fibres. Single fibres were the object of a study by de Yong et al. reported at the Oxford symposium, 1977 [251]. Adopting a method developed by Mark et al. [252], they measured the torsional response of fibres to changes of the relative humidity under minimal tensile load. Typically, fibres twist ("unwind") by about 200° when the relative humidity is increased from 20 to 80%, and the twist exhibits a hysteresis loop on reducing the relative humidity similar to that shown by the moisture expansion of sheets made from the fibres. The authors tried to relate the torsional behaviour of fibres to the dimensional stability of paper (see below), but although such a relationship can perhaps not be ruled out it is unlikely to be very significant.

At the same symposium Kadoya and his co-workers [253] compared the effect of increasing relative humidity on the time-to-failure and the strain-to-failure of handsheets made from bleached softwood kraft pulp under static and dynamic conditions and found that under static conditions both criteria vary in parallel over the entire humidity range but under dynamic conditions only up to 80% r.h., where the strain-to-failure continues to increase but the time-to-failure remains constant. They showed that at 80% r.h. the fracture mechanism changes from bond-breaking to fibres sliding out of the network under load and related this to a maximum of the energy absorption and the impulse at 80% r.h. (their own experiments) and to the steady increase of the mechanical loss factor of paper up to 80% r.h. as observed by Kubát and Lindbergson [254] in experiments using a torsion pendulum technique.

In an earlier study of the effect of the relative humidity on the relaxation of paper, Kubat [255] loaded strips of paper to the same value ($f_0$ = 2000 g) and determined the equilibrium load after complete relaxation ($f_\infty$) at different humidities. He found a nearly linear decrease of $f_\infty/f_0$ from 0.65 at 0% to 0.15 at 60% r.h., indicating a reduction of the energy of activation for the transition of a cellulose-cellulose hydrogen bond to a cellulose-water hydrogen bond. The relaxation time (after long periods of time) decreased almost hyperbolically with increasing relative humidity. In the same year, Kubát [256] published a more practical paper in which he treated four types of paper with nine plasticising agents in the form of aqueous solutions of different concentrations and determined their effect on the stiffness after Clark [257]. He found a close negative relationship between the dimensional changes caused by the treatment and the stiffness. The relationship was independent of the chemical nature of the plasticisers, which included water. The conclusion seems to be that the true plasticiser for paper is water and that the chemicals used for plasticising hold the moisture in the sheet because of their hygroscopic nature, e.g. glycerol, ammonium lactate, trymethylbenzyl ammonium chloride, ethanol amine hydrochloride, and others. This, however, was disputed by McPherson [258] who, at least for glycerol, claims a specific lubrication mechanism to be responsible for the plasticising effect.

Compared with the changes of mechanical properties of paper with the relative humidity, the dimensional changes of paper are small, but they are nonetheless very important, particularly to the printer. Perhaps, the first systematic investigation was carried out by Larocque in 1936 [259]. He measured the rate at which dry strips of paper expanded when exposed to 20 and to 50% r.h. at 25°C, both in the absence and in the presence of air, and found that, in the absence of air, the strip approaches the final length in a perfectly exponential fashion as shown in Fig. 1.10 for the increase of the specific volume with beating time and as suggested on p. 64 for the thickness swelling of paper by liquid water. In the presence of air, the initial rate of expansion is retarded and depends very strongly on the velocity of the air stream. The rates of expansion in the absence of air are different for different papers and vary with the humidity to which they are exposed. For example, the "half-value" times for glassine at 20 and at 50% r.h. were 10 and 55 min, respectively, but for newsprint the figures were only 0.5 and 0.6 min. This is explained by a time-consuming swelling process at higher humidity, which is more pronounced for glassine than for newsprint. Larocque then compared the moisture expansions from 25 to 65% r.h. at 25°C for 21 papers in the cross- and machine-directions and found values ranging from 0.3% for newsprint to 0.95 for glassine in the cross-direction and a cross-to-machine direction ratio ranging from 2 to 6, approximately. The wet expansion values were found to be typically 2—3 times the 25—65% r.h. expansion in the machine direction and 4—6 times that in the cross-direction. The author compared the hysteresis of the moisture expansion with that of the water vapour adsorption and found that they were opposite in the machine direction (lower expansion on the desorption branch in spite of higher moisture content). In the cross-direction, both hysteresis loops were in the same direction for papers containing groundwood but not for the others. If one adds to this that Larocque also observed the stabilising effect of repeated humidity cycles and introduced the term "conditioning" for it, one has to say that, phenomenologically speaking, not much really new has been added since his pioneering investigation. Even his explanation of the mechanism of the "negative" hysteresis effect anticipates in essence our present understanding: The natural tendency of paper to expand when adsorbing water is opposed by the release of "dried-in" strains and the effect in the machine direction is greater (the strip after several cycles finishes up shorter than it was originally) than in the cross-direction (where it sometimes shows a net increase in length). Glassine strips in the machine direction taken from a freshly supercalendered reel and exposed to high relative humidity actually *shrink* by several percent, which has caused great difficulties when trying to make flat bags, for example for men's shirts, by combining glassine with cellulose film and using an aqueous glue for joining the edges (unpublished work by the author).

Brecht [260, 261] and Brecht and Hildenbrandt [262] added considerable experimental material about the effects of beating, furnish, calendering, mode of drying etc., on the moisture expansion. Brecht and Räderer [263] studied in some detail the rate of the dimensional change and Hudson and co-workers [264,265] investigated the possibility to improve the dimensional stability by wetting and stretching machine-made paper in the cross-direction and then drying it under CD tension. For the same purpose, a mechanical conditioning procedure by repeated loading-unloading cycles was suggested by Poppel et al. [266].

It was said before that Larocque already had the right idea about the difference of MD and CD moisture expansions. Our understanding of the whole complex would obviously benefit from an answer to the question what precisely a "dried-in" or, some say, a "frozen-in" tension is. The answer is found in the work by Broens [165,166] discussed on pp. 51/52. Dried-in tensions are hydrogen bonds between fibres which are not in the state of minimum potential energy, brought about by the fact that there was not enough time during the drying process for a particular OH-group to find the ideal partner and by external forces distorting bonds about to be formed. As was said there, these "unhappy" bonds gradually part in the course of time (curing), causing the paper to move. The encouragement to part given by water is the result of its catalytic effect, mentioned on p. 68. It reduces the activation energy of the dissociation reaction by offering alternative bonding opportunities and, therefore, shortens the time to reach equilibrium (minimum potential energy). This is conditioning.

Additional insight is gained from the effect of shrinkage during drying on the moisture expansion afterwards. The shrinkage (or lack of it) is an outward indication of dried-in tensions. Swollen fibres want to shrink laterally on drying. The fact that the wet web as whole shrinks, given the opportunity to do so, is by no means obvious, as Corte and Herdman pointed out [164] who said that the "first satisfactory explanation was probably given by Page and Tydeman" [162]. This is referred to on p. 50 and is ultimately caused by the capillary forces of eqn (1.2). These are volume forces or, rather, pressures which act uniformly in all directions, the same way as the pressure exerted by a piston on an enclosed volume of water. They act inside fibres to pull them together, mostly laterally, and they act between fibres to pull them together, mostly in the z-direction. The two actions combine to contract the web as a whole in all directions. If prevented from contraction in one or two directions, as on paper machines, the web will try to do it as soon as energetically possible, and to an extent that is controlled by the severity of the preventive constraints. Newman [267] gave an account, based on extensive experimentation, of the influence of machine variables controlling the shrinkage of the web, on the moisture and wet expansion of paper. It must be emphasised again that even "ideally" conditioned paper would still exhibit moisture expansion. The dimensional increases under swelling forces exerted by water are an intrinsic feature of the cellulose-water relationship. Their extent depends on the ease with which the web yields to these forces. This is much greater in the z-direction with not much to "push apart" (paper has a layered structure!) than in the plane of the sheet where the network structure restricts the movement. The thickness expansion is therefore much greater than the lateral expansion, typically 30% (Rance [268] quotes a case where it is much greater). In the plane of the sheet, the microcompressions mentioned on p. 50 yield as easily to swelling forces as they do to external mechanical forces. Their anisotropic orientation in machine-made papers, a combined effect of the fibre orientation and of anisotropic shrinkage constraints, as was said before, explains the difference of the MD and CD moisture expansion already observed by Larocque [259].

If the moisture expansion is different on the two sides of the sheet, the paper curls. One contributing factor, that is virtually always present, is the more pronounced fibre orientation on the wire side compared with the top side. Other factors are uneven fines and filler distributions, and uneven drying through the thickness of the web. In fact, anything that causes or increases two-sidedness of the web will affect the sensitivity of the two sides of the paper to moisture differently and therefore increase curl. Notable contributions to the problem of curl were made by Gallay and his co-workers [269—271] and by Brecht et al. [272]. Hendry [273] and Hendry and Newman [274] gave an account, based on extensive laboratory and machine experimentation, of practically all the machine variables, including the size press, on curl and showed that it can be controlled at the slice (by controlling the fibre orientation) and by the contribution of each drying cylinder to the total drying capacity.

Compared with the mechanical properties and the dimensional stability, the effect of water on other usage properties is less important and has therefore attracted less attention. When Chapman [275] published the first article on his printing smoothness tester (measuring the area of the sample in contact with one face of a glass prism pressed against it), he found that the smoothness increases with increasing relative humidity, presumably because of the increased compressibility of the paper. This was confirmed by Brecht and Nagler [276] who studied the effect of humidifying methods on the smoothness of paper. They found in addition that the smoothness after Bekk [277] decreases with increasing humidity but give no reason for it. Now, in the Bekk instrument, the sample is clamped with a pressure of 0.1 kg $cm^{-2}$ and the air leaking between the paper and the clamp face when a suction is applied is measured (a principle similar to that of the Bendtsen tester). The pressure in the Chapman tester is much higher, typically 28 kg $cm^{-2}$. This explains the opposite effects of the humidity on the results of the two tests: Paper swells with the uptake of water and surface irregularities are enhanced. At low pressures much of the enhanced roughness persists but at a high pressure the increased compressibility helps to literally "iron" the paper surface flat. Indeed, Brecht and Nagler found that everything that increases the gain of the Chapman smoothness increases the loss of the Bekk smoothness. The increased compressibility of paper with increasing moisture has an effect on the calendering process, in particular on the onset of "plastic flow". Observations often concern groundwood or groundwood containing papers in which temperature effects on the softening of non-cellulosic furnish components occur simultaneously. The subject was investigated by Jackson and Gavelin [279] and by Colley and Peel [280] and discussed at the Oxford symposium, 1977 by Lyne [281].

Stoewer [278] reported that the air permeability increases slightly with increasing relative humidity at Schopper-Riegler values up to 70 but decreases markedly at higher values. In other words, relatively open papers become more open with moisture, dense papers become more closed. This at least does not disagree with Corte's results [170] that the pore size distribution of a 62° SR handsheet was shifted towards larger pore sizes by the uptake of water.

### *1.5.3 How to reduce adverse effects of water*

The dimensional instability of paper through the action of water can be reduced by measures mentioned in the previous sub-section, that is, by humidity or mechanical cycling or, in the case of curl, judicious drying. Essentially, these measures are aimed at achieving some form of stability or uniformity. Their effectiveness is by nature limited, as they do not fundamentally change the cellulose-water relationship. A greater reduction of adverse effects of water can be expected from measures which either make it more difficult for the water to react with cellulose or prevent the water from reaching the cellulose or the paper surface in the first place. They will be discussed in that order.

The simplest method to impart some measure of water resistance to paper is to overdry it. Such paper is less wettable by water and the change to its surface that has apparently occurred is often vaguely called hornification. It is not over-drying as such that causes the effect, however, but the fact that it is done at high temperature. Acetal formation, i.e. a genuine condensation reaction, has been suggested as a possible mechanism [282]. Back [283,284] heated kraft paper to 190°C and found that after 20 min the wet tensile strength had quadrupled and was 30% of the dry strength. The fact that the increased wet strength is accompanied by increased dimensional stability is taken as proof that internal dehydration and auto-cross-linking between cellulose chains occurs. Treatment with oxidising chemicals such as potassium periodate, which promote the formation of carbonyl groups, accelerates the reaction. Back concludes that one possible reaction mechanism is the forming of hemiacetal bridges. Indeed, treatment with reducing chemicals such as sodium borohydride, which remove any existing carbonyl groups, retard the reaction [285]. Desai [286] observed an increase of the wet strength of paper after exposure to UV light and attributes this to photochemical auto-cross-linking.

"Ether-like" cross-links (i.e. acetals) were suggested by Mark [287] as early as 1938 as the operative chemical reaction in the manufacture of vegetable parchment and vulcanised fibres. The resistance to water, which distinguishes vegetable parchment from another greaseproof wrapping paper, namely, glassine, is so high that one of the tests suggested to distinguish between the two is the "chew test": A piece of glassine chewed long enough is turned into a "pap", a piece of vegetable parchment remains intact [288]. The precise nature of the parchmentising reaction is still a matter of dispute although it has been made for well over a hundred years and the first patent specification dates from 1889. There is agreement that paper swells considerably in sulphuric acid and in zinc chloride solution of commercially used concentrations. Some of the cellulose is said to be removed from the fibres and dispersed as a gelatinous substance between them were it forms a skin on drying. Bucher [289] used the microscope to follow the parchmentising process of very thin (0.5 g $m^{-2}$) sheets and found that the crystal lattice of the cellulose had not changed to that of regenerated cellulose, which one would expect if cellulose is truly dissolved and re-precipitated. This is disputed by Skalicky and Vondrakova [290,291] who claim that some regenerated cellulose can be found by X-ray diffraction (indicating molecular solution), provided that the drying is done at low temperature. If dried at high temperature the regenerated cellulose is reverted back to native cellulose. They support their claim by pointing out that, if regenerated cellulose is heated in polar liquids to 130—300°C, it is partly converted to native cellulose. Sakurada and Fuchino [292] found no regenerated cellulose in vegetable parchment but did find it when cellulose was completely dissolved in sulphuric acid and recovered. They concluded that in parchmentising only the amorphous cellulose is dissolved so that the crystal lattice remains unchanged.

Formaldehyde is the traditional cross-linking agent for polymers and apart from being a component of several wet-strengthening chemicals, it was tried on its own with paper by a number of people, for example, Stamm and his co-workers [293,294]. Paper containing about 1.25% zinc chloride as a catalyst was exposed to formaldehyde vapour and was found to show reduced swelling in water, particularly when the formaldehyde was forced to pass through the sheet. The square of the reduction of swelling increased linearly with the duration of treatment and also with the bound formaldehyde content of the paper. Typically, 0.35% bound formaldehyde caused the swelling in water to be halved. The swelling in pyridine and in 2% sodium hydroxide solution was similarly reduced. This was taken to be "good evidence that cross-linking had occurred". Hydrochloric acid is known to hydrolyse acetal links and indeed, when the cross-linked papers were soaked for a day in 10% hydrochloric acid solution the reduction of the swelling in water dropped from 73 to 44%, for example, indicating that acetal linkages between formaldehyde and cellulose had in fact been formed. Similar results were obtained when sheets were soaked in a 20% formalin solution containing 1% zinc chloride, and then dried. With hydrochloric acid as catalyst a reduction in swelling of 60% could be achieved but with formic acid as catalyst it was only 40%. Wet strength values increased slightly but pronounced embrittlement was the price to pay for substantial reductions in swelling and this loss of dry strength, particularly folding endurance, was independent of the method of treatment and of the catalyst and depended only on the reduction of swelling in water achieved.

Caulfield and Weatherwax [295,296] stressed the difference between wet strength and wet stiffness. Wet strength is caused by a number of polymers which re-enforce the bonds between the fibres somehow without penetrating the fibre wall. Such papers therefore still exhibit a low modulus when wet. In order to increase wet stiffness, the cross-linking agent must penetrate the fibre-wall and generate cross-links inside as well as between the fibres. Cross-linking agents at the inter-fibre bond sites are most effective when the fibres are already so close to each other that only short distances have to be bridged, that is, when the paper is dry. But then the fibres are also in a collapsed state and the chance for a cross-linking agent to penetrate the fibre wall is much reduced. It is greater when the fibres are in a swollen state, but then the distances to be bridged by cross-links are much greater. Caulfield and Weatherwax overcame this contradiction by exposing moist paper to a mixture of formaldehyde vapour, sulphur dioxide and dry air. The first two react with water to form a strong acid ($HOCH_2SO_3H$), which the authors believe to be the active agent in this cross-linking process, forming a hemiacetal with the cellulose which turns into an acetal linkage as the complex decomposes. It proceeds while the paper is being dried. A wet stiffness of 90% of the dry stiffness could be obtained. The function of the cross-links in this process is not so much to re-enforce between-fibre bonds as to protect hydrogen-bonded hydroxyl groups between and inside the fibres from the advances of competitive water molecules. This is concluded from the reduction of the fibre saturation point (determined by the polymer exclusion method) from, for example, 78% (dry basis) of untreated to 17% of cross-linked paper containing 1.6% bound formaldehyde. The negative-exponential relationship between the (sonic) modulus of elasticity and the moisture content as shown in Fig. 1.6 holds for the moisture range up to the fibre saturation point, in accordance with the concept of hydrogen bonding in paper.

One of the oldest wet-strengthening methods used commercially was to tub-size paper with gelatine or animal glue and to treat it subsequently with formaldehyde. This method obviously affects only the surface of the paper and cannot be used if the paper is to remain absorbent such as tissue or plastic base for decorative laminates. "Internal" wet-strenghtening requires the addition of suitable chemicals in the stock preparation phase. Such chemicals, specifically developed for the paper industry, started to appear on the market during the early 1940s and include phenol-formaldehyde resins, urea-formaldehyde resins, melamine formaldehyde, polyamide-polyamine-epichlorohydrin resins and polyethyleneimines. It was said before that these (usually fairly high-molecular) compounds do not penetrate the fibre walls but act as re-enforcing agents at the fibre-to-fibre bond sites. The precise mechanism of wet-strenghtening is still not quite clear. Stannett [297] cured four urea-formaldehyde treated papers from different types of fibres including glass fibres at different temperatures. The wet tensile strength was measured as a function of the time and the initial slopes of the resulting curves were considered as "reaction rates". Their logarithms were plotted against the reciprocal temperature (Arrhenius plot) and from the slopes of the resulting straight lines activation energies of 12—13 kcal $mole^{-1}$ were found, regardless of the type of fibre, of the mode of addition (saturation and beater addition) and of the amounts of resin added. This procedure (Arrhenius plot based on mechanical properties), which incidentally was also used by Stamm [293], may seem somewhat odd to the purist, but it shows that whatever happens is independent of the fibres and happens to the resin itself. Similarly, the loss of wet strength by hydrolysis when papers were soaked in water at various temperatures was shown to be "associated with the resin itself and not with the fibre and resin combination". The wet strengthening mechanism, at least for the formaldehyde-based resin, therefore, seems to be their polycondensation and their ability to "waterproof" (Stannett) the existing fibre-to-fibre bonds. Russell et al. [298] applied a melamine-formaldehyde resin and a polyamide resin to very thin sheets and found that the load-elongation curves of the fibres had not changed, nor had the area and the dry shear resistance of the fibre-to-fibre contacts; but the shear resistance in the wet state was much increased. The authors hypothesise that polar forces are responsible. No condensation mechanism can be operative in the case of polyethyleneimines because they do not show any curing effect during subsequent heating or storage and reach their full wet-strengthening potential while the paper is dried on the paper machine. Sarkanen [299] suggests electrostatic interaction between the amino groups of polyethyleneimine and cellulose (i.e. a kind of ionic bond formation) while Trout [300] advocates polymer bridges between carboxyl groups of adjacent fibres. A particular feature of this polymer is that it does not reduce the porosity or absorbency of the paper. It shares this property with polyamide-epoxide resins whose wet-strengthening mechanism could be similar to that of polyethyleneimines because of their cationic nature, although they show curing effects.

Ionic bond formation may or may not be the operative wet-strengthening principle for the two last-mentioned polymers but it certainly is, according to Allan and his co-workers [301,302] for chitosan, a polymer similar to cellulose but with amino groups replacing some of the secondary alcohol groups, and occurring naturally in the shells of crabs, shrimps and other crustaceans. No matter how it is applied but apparently best when it is sprayed on the sheet, this material increases both the dry and the wet tensile strengths considerably at very low addition levels. For example, the dry tensile strength of unbeaten kraft pulp could be raised to 90% of its value when fully beaten by the addition of 1% chitosan. The wet strength approaches 30% of the dry strength and increases with the molecular weight (i.e. chain length) of the chitosan.

An interesting effect is observed when paper is soaked in aqueous solutions of a zirconium compound, for example, ammonium zirconyl carbonate, $(NH_4)_2ZrO(CO_3)_2$, or when it is made from a fibre suspension containing this compound. For example, a sheet from moderately beaten (30° SR) softwood sulphite pulp containing 2% $ZrO_2$ had a dry breaking length of about 5000 m compared with 3600 m for the untreated sheet, but a wet breaking length of 1000 m compared with 50 m. In other words, the dry : wet tensile strength ratio had decreased from 72 : 1 to 5 : 1. Whether this effect was due to some form of ionic bonding was not established (unpublished work).

There remain the methods aimed at preventing water from reaching the cellulose surface of paper. They are, strictly speaking, outside the subject cellulose-water interactions but are briefly included here because of their technical importance. By far the oldest method to mitigate adverse effects of water on paper is sizing. The Chinese used starch from wheat or rice as early as 768. This provided sufficient ink holdout for their writing with a squirrel hair brush and for their delicate wood block printing. Early European paper had to be adapted to writing with a quill pen. This was achieved by dipping the sheets in a solution of gelatine made from the hoofs, hides and horns of animals. The resulting relatively hard surface required a heavier impression during printing and this was provided by Gutenberg's invention of the printing press [303]. This method of sizing formed a water-resistant film on the paper and is still occasionally used today. The more common method, however, is rosin (colophony) sizing introduced in 1807 by its inventor, Moritz Friedrich Illig [303]. Unlike animal glue sizing it requires the addition of aluminium sulphate (the papermaker's "alum") which precipitates aluminium rosinate in particulate form on the surface of the paper. Opinions about the mechanism vary. Some possibilities were reviewed by Petrow [304]. The most likely primary mechanism which causes a drop of water to lie on the surface without spreading is perhaps simply the fact that water, because of its high surface tension, bridges the hydrophobic rosinate particles without touching the cellulose surface between them. Exactly the same effect causes hydrophobic objects to float on a water surface although their density may be much greater than water *. Such a mechanism is considered possible by

---

* A well-known party trick is to float a darning needle on water. It helps if the water formes a convex surface on the filled glass and if the needle is slightly greasy.

Reaville and Hine [305]. It is unlikely, however, for some of the new sizing agents such as alkyl ketene dimers which do not form particulate precipitates. The mechanism may be the formation of a very thin hydrophobic film and something similar probably happens in "self-sizing", observed by Swanson [306], when absorbent papers lose their absorbency in the course of time by distillation of residual hydrophobic substances from the interior of the sheet to the surface.

A much more drastic way to protect the paper surface from water is to coat it with a water-repellent substance. One of the first was microcrystalline wax, widely used for making breadwrap paper. Polyethylene extrusion coatings and PVC/PVDC emulsion coatings are even more effective and a recent interesting development is the encapsulation of paper with polyethylene by Marchessault and his co-workers [307,308], but this takes us finally away from the subject of cellulose-water interactions.

### *1.5.4 Recycling*

We are very much back with the subject when we look at recycling. The practice is not new. The re-use of broke is probably as old as the paper machine, since the amount of paper piling up when the web broke (hence presumably the name) was so large that no papermaker could afford to waste it. The incentive was to save money. Even the re-use of waste paper is not new. Some of the older readers, at least from Europe, will remember the weekly rag-and-bone-man who also accepted scrap-iron and waste paper. It was used by manufacturers of cheap board and also in furnishes for lower-grade writing and printing papers. The incentive was again cost saving. What is new is that recycled fibres have been "elevated to the status of a raw material worthy of fundamental research in its own right", as Grant expressed it, commenting on the Fundamental Research Symposium, Oxford, 1977 [309]. The incentive transcends cost saving and, with the growing realisation of the fact that the natural resources on our planet are falling behind the increasing demand, is shifted towards conservation. The size of the problem is seen from some statistics. In 1974, Pulp and Paper International published the first waste-paper statistics for Western Europe: 11.6 million tons were used in 1973 for the production of 40.0 million tons paper and board, a recovery rate of 29% [310]. In 1976, the figures were 12.2 million tons waste paper and 38.5 million tons paper and board, a recovery rate of 32% [311].

The literature on the subject before 1970 is rather limited and much of it was concerned with deinking. Even after the interest in recycling and with it the number of publications began to increase many of them were about economic aspects or about extrinsic problems (as far as cellulose-water relationships are concerned), such as the utilisation of "urban fibres" (how to recover fibres from municipal sewage where they make up ca. 40% of the solids), or the removal of contraries. Gradually, recycling became a subject of laboratory investigations. These can be divided into attempts to characterise the new raw materials with the aim to evaluate and assess the usefulness of secondary fibres from different sources, and laboratory simulations of recycling aimed at establishing how the properties of fibres and of paper made from them change as a result of recycling.

Kleinau [312] examined a number of waste papers after deinking and found them "fast beating", compared with the virgin fibres. He also found that strength losses can be recovered by refining, albeit at the expense of freeness and generally comes to the conclusion that the use of deinked waste paper "can be a real asset to improved formation". Raymond [313] described production trials making offset printing with 30 and 55% bleached urban fibres in the furnish. The physical properties were much improved by bleaching and the conclusion was that a pulp can be produced from urban refuse "that has physical and optical properties in the range of pulps generally accepted by the paper industry". The freeness was lower than for virgin pulps but the main disadvantage was contamination with foreign material that prevented its use for many grades. Szwarcsztajn and Przybysz [314] blamed hornified fines in waste paper which, in contrast to hornified fibres, recover none of their usefulness during refining, for the increased wetness of pulps from secondary fibres and for the loss of strength of paper made from them. They used fibre fractionation as a method to characterise waste paper and claim that the removal of fines before refining improves the strength of paper made from secondary fibres. On the whole, the various methods of characterisation or examination show that secondary fibres produce weaker papers and that any recovery of the strength loss is accompanied by an increase in the drainage resistance.

This is more or less the common result of laboratory simulations of recycling which have been reported regularly by a number of authors in the last seven years or so [315—321]. The experimental procedure was usually that handsheets were made and tested, then re-disintegrated and then with or without further treatment made into handsheets again. The treatment could be a certain amount of beating or the removal of fines, for example, and the procedure was often repeated several times. A detailed comparison, which will not be carried out here, is sometimes made difficult by the fact that different experimenters select different standards by which they assess the effects of their repulping cycles on the sheet properties. The intermediate repulped fibres are either beaten to the same wetness or to the same tensile strength, or they are made into sheets of the same density; or they are not treated at all. This causes some seemingly conflicting results. Some of the methods were recently reviewed and compared by Cardwell and Alexander [322]. Occasionally a detail of observation highlights one aspect of the problem. For example, McKee [315] concludes from his experiments that "the loss in bonding potential appears to overshadow the loss in fibre strength as they relate to sheet properties". Bovin et al. [318] found that the "beatability", expressed as the increase of the breaking length per minute beating time, is increased from 0.1 km $min^{-1}$ for virgin pulp to 0.3 km $min^{-1}$ for recycled pulp, but "the production of fines is unfortunately enhanced even more".

This empirical approach, using conventional disintegrating, beating and sheet-making equipment, is somehow reminiscent of the time when standardised beater runs (for which much of the equipment was in fact designed) were virtually the only basis for the assessment and comparison of the paper-making potential of pulps. It is certainly a useful and probably necessary start and it certainly reveals a very important aspect of the cellulose-water relationship, but the results of laboratory recycling experiments, in spite of differences in detail, are discouragingly predictable as the author pointed out when summarising the Paper Physics Conference, Ellenville, 1975: "Once made into paper, the mechanical strength of the redispersed pulp is reduced and, in trying to revive the original strength, the fines content and hence the drainage resistance is increased" [323]. The all-important question is what causes fibres that had been made into paper to be different when they are redispersed, from what they were in the virgin pulp, remembering that in commercial pulps they are already in a once-dried state. This question cannot be answered by laboratory repulping cycles alone but requires more sophisticated experimentation. An outward sign of the change suffered by beaten and dried fibres is that they have lost their ability to swell to the state in which they were after they had been beaten. This could be investigated in greater detail by the solute exclusion technique referred to on p. 40. Treiber and Abrahamson [324] used the non-solvent water method for the determination of the uptake of water by a dissolving pulp that had been dried under different conditions. They used a very high-molecular dextran fraction with molecules of 500 Å diameter and found that the never-dried control sample took up 0.92 g water per g cellulose. Drying at 125°C reduced this figure to 0.47 g $g^{-1}$ but drying at 25°C only to 0.58 g $g^{-1}$, and drying at 25°C after a surfactant had been added produced a value of 0.74 g $g^{-1}$. The determination of complete solute exclusion curves appears to offer the possibility to follow in detail the partly irreversible closure of interlamellar pores of different pulps when made into different types of paper. Such irreversible closures of pores could incidentally be responsible for the low yield of milk from hay as fodder compared with that from the same amount of dry cellulose in the form of fresh grass: The pores are not opened up by the ruminants in the cow's stomach and part of the cellulose remains undigested.

More important than the processes inside the fibres are the irreversible processes on the fibre surface as suggested by McKee's observation mentioned before [315]. Here, a suitable investigational method is more difficult to envisage but, as Treiber and Abrahamson [324] suggested, a treatment with weak alkali could be a possibility.

The detailed investigation of these irreversible changes does not automatically answer the question what they are. "Hornification" is the name often given to them, as was mentioned on p. 72, but it is no more than a different name. Acetal formation or hemiacetal linkages were mentioned before as possible chemical mechanisms suggested by several research workers, particularly for the parchmentising process. This would be a true chemical reaction and to establish whether such a reaction actually occurs when fibres are beaten and made into paper is certainly worth some research effort. At this point a little speculation may be permitted. It is based on the observation that relatively little mechanical treatment of secondary fibres is required to recover much of their former bonding potential. This brings to mind a favourite chemistry lecture demonstration. A rod of very pure iron is suspended in the centre by a steel wire or a nylon string and immersed horizontally in a tank containing dilute hydrochloric acid. Nothing happens. This is known as "passivity". If the rod is struck with a hammer the bell-like ring sounds like a starting signal for a lively reaction in the form of hydrogen bubbles developing on the surface of the rod and rising in the liquid. The passive state is one of non-equilibrium, but the activation energy is too high for the reaction to commence. The mechanical deformations caused by the hammer blow are sufficient to overcome the activation barrier and the reaction starts. Could not a similar mechanism be operative in recycling? The fibre surfaces as well as their insides had been loosened by beating but on drying they were closed again by hydrogen bonds. These are not totally irreversible but the activation barrier is too high for water molecules to open them again. The deformation of the fibres during mechanical treatment could be sufficient, however, to lower the barrier to the advancing water molecules, a kind of mechanical catalysis. Would it be too much speculation to try to think of chemical or electrochemical means to "catalyse", i.e. to accelerate, the OH-dissociation reaction without excessive fines production?

Even if the mechanism or mechanisms of "hornification" were known it would still leave the problem of what to do about them. De Ruvo and his co-workers [325] found that a partial acetylation (3—4% by weight) improves the recyclability of pulp, presumably because the acetate groups prevent hydroxyl groups in their neighbourhood from recombination. This is perhaps not a realistic proposition for making recyclable paper but the ultimate aim of research in this field must be to find ways and means to reduce the damage done to the fibres as much as possible, that is, to make paper with recycling in mind. This, of course, clashes with the efforts of generations of papermakers to increase the utility of many grades of paper by increasing their resistance to water. The vital, yet deadly, affinity between cellulose and water thus represents a true conflict of interests, and to find the optimum balance remains a continuing challenge.

## 1.6 CONCLUDING REMARKS

"Fibre-Water Interactions in Paper-Making" was the theme of the sixth Fundamental Research Symposium held at Oxford, September 1977. Fifty-four papers and prepared contributions were presented, but even so the field was far from completely covered. Indeed, the range of the subject is such that every section and sub-section of this chapter could form, and in most cases has formed, the theme of a full-size conference. Even approximate completeness could hardly be contemplated, let alone achieved. The unfortunate result is that much had to be left out, amongst it contributions by research workers who have published significant articles in the field. Others had to be cited less fully than they deserve. The omissions do not reflect the author's estimation of them but were dictated by the available space and by an attempt to maintain a measure of continuity. This is the role of the fibre morphology in the cellulose-water relationship, not in any specific botanical but in a general sense. It is the true variable in many of the manifestations of this relationship. It is an important fact of papermaking life that we do not find cellulose in nature as we find clay or chalk or aluminium in deposits from which they can be extracted in more or less identical forms. Cellulose to the papermaker is not a chemical and the cellulose-water relationship is really a fibre-water relationship. When the author in 1955 addressed the German pulp and paper association Zellcheming at their fiftieth annual meeting and presented what in his opinion was the experimental proof of hydrogen bonding in paper [158], the presentation ended with these words: "The special properties of cellulose fibres which enable us to make paper at all are by no means obvious . . . . If somebody had never seen a piece of paper and would have to be convinced of its existence solely by physical argument he would perhaps summarise the present situation as follows: "Whilst it has not yet been one hundred percent proved that paper really exists, its existence is highly probable"". The first half of this statement is as true now as it was then. As for the second half, we might say that the probability is now almost imperceptibly different from unity.

## 1.7 REFERENCES


1 D.A.I. Goring, in Fibre-Water Interactions in Paper-Making, Transactions of the 6th Fundamental Research Symposium, held at Oxford, 1977, ed. Fund. Res. Comm., Tech. Div. Brit. Paper Board Ind. Fed., London, 1978, p. 43.

2 W.C. Röntgen, Ann. Phys., 45 (1892) 91—97.

3 K.E. Bett and J.B. Cappi, Nature, 207 (1965) 620—621.

4 J.D. Bernal and R.H. Fowler, J. Chem. Phys., 1(8) (1933) 515—548.

5 J. Morgan and B.E. Warren, J. Chem. Phys., 6(11) (1938) 666—673.

6 G.W. Brady and W.J. Romanon, J. Chem. Phys., 32 (1960) 306.
7 M.D. Danford and H.A. Levy, J. Am. Chem. Soc., 84(20) (1962) 3965—3967.
8 A. Eucken, Nachr. Akad. Wiss. Göttingen, (1946) 38—48.
9 A. Eucken, Z. Elektrochem., 52(6) (1948) 255—269.
10 P.W. Bridgman, Proc. Nat. Acad. Sci., 47(13) (1912) 441—558; 48(9) (1913) 310—362.
11 E.H. Amagat, Ann. Chim. Phys., 29 (1893) 505—574.
12 F.D. Rossini, Bur. Stand. J. Res., 7 (1931) 47—55.
13 E. Bose, Phys. Z., 8(15) (1907) 513—517; 9(21) (1908) 708—713.
14 G.E. Stewart, Koll.-Z., (1934) 130—135.
15 J. Frenkel, Kinetic Theory of Liquids, Oxford University Press, Oxford, 1947, 102 ff.
16 H.S. Frank and W.-Y. Wen, Disc. Faraday Soc., 24 (1958) 133—140.
17 H.S. Frank, Proc. Roy. Soc. A, 247 (1958) 481—492.
18 W.H. Zachariasen, J. Am. Chem. Soc., 53 (1931) 2123—2130.
19 B.T. Darling and D.M. Dennison, Phys. Rev., 57 (1940) 128—139.
19a D'Ans-Lax, Taschenbuch für Chemiker und Physiker, 3rd ed, Vol. I, Springer, Berlin, 1967, p. 600.
20 W.H. Barnes, Proc. Roy. Soc. (London) A, 125 (1929) 670—693.
21 C.H. Collie, J.B. Hasted and D.M. Ritson, Proc. Phys. Soc., 60 (1948) 71—82.
22 G. Némethy and H.A. Scheraga, J. Chem. Phys., 36(12) (1962) 3382—3400; 3401—3417.
23 W. Luck, Fortschr. Chem. Forsch., 4 (1964) 653—781.
24 W. Luck, Ber. Bunsenges., 67(2) (1963) 186—189.
25 W.A.P. Luck, Ber. Bunsenges., 69(7) (1965) 626—637.
26 G.H. Haggis, J.B. Hasted and J. Buchanan, J. Chem. Phys., 20(9) (1952) 1452—1465.
27 B.R. Lentz, A.T. Hagler and H.A. Scheraga, J. Phys. Chem., 78(15) (1974) 1531—1550.
28 T.T. Wall and D.F. Hornig, J. Chem. Phys., 43(6) (1965) 2079—2087.
29 D.P. Stevenson, J. Phys. Chem., 69(7) (1965) 2145—2152.
30 M. Falk and T.A. Ford, Can. J. Chem., 44 (1966) 1699—1707.
31 M.V. Ramiah and D.A.I. Goring, J. Polym. Sci. C, 11 (1965) 27—48.
32 M. Wahba and K. Aziz, J. Text. Inst., 53(6) (1962) T291—307.
33 G.F. Davidson, J. Text. Inst., 18 (1927) T175—186.
34 D.A.I. Goring, l.c. [1], 43—59.
35 C.E. Boesen, Cellul. Chem. Technol., 4 (1970) 149—164.
36 D.F. Caulfield, l.c. [1], 63—76.
37 J.E. Stone and A.M. Scallan, Tappi, 50(10) (1967) 496—501.
38 A.J. Stamm and W.K. Loughborough, J. Phys. Chem., 39 (1935) 121—132.
39 K.H. Meyer and L. Misch, Ber., 70B(2) (1937) 266—274.
40 H. Kiessig, Z. Elektrochem., 54(4) (1950) 320—325.
41 G. Jayme and D. Froundjian, Cellulosechemie, 18(1) (1940) 9—12.
42 O. Kress and H. Bialkowsky, Pap. Trade J., 93 (20) (1931) 35—44.
43 J. d'A. Clark, in Ott, Spurling and Grafflin (Eds.), Cellulose and Cellulose Derivatives, Interscience, New York, 1954, Ch. VIII.
44 B. Steenberg, Proceedings of the sixth EUCEPA symposium, Stockholm, 1962 8 : 1—8 : 7.
45 A.A. Robertson and S.G. Mason, Pulp Pap. Mag. Can., 50(13) (1949) 103—110.
46 H. Corte, Das Papier, 6 (1952) 1—8; Tappi, 35(3) (1952) 124—128.
47 A.A. Robertson, in F. Bolam (Ed.), Fundamentals of Papermaking Fibres, Transactions of the Fundamental Research Symposium, held at Cambridge, 1957, ed. F. Bolam, Tech. Sect. Brit. Paper Board Makers Ass., Kenley, 1958, pp. 411—417.
48 H.W. Emerton, Fundamentals of the Beating Process, The British Paper and Board Industry Research Association, Kenley, Surrey, 1957.

49 D.H. Page and J.H. De Grâce, Tappi, 50(10) (1967) 489—495.
50 J.E. Stone and A.M. Scallan, Pulp Pap. Mag. Can., 66(8) (1965) T407—414.
51 H. Corte, Das Papier, 8(9/10) (1954) 163—172.
52 M.W. Kane, Pulp Pap. Mag. Can., 60(10) (1959) T308—315.
53 L. Nordman and P. Laininen, EUCEPA Conference, Venice, 1964, pp. 51—63.
54 R. Meyer, K.E. Almin and B. Steenberg, Br. J. Appl. Phys., 17 (1966) 409—416.
55 J.C. Woolwage, Tech. Assoc. Pap., 22 (1939) 578—594.
56 A. Erspamer, Tech. Assoc. Pap., 23 (1940) 132—137.
57 S.G. Mason, Pulp Pap. Mag. Can., 49(3) (1948) 99—104.
58 A.A. Robertson and S.G. Mason, Tappi, 40(5) (1957) 326—334.
59 a O.L. Forgacs, A.A. Robertson and S.G. Mason, l.c. [47], 447—473.
59 b O.L. Forgacs, A.A. Robertson and S.G. Mason, Pulp Pap. Mag. Can., 59(5) (1958) 117—128.
60 D. Wahren, Sven. Papperstidn., 67(13) (1964) 536—539.
61 B. Steenberg, N. Thalén and D. Wahren, in Consolidation of the Paper Web, Transactions of the Fundamental Research Symposium, held at Cambridge, 1965, ed. F. Bolam, Tech. Sect. Brit. Paper Board Makers' Ass., London 1966, p. 177—186.
62 F.D. Broadbent and H.A. Harrison, Proc. Tech. Section, Paper Makers' Ass. of Great Britain and Ireland, 21 (1940) 213—223.
63 S.G. Mason, Pulp Pap. Mag. Can., 51(9) (1950) 94—98.
64 A.A. Robertson and S.G. Mason, Pulp Pap. Mag. Can., 55 (1954) 263—269.
65 A.A. Robertson and S.G. Mason, l.c. [47], 447—473.
66 O. Huber and J. Weigl, Das Papier, 24(10A) (1970) 823—832.
67 D.A.I. Goring and S.G. Mason, Can. J. Res., 28B (1950) 307—322, 323—338 and (with G.J. Biefer) 339—344.
68 J. Melzer, Das Papier, 26(7) (1972) 305—332.
69 A. Coehn, Ann. Phys., 64 (1898) 217—232.
70 T.E. Chripunowa, W.I. Jurjew and S.S. Pozin, Probl. Fiz. Chim. Mech. Wolok. Poristych Dipers. Struktur. Mater., Mater. Conf., Riga, (1965) 473—479.
71 L. Michaelis, Die Wasserstoffionenkonzentration, 2nd ed, Springer, Berlin, 1922.
72 M.Y. Chang and A.A. Robertson, Pulp Pap. Mag. Can., 68(9) (1967) T438—444.
73 G.J. Biefer and S.G. Mason, J. Colloid Sci., 9 (1954) 20—35.
74 H.R. Kruyt (Ed.), Colloid Science, Elsevier, Amsterdam, 1952, p. 229.
75 a H. Corte, Das Papier, 29(12) (1975) 541—545.
75 b H. Corte, l.c. [1], 107—114.
76 J.Th.G. Overbeek, l.c. [1], 85—103.
77 A.R. Urquhart, J. Text. Inst., 20 (1924) T125—132.
78 W.B. Campbell, Bull. For. Ser., Dept. Int. Can., Ottawa, 84 (1933) 20.
79 E. Poppel and I. Bicu, Das Papier, 26(4) (1972) 162—173.
80 Th. Hennig, Wochenbl. Papierfabr., 94(23/24) (1966) 862—872.
81 D. Horn and J. Melzer, l.c. [1], 135—148.
82 C.A. King and D.G. Williams, Tappi, 58(9) (1975) 138—141.
83 K.W. Britt, A.G. Dillon and L.A. Evans, Tappi, 60(7) (1977) 102—104.
84 T. Lindström and C. Söremark, J. Colloid Interface Sci., 55 (1976) 305—312.
85 S.T. Han, The Status of the Sheet-Forming Process, IPC, Appleton, WI, U.S.A., 1965.
86 J.D. Parker, The Sheet-Forming Process, STAP No. 9, T.A.P.P.I., Atlanta, GA, U.S.A., 1972.
87 T.A. Arnevik and T. Helle, l.c. [1], 315—353.
88 F.M. Williams, Pap. Trade J., 107(16) (1938) 37—38.
89 H. Corte, Das Papier, 8(11/12) (1954) 205—209.
90 W.L. Ingmanson, Tappi, 47(12) (1964) 742—750.
91 R.W. Nelson, Tappi, 47(12) (1964) 752—764.

92 S.T. Han, Tappi, 47(12) (1964) 782—787.
93 S.T. Han and W.L. Ingmanson, Tappi, 50(4) (1967) 176—180.
94 N.L. Chang, Tappi, 52 (1969) 923—929.
95 P. Mauranen, N. Ryti and P. Aaltonen, Pap. Puu, 53 (1971) 607—616.
96 W.E. Bennett, Proc. Tech. Sect., Br. Pap. Board Maker's Assoc., 36 (1955) 359—374.
97 D.A. Gill, The British Paper and Board Industry Research Assoc., Report RA/T/32, January 1953, p. 40.
98 A.H. Nissan, Tappi, 37 (1954) 597—608.
99 P.E. Wrist, Pulp Pap. Mag. Can. 65 (1964) T284—296.
100 H.D. Turner, J.P. Hohf and S.L. Schwartz, Pap. Trade J., 127(9) (1948) 43—50.
101 C.A. Asklöf, K.O. Larsson and P.B. Wahlström, Pulp Pap. Mag. Can., 65(8) (1964) T339—347.
102 P.B. Wahlstrom, Pulp Pap. Mag. Can., 70(10) (1969) T349—369.
103 P.B. Wahlström, Pulp Pap. Mag. Can., 61(8) (1960) T379—401; (9) T418—451.
104 J.S. Sweet, Pulp Pap. Mag. Can., 62(7) (1961) T367—371.
105 R.E. Morrison, Proc. Tech. Section, C.P.P.A., (1969) T35—39.
106 H.H. Heller and C.G. Tewksbury, Tappi, 55(6) (1972) 893—896.
107 Chr. Schiel, Das Papier, 29(4) (1975) 137—145.
108 H.H. Heller, M.A. MacGregor and W.C. Bliesner, Pap. Technol. Ind., 16(3) (1975) 154—163.
109 T.K. Sherwood, H.S. Gardner and R.P. Whitney, Pap. Trade J., 106(24) (1938) 29—35.
110 E. Cowan and B. Cowan, Tech. Assoc. Pap., 20(1) (1937) 271—277.
111 A.H. Nissan, Chem. Ind., (1956) 198—211.
112 S.T. Han, Pulp Pap. Mag. Can., 65(12) (1964) T537—549.
113 S.T. Han, Tappi, 53(6) (1970) 1034—1046.
114 B.W. Attwood and S.F. Smith, Proc. Tech. Sect., Br. Paper Board Makers' Ass., 31(3) (1950) 577—609.
115 S.F. Smith and B.W. Attwood, Tappi, 36(11) (1953) 481—490.
116 A.C. Dreshfield, Chem. Eng. Progr., 53(4) (1957) 174—180.
117 G. Gavelin, Paper and Paperboard Drying, Lockwood Trade Journal Co., Inc., New York, 1964, p. 4.
118 J.R. Bell, W.F.E. Robinson and A.H. Nissan, Tappi, 40(7) (1957) 558—564.
119 L.A. Kirk, F.L. Hudson and J.W.S. Hearle, Pap. Technol., 4(3) (1963) T75—81.
120 Ju Ching Chu and Weng Ling Kuo, Tappi, 50(8) (1967) 405—415.
121 P.D. Martin and H. Corte, Verfahrenstechnik, 6(6) (1972) 178—181.
122 E.U. Schlünder, l.c. [1], (1977) 443—455.
123 G. Champetier, Ann. Chim., 20 (1933) 5—96.
124 F.A.H. Schreinemakers, Z. Phys. Chem., 11 (1893) 75—109.
125 A. König, Landw. Jahrb., 11 (1882) 1—56.
126 J. Tankard, J. Text. Inst., 28(9) (1937) T263—292.
127 P.H. Hermans, Physics and Chemistry of Cellulose Fibres, Elsevier Amsterdam, 1949, p. 188.
128 G. Jayme, Wochenbl. Papierfabr., (6) (1944) 187—194.
129 A. Matthes, Koll. Z., 84(2/3) (1944) 79—94.
130 T.R. Le Compte, Le Papier, (11) 34 (1931) 1193—1194; 1197—1198.
131 Zellcheming Merkblatt IV/33/57.
132 A.G. Assaf, R.H. Haas and C.B. Purves, J. Am. Chem. Soc., 66 (1944) 59—65; 66—73.
133 P.S. Minhas and A.A. Robertson, Text. Res. J., 37(5) (1967) 400—408.
134 V.J. Frilette, J. Hanle and H. Mark, J. Am. Chem. Soc., 70 (1948) 1107—1113.
135 Y. Sumi, R.D. Hale, J.A. Meyer, B. Leopold and B.G. Rånby, Empire State Paper Research Institute, Contribution No. 25, (1962).

136 J.L. Neal, M.G. Cohen and D.A.I. Goring, Cellulose Chem. Technol. 4(12) (1970) 383—387.
137 H. Tarkow, Tappi, 33(12) (1950) 595—599.
138 B.G. Rånby, Das Papier, 18(10a) (1964) 593—600.
139 S.P. Rowland, ACS Symp. Ser., 49 (1977) 20—45.
140 J.E. Stone and A.M. Scallan, Tappi, 50(10) (1967) 496—501.
141 J.E. Stone and A.M. Scallan, Pulp Pap. Mag. Can., 69 (1968) T288—293.
142 F.F, Morehead, Text. Res. J., 17(2) (1947) 96—98.
143 W.W. Barkas and R. Hallan, Proc. Tech. Sect. Br. Pap. Board Makers' Assoc., 34(2) (1953) 289—304.
144 A.A. Robertson, Tappi, 48(10) (1965) 568—573.
145 A.M. Scallan, l.c. [1], 9—27.
146 J.E. Stone, A.M. Scallan and B. Abrahamson, Sven. Papperstidn., 71(19) (1968) 687—694.
147 J.J. Thomson, Application of Dynamics to Physics and Chemistry, Chapter XI, Evaporation, London, 1888.
148 A.W. Porter, Proc. Roy. Soc. (London) A, 79 (1907) 519—528.
149 W.W. Barkas, Swelling Stresses in Gels, D.S.I.R., Forest Products Research Special Report No. 6, H.M.S.O., London, 1945.
150 J.A. Van den Akker, Tappi, 35(1) (1952) 13—15.
151 H.W. Emerton, l.c. [47], 431—434.
152 H.W. Giertz, l.c. [47], 389—417.
153 G. Jayme and G. Hunger, in F. Bolam (Ed.), The Formation and Structure of Paper, Transactions of the Fundamental Research Symposium held at Oxford, 1961, Tech. Sect. Br. Pap. Board Makers' Assoc., London, 1962, pp. 135—170.
154 S. Asunmaa and B. Steenberg, Sven Papperstidn., 61(18B) (1958) 686—695.
155 Fan Yeh, Hou Han Shu (History of the Later Han Dynasty), Biographies, Ch. 68 (445, approx.).
156 A. Wincziakiewics, Przegl. Papier., 16(2) (1960) 57—58.
157 J.W. Swanson, Tappi, 39(5) (1956) 257—267.
158 H. Corte and H. Schaschek, Das Papier, 9 (21/22) (1955) 519—530).
159 G. Broughton and J.P. Wang, Tappi, 37(2) (1954) 72—79.
160 L. Nordman, Ch. Gustafsson and G. Olofsson, Pap. Puu, 36(8) (1954) 315—320.
161 A.H. Nissan, Nature, 175 (1955) 424.
162 D.H. Page and P.A. Tydeman, l.c. [153], 397—413.
163 P.A. Tydeman, D.R. Wembridge and D.H. Page, l.c. [61], 119—144.
164 H. Corte and P.T. Herdman, Das Papier, 29(7) (1975) 288—295.
165 O. Broens, Das Papier, 11(5/6) (1957) 89—97.
166 H. Corte, H. Schaschek and O. Boens, Tappi, 40(6) (1957) 441—447.
167 W. Vollmer, Chem. -Ing. -Tech., 26(2) (1954) 90—94.
168 M. Knudsen, Ann. Phys., 28 (1909) 75—130.
169 H. Schaschek, Chem.-Ing. -Tech., 28(11) (1956) 698—702.
170 H. Corte, l.c. [47], 301—331.
171 K. Kammermeyer, Ind. Eng. Chem., 50(4) (1958) 697—702.
172 R.R. Rounsley, Tappi, 47(2) (1964) 95—98.
173 A.T. Ahlen, Tappi, 53(7) (1970) 1320—1326.
174 J.A. Bristow, Sven. Papperstidn., 68(23) (1968) 834—838.
175 The French patent was granted in about 1955 but the author has been unable to locate it and cannot quote the number.
176 S.E. Sheppard and P.T. Newsome, Ind. Eng. Chem., 26(3) (1934) 285—290.
177 H. Corte and W. Schoch, Neue Verpack., 8(10) (1955) 489—496.
178 M. Wahba and S. Nashed, J. Text. Inst., 48(1) (1957) T1—20.

179 S.L. Gupta and R.K.S. Bhatia, Text. Res. J., 41(2) (1971) 97—100.
180 T. Lukanoff and J. Baudisch, Zellst. Pap., 16(3) (1967) 88—93.
181 V.L. Simril and S. Smith, Ind. Eng. Chem., 34(2) (1942) 226—230.
182 J. McGavack and W.A. Patrick, J. Am. Chem. Soc., (1920) 946—978.
183 H.A. Hamm and W.A Patrick, Text. Res. J., 6(9) (1936) 401—409.
184 D. Arnhold and K. Lautenschläger, Zellst. Pap., 24(3) (1975) 79—82; (4) 111—112.
185 A.R. Urquhart, J. Text. Inst., 20 (1929) T117—124.
186 W.W. Barkas, The Swelling of Wood under Stress, Meddelande 61, Svenska Träforskningsinstitutet, 1950, pp. 50—57.
187 W.W. Barkas, The Swelling of Wood under Stress, D.S.I.R. Forest Products Research, H.M.S.O., London, 1949.
188 W.W. Barkas, in R. Meredith (Ed.), Mechanical Properties of Wood and Paper, North Holland, Amsterdam, 1953, Part A.
189 A.W. Porter, l.c. [148], 520—522.
190 K.R.G. Treloar, Trans. Faraday Soc., 48 (1952) 567—576; 49 (1953) 816—823.
191 J. Kubát and L. Nyborg, Sven. Papperstidn., 65(18) (1962) 698—702.
192 F.T. Peierce, J. Text. Inst., 20 (1929) T133—150.
193 S.E. Scheppard and P.T. Newsome, Ind. Eng. Chem., 26(3) (1934) 285—290.
194 J.D. Babbit, Can. J. Res., 20A(9) (1942) 143—172.
195 S. Brunauer, P.H. Emmet and E.J. Teller, Am. Chem. Soc., 60(1) (1938) 309—319.
196 P.K. Christensen and H.W. Giertz, l.c. [61], p. 72.
197 E.A. Filby and O. Mass, Can. J. Res., 7(2) (1932) 162—177.
198 B. Alinče, Sven. Pappaerstidn., 65(6) (1962) 216—221.
199 P.H. Herman, Contributions to the Physics of Cellulose Fibres, Elsevier, Amsterdam, 1946.
200 A.J. Stamm and R.M. Seborg, J. Phys. Chem., 39 (1935) 133—142.
201 C.G. Mills, Sven. Papperstidn., 65(5) (1962) 182—184.
202 J.H. Shipley, W.B. Campbell and O. Maass, Can. J. Res., 17B(1) (1939) 40—50.
203 M. Takeda, J. Bull. Chem. Soc. Jpn., 24 (1951) 169—173.
204 C.C. Houtz and D.A. McLean, J. Phys. Chem., 45 (1941) 111—127.
205 S.M. Neale and W.A. Stringfellow, Trans. Faraday Soc., 37 (1941) 525—532.
206 M. Wahba and S. Nashed, J. Text. Inst., 48(1) (1957) T1—20.
207 M. Wahba, S. Nashed and K. Aziz, J. Text. Inst., 49(11) (1958) T519—531.
208 G.H. Argue and O. Maass, Can. J. Res., 12(4) (1935) 564—574.
209 J.C. Guthrie, J. Text. Inst., 40(8) (1949) T489—504.
210 M. Wahba, J. Phys. Coll. Chem., 54(11) (1950) 1148—1160.
211 S.R.S. Iyer and N.T. Baddi, Coll. Chem. Technol., 3(6) (1969) 561—566.
212 N.K. Adam, The Physics and Chemistry of Surfaces, 3rd ed, Oxford University Press, London, 1941, p. 185.
213 K. Borgin, Nor. Skogind., 13(3) (1959) 8—92; (11) 429—442; 14(11) (1960) 485—495; 15(9) (1961) 384—391.
214 P. Luner and M. Sandell, J. Polym. Sci., Part C, 38 (1969) 115—142.
215 C.D. Knight, Tappi, 33(2) (1950) 59—66.
216 R. Lucas, Kolloid Z., 23 (1918) 15—22.
216 a R.L. Peek and D.A. McLean, Ind. Eng. Chem. (Analytical Edition), 6 (1934) 85—90.
217 J. Polčin, Sbornik VUPP, 5 (1960) 185—209.
218 L. Kuniak, Sven. Papperstidn., 65(19) (1962) 760—766.
219 P.K. Chatterjee, Sven. Papperstidn., 74(17) (1971) 503—508.
220 P. Klemm, Handbuch der Papierkunde, 3rd ed, Grieben, Leipzig, 1923.
221 J.A. Bristow, Sven. Papperstidn., 71(2) (1968) 33—39.
222 R.M. Cobb and D.V. Lowe, Tech. Assoc. Pap., (1934) 213—216.

223 N.O. Clark, W. Windle and K.M. Beazlay, Tappi, 52(11) (1969) 2191—2202.
224 W. Windle, K.M. Beazlay and M. Climpson, Tappi, 53(12) (1970) 2232—2236.
225 L. Österberg, O. Brauns and E. Back, Sven. Papperstidn.,63(19) (1960) 658—664.
226 J.A. Bristow, Pap. Puu, 50(11) (1968) 639—646.
227 W. Wink and J.A. Van den Akker, Tappi, 40(7) (1957) 528—536.
228 J.A. Van den Akker and W. Wink, Tappi, 52(12) (1969) 2406—2410.
229 J. Verhoeff, J.A. Hart and W. Gallay, Pulp Pap. Mag. Can., 64(12) (1963) T509—516.
230 R.W. Hoyland and R. Field, Pap. Tech. Ind., 17(10) (1976) 213, 215—219; (12) 291—293, 298, 299, 304—306; 18(1) (1977) 7—9.
231 J.A. Bristow, Sven. Papperstidn., 74(20) (1971) 645—652.
232 J.A. Bristow, Sven. Papperstidn., 75(21) (1972) 847—852.
233 R.W. Hoyland, l.c. [1], 557—577.
234 H. Schubert, l.c. [1], 537—553.
235 B.W. Scribner and F.T. Carson, in J. Newell Stephenson (Ed.), Pulp and Paper Manufacture, Vol. 3, McGraw-Hill, New York, Toronto, London, 1953, p. 877.
236 P.L. Houston, F.T. Carson and R.S. Kirkwood, Paper Trade J., 76(15) (1923) 235—251.
237 R. Korn and F. Burgstaller, "Papier- und Zellstoffprüfung", 2nd ed, Springer, Berlin, Göttingen, Heidelberg 1953, p. 127.
238 Ibid., p. 128.
239 O. Andersson and E. Berkyto, Sven. Papperstidn., 54(13) (1951) 441—444.
240 H.G. Higgins and V. Balodis, in C.J. Osborn (Ed.), Fracture, University of Melbourne, 1965, pp. 201—220.
241 J.K. Carver and D.L. Taylor, Tappi, 48(3) (1965) 142—147.
242 W.P. Riemen and S.P. Kurath, Tappi, 47(10) (1964) 629—633.
243 R.E. Benson, Tappi, 54(5) (1971) 699—703.
244 A.H. Nissan, l.c. [1], 603—623.
245 A.H. Nissan, Text. Res. J., 25(9) (1955) 780—788.
246 A.H. Nissan and S.S. Sternstein, Tappi, 47(1) (1964) 1—6.
247 A.H. Nissan, in R.H. Marchessault and C. Skaar (Eds.(, Surfaces and Coatings Related to Paper and Wood, Syracuse University Press, 1967, 221—265.
248 H.W. Starkweather, Jr., Macromolecules, 8(4) (1975) 476—479.
249 N.L. Salmén and E.L. Back, l.c. [1], 624—625.
250 A. de Ruvo, R. Lundberg, S. Martin-Löf and C. Söremark, in F. Bolam (Ed.), The Fundamental Properties of Paper Related to Its Uses, Transactions of the 5th Fundamental Research Symposium, held in Cambridge, 1973, Tech. Div. Brit. Paper Board Ind. Fed., publ. Benn, London, 1976, pp. 785—806.
251 J. de Yong, H.G. Higgins and G.M. Irvine, l.c. [1], 583—600.
252 R.E. Mark, J.L. Thorpe, A.J. Angello, R.W. Perkins and P.P. Gillis, J. Polym. Sci. Part C, 36 (1971) 177—195.
253 M. Kimura, M. Usuda and T. Kadoya, l.c. [1], 665—675.
254 J. Kubát and B. Lindbergson, Sven. Papperstidn., 68(21) (1965) 743—756.
255 J. Kubát, Sven. Papperstidn., 56(17) (1953) 670—675.
256 J. Kubát, Sven. Papperstidn., 56(1) (1953) 8—10.
257 J. d'A. Clark, Pap. Trade J., 100(13) (1935) 41—44.
258 H.W. McPherson, Jr., Tappi, 33(11) (1950) 531—537.
259 G. Larocque, Pulp Pap. Mag. Can., 37(4) (1936) 199—209.
260 W. Brecht, l.c. [47], 241—281.
261 W. Brecht, Das Papier, 13(9/10) (1959) 201—207.
262 W. Brecht and W. Hildenbrandt, Das Papier, 14(10A) (1960) 610—624.
263 W. Brecht and P. Räderer, Das Papier, 18(10A) (1964) 652—657.

264 F.L. Hudson and H.J. Heinsius, Proc. Tech. Sect., Br. Pap. Board Makers' Assoc., 40(1) (1959) 31—47.
265 D.R. Coles and F.L. Hudson, Paper Technol., 2(5) (1961) T183—189.
266 E. Poppel, I. Bicu and Z. Ladó, Cellul. Chem. Technol., 2(4) (1975) 413—423.
267 J.A.S. Newman, Das Papier, 17(10) (1963) 491—496.
268 H.F. Rance, Tappi, 37(12) (1954) 640—654.
269 P. Glynn, H.W.H. Jones and W. Gallay, Pulp Pap. Mag. Can., 60(10) (1959) T316—323.
270 P. Glynn, H.W.H. Jones and W. Gallay, Pulp Pap. Mag. Can., 62(1) (1961) T38—48.
271 P. Glynn and W. Gallay, Pulp Pap. Mag. Can., 63(8) (1962) T418—424.
272 W. Brecht, P. Räderer and W. Weitzel, Das Papier, (11/12) 13 (1959) 237—244.
273 I.F. Hendry, Das Papier, 16 (1962) 588—595.
274 I.F. Hendry and J.A.S. Newman, Pap. Technol., 4(4) (1963) T121—128.
275 S.M. Chapman, Pulp Pap. Mag. Can., 48(3) (1947) 140—150.
276 W. Brecht and E. Nagler, Das Papier, 20(10A) (1966) 771—778.
277 Bekk, Z. Dtschl. Buchdr., 42 (1930) 755.
278 W. Stoewer, Wochenbl. Papierfabr., 64(4) (1933) 57—59.
279 M. Jackson and G. Gavelin, Sven. Papperstidn., 69(5) (1966) 131—138.
280 J. Colley and J.D. Peel, Pap. Technol., 13(10) (1972) 350—357.
281 M.B. Lyne, l.c. [1], 635—659.
282 E.L. Taylor, private communication, (1965).
283 E. Back, Sven. Papperstidn., 70(3) (1967) 84—90.
284 E.L Back, Das Papier, 27(10A) (1973) 475—483.
285 L. Klinga, Medd. Wallboardind. Centrallab., 28B (1962).
286 R.L. Desai, Can Dept. For. Bimo Res. Notes, 24(3) (1968) 25—26.
287 H. Mark, Wochenbl. Papierfabr., 69(49) (1938) 1031—1034.
288 R. Korn and F. Burgstaller, l.c. [237], 360.
289 H. Bucher, Das Papier, 11(7/8) (1957) 125—133.
290 C. Skalicky and M. Vondrakova, Pap. Cellul., 9 (1957) 194—198.
291 C. Skalicky and M. Vondrakova, Sb. VÚPP, 3 (1958) 97—116.
292 I. Sakurada and K. Fuchino, Cellul. Ind. (Tokyo), 9 (1933) 110—115.
293 A.J. Stamm, Tappi, 42(1) (1959) 44—50.
294 W.E. Cohen, A.J. Stamm and D.J. Fahey, Tappi, (12) 42 (1959) 934—940.
295 D.F. Caulfield and R.C. Weatherwax, International Paper Physics Conference, TAPPI, Ellenville, NY, 1975, pp. 51—61.
296 D.F. Caulfield and R.C. Weatherwax, l.c. [1], 735—753.
297 V.T. Stannet, in F. Weidner (Ed.), Wet Strength in Paper and Paperboard, TAPPI monograph No. 29, 1965, pp. 85—104.
298 J. Russell, O.J. Kallmes and C.H. Mayhood, Tappi, 47(1) (1964) 22—25.
299 K.V. Sarkanen, l.c. [297], 38—49.
300 P.E. Trout, Tappi, 34(12) 1951) 539—544.
301 G.G. Allen, G.D. Crosby and K.V. Sarkanen, l.c. [295], 109—114.
302 G.G. Allen, J.R. Fox, G.D. Crosby and K.V. Sarkanen, l.c. [1], 359—787.
303 D. Hunter, Papermaking, Pleiades Books, London, 1947.
304 A.P. Petrov, Zellst. Pap., 22(6) (1973) 163—172.
305 E.T. Reaville and W.R. Hine, Jr., Tappi, 50(6) (1967) 262—270.
306 J.W. Swanson and S. Cordingly, Tappi, 42(10) (1959) 812—818.
307 B. Fisa, J.F. Revol and R.H. Marchessault, l.c. [295], 115—121.
308 R.H. Marchessault, M. Dubé, J. St. Pierre and J.F. Revol, l.c. [1], 789—807.
309 R. Grant, Pulp Pap. Int., 20(1) (1978) 58.
310 Pulp Pap. Int., 16(8) (1974) 14.
311 Pulp Papl Int., 19(8) (1977) 32.

312 J. Kleinau, Tappi, 49(7) (1966) 90A—93A.
313 D.R. Raymond, T.A.P.P.I. Secondary Fibres/Testing Conference, Madison, 1975, pp. 3—7.
314 E. Szwarcsztajn and K. Przyhysz., Zellst. Pap., 25(9) (1976) 271—275.
315 R.C. McKee, Pap. Trade J., (1971) (21) 34—40.
316 D. Wahren and B. Berg, Sven. Paperstidn., 75(4) (1972) 125—130.
317 H. Cildir and P. Howarth, Pap. Technol., 13(5) (1972) T149—151.
318 A. Bovin, N. Hartler and A. Teder, Pap. Technol., 14(5) (1973) T169—172.
319 E. Swarcsztajn and K. Przybysz, Zellst. Pap., 23(7) (1974) 203—207.
320 R.A. Horn, Pap. Trade J., 159(7/8) (1975) 78—82.
321 E. Schwarcsztajn and K. Przybysz, Cellul. Chem. Technol., 10(6) (1976) 737—749.
322 R.D. Cardwell and S.D. Alexander, Appita, 30 (1977) (4) 327—333.
323 H. Corte, l.c. [295], Proceedings, p. 52.
324 E. Treiber and B. Abrahamson, Holzforsch. Holzverwert., 24(3) (1972) 54—55.
325 E. Ehrnrooth, M. Htun and A. de Ruvo, l.c. [1], 899—911.

*Chapter 2*

# THE FIBROUS RAW MATERIALS OF PAPER

H.W. Emerton

## 2.1 INTRODUCTION

For the purpose of this chapter, paper may be defined as a sheet material essentially made up of interlacing cellulose fibres of natural origin, made by depositing them from an aqueous suspension. The definition is to be taken to include board, a term reserved technically for paper above a certain substance (weight per unit area) and a certain rigidity. The manifold properties of paper are achieved by selection of the type of fibre, their subsequent treatment, the additives employed and by the way in which the paper machine is operated when the sheet is formed, consolidated and dried. In addition, paper may be modified by numerous subsequent conversion processes but these are outside the scope of this work. The treatment of fibres in suspension is dealt with in Chapter 3 and the effects of forming, consolidation and drying subsequently. In this chapter, we are concerned with the raw materials of paper.

The world's total annual production of paper and paperboard is exceeded by that of few other manufactured products. It is estimated to have amounted in 1974 to 152.4 million metric tons and to be increasing exponentially at more than 5% per annum [1]. This production may be compared with that of raw steel, some seven times more dense, which in the same year was 707.7 million metric tons. To sustain this enormous output, the living plants from which the fibrous material is obtained must be available in extensive, concentrated and botanically homogeneous stands and must be capable of yielding a high percentage of useful fibre. The latter requirement is essential not only to keep low the cost of manufacturing the pulp itself but also to minimise the cost and environmental impact of disposing of what would otherwise be embarrassingly large quantities of waste material.

Of more than one third of a million species of plant that have been described it is the trees that yield the highest proportion of their weight as fibres particularly suitable for papermaking; and of these pre-eminently the conifers. However, the cellulose in trees is always heavily encrusted with lignin which must be removed. For many centuries papermaking was based on waste linen, in the preparation of which from the flax plant the lignin had been eliminated, and on rags made from cotton, which is almost pure cellulose. From time to time a shortage of fibres from these sources arose but it was the rapid growth of industry in the western world in the late 18th and early 19th centuries and the increase in population, communications and literacy to which this directly or indirectly gave rise, together with the invention to meet their needs of the paper machine with its greatly increased productive capacity, that provided the strong incentive to find ways of processing wood to provide fibres for papermaking. Today more than 90% of the virgin fibre used for this purpose is derived from trees.

## 2.2 WOOD — SOURCES

Trees grow in forest communities over large tracts, amounting in total to about $3.9 \times 10^9$ hectares or about 30% of the earth's land surface. Of this area it is estimated that at present very approximately half is productive; rather more than half of the wood extracted is for industrial purposes, the remainder being fuel wood [2]. There are, however, striking regional variations as Tables 2.1—2.4 reveal. Thus considering extractions for industrial purposes only, as a proportion of the world's total industrial removals and in the context of the world's total forest area, N. America removes one third (Table 2.4) from less than one fifth of the area (Table 2.2). On the other hand, Central and S. America account for only 4% of the removals notwithstanding that this region contains almost a quarter of the area. Again whereas N. America and Africa have comparable forest areas (Table 2.2), the industrial removals are in the ratio of 18 to 1. More than four fifths of the world's total for industrial use is extracted from the forests of N. America, U.S.S.R.

TABLE 2.1

Distribution and regional composition of world's forests — based on ref. 3

| Region | Forest area as proportion of world's total (%) | Proportion of region's forest area that is productive (%) | Proportion of region's forest area that is | |
|---|---|---|---|---|
| | | | Conifer (%) | Broad-leaf (%) |
| N. America | 19 | 41 | 63 | 37 |
| Central and S. America | 24 | 39 | 5 | 95 |
| Africa | 19 | 40 | 1 | 99 |
| Europe | 4 | 88 | 58 | 42 |
| U.S.S.R. | 19 | 95 | 76 | 24 |
| Asia | 13 | 66 | 18 | 82 |
| Pacific area | 2 | 52 | 5 | 95 |

TABLE 2.2

Total forest area: ranking of regions — based on ref. 3

| Region | Proportion of world forest area (%) |
|---|---|
| Central and S. America | 24 |
| U.S.S.R. | 19 |
| N. America | 19 |
| Africa | 19 |
| Asia | 13 |
| Europe | 4 |
| Pacific area | 2 |

TABLE 2.3

Removals from world's forests — based on ref. 3

| Region | Proportion of world's total removals | | | Proportion of region's removals for industrial use that is | |
|---|---|---|---|---|---|
| | Industrial use (%) | Fuel (%) | Total (%) | Conifer (%) | Broadleaf (%) |
| N. America | 18 | 3 | 21 | 81 | 19 |
| Central and S. America | 2 | 10 | 12 | 43 | 57 |
| Africa | 1 | 9 | 10 | 11 | 89 |
| Europe | 12 | 5 | 17 | 77 | 23 |
| U.S.S.R. | 14 | 5 | 19 | 88 | 12 |
| Asia | 6 | 14 | 20 | 50 | 50 |
| Pacific area | 1 | Neg. | 1 | 40 | 60 |
| Total | 54 | 46 | (100) | Overall 75% | 25% |

TABLE 2.4

Removals for industrial usage: ranking of regions — based on ref. 3

| | Ratio $\frac{\text{industrial removals}}{\text{world total removals}}$ | Percentage of total industrial usage (%) |
|---|---|---|
| N. America | 0.18 | 33 |
| U.S.S.R. | 0.14 | 26 |
| Europe | 0.12 | 22 |
| Asia | 0.06 | 11 |
| Central and S. America | 0.02 | 4 |
| Africa | 0.01 | 2 |
| Pacific area | 0.01 | 2 |
| Total | 0.54 | (100) |

and Europe (Table 2.4) which comprise little more than two fifths of the forest area (Table 2.1).

In commercial terms about half the value of the world's extractions is represented by lumber and one third by pulp; fuel accounts for 7%, the balance being made up of a number of minor uses including so-called naval stores. Thus in regions in which the wood usage is predominantly for fuel, such as Central and S. America, Africa and Asia (Table 2.3), the annual value of the forest products is correspondingly lower.

Forests bestow other benefits than the immediate economic ones. They are generally held to moderate the local climate and the weather by reducing the force of the wind, moderating the air temperature and increasing its relative humidity. In addition the leaves, while on the tree, reduce the erosive power of failing rain and, when fallen, form a spongy soil which releases water to drainage more slowly than would otherwise be the case. At the same time the roots bind the soil together and so further reduce erosion.

## 2.3 WOOD — AS TREES

Trees, even forest trees, show great diversity of form in their branching habits, their bark, their root systems, their flowers and fruits and autumn colouration; not least in their size and shape. They have exerted a strange spiritual influence over man throughout the ages: as objects of worship (see for example The Golden Bough [4]) or of simple admiration. The largest animals the earth has known, the dinosaurs, having evolved to their great size some 150 million years ago, became extinct. Trees as we know them were coeval in origin with the dinosaurs but at no earlier stage have they approached the size of some of the largest conifers of the present time. The coniferous genus *Sequoia* as represented by the Californian redwood, *S. sempervirens*, has produced the tallest trees ever known: well over 100 metres. Another species, *S. gigantea*, the most massive: the weight of the largest known living specimen — the General Sherman tree in the Sequoia National Park — is estimated to be no less than 6,000 metric tons. (The name *Sequoia* commorates a Red Indian chief of remarkable intellectual powers: he is said to have invented a phonetic alphabet and to have taught his tribe to read.) Sequoias were also thought to be the longest lived but this distinction is now known to belong to the bristlecone pine (*P. aristata*), some extant examples of which are known to be more than 4,500 years old. To put this in perspective: there are trees still living that were several centuries old in the time of Abraham (ca. 1500 B.C.). This long life is also approached by *Sequoia sempervirens.*

Another genus of the family that includes the sequoias has produced the Big Tree of Tule, in southern Mexico. Of this bald or swamp cypress (*Taxodium distichum* var. *mucronatum*), it has been said that "probably no tree in the world has been measured so often, over so long a period". It is thought almost certainly to be the tree described in 1590 by a Spanish writer and appears still to be in full vigour. It is not its height that excites comment and wonder but the diameter of its trunk: this was reported in 1968 to girth, at a height of one metre, 34.25 m (112 feet) with the tape taut [5].

At the other end of the size spectrum is the dwarfing of forest trees caused by adverse growing conditions. This is deliberately brought about by man in the Japanese garden art of Bonsai but it can occur fortuitously in nature. Extreme examples were noted, for example, on Vancouver Island where two specimens of Sitka spruce, both 30 cm or less in height and in the one case 19 mm and in the other 25 mm in diameter, were found from a count of the annual rings to be 86 and 98 years old, respectively [6].

Despite the remarkable longevity of some bristlecone pines, in general an individual tree faces a wide range of threats, hazards and enemies that may prevent the attainment of its full stature. Storm winds may fell it, severe frost blast it, lightning rive or fire consume it; it may be gnawed by animals or hewn by man, bored by insects or rotted by fungi; it may succumb to disease, encounter a deficiency of water or of essential nutrients or suffer the effects of atmospheric pollution. All these are capable of stunting a tree's growth, if not of ending its life prematurely.

The three principal parts of a tree are well-known: the *crown* of leaves, supported by the *stem* which in turn is anchored by the *roots*. As a rule they are visually obvious and, to a large extent, functionally distinct but structurally they are continuous.

The crown serves to expose the leaves to sun and air as advantageously as possible. In addition it usually bears the organs of sexual reproduction. The leaves, referred to in the case of conifers as needles, take part in the two major physiological processes of photosynthesis and transpiration.

The ability to convert light energy to chemical energy is a property unique to plants and one on which all life is, directly or indirectly, dependent. No other synthesising process upon earth approaches this in scale. Under the mediation of the green pigment chlorophyll, acting in the presence of sunlight, carbohydrates are built up from carbon dioxide absorbed from the atmosphere through the pores (stomata) of the leaves and the water drawn up from the roots. Although the process is augmented by small openings, known as lenticels, in other parts of the plant, more than nine tenths of the dry weight of a tree is derived from the atmosphere by photosynthesis in the leaves. For deciduous trees growing in temperate climates, the life of a leaf on the tree is only about five months; it is therefore perhaps not surprising that the transport of elaborated foodstuffs from the leaf to the rest of the tree is under way even before the leaf is half grown.

Although the purpose and mechanism of photosynthesis are well understood, the same cannot be said of the phenomena of translocation of water up the tree and its transpiration through the leaves. Water is continuously absorbed through the fibrous roots, drawn up through the stem and, as vapour, given off by the leaves into the atmosphere. Considerable quantities are discharged in this way: a value has been cited for a mature hardwood tree on a summer's day of a few hundred litres an hour; that is, a quantity approaching half a ton an hour: small wonder that trees can have a significant effect on local climate. To sustain such levels, flow-rates of up to 40 metres per hour are required. The purpose and mechanism of this transpiration has puzzled plant physiologists of many generations. Considerable quantities of water are required for the growth and metabolism of the cells; there is little doubt that transpiration can moderate the temperature of leaves exposed to hot sun; and water is the vehicle for the translocation of essential nutrients from the soil into and through the vascular system: but when all these have been taken into account, the water economy of a tree appears distinctly inefficient.

Likewise, the precise mechanism by which it is possible for a column of water to be pulled at a considerable rate to the top of a tree more than a hundred metres tall has never been unequivocally explained. Upward pressure from the roots, osmotic pressure and the vital processes of the living cells have all been considered and found, severally and collectively, inadequate to account fully for the phenomenon. At the present time, the most favoured explanation is that known as the cohesion-tension theory, first proposed in 1895 [7]. If a simple laboratory apparatus is set up, consisting of a vertical glass tube filled with water with its lower end submerged in a reservoir of water and its upper end embedded in a well-soaked sponge, and a current of air is directed onto the sponge so as to encourage evaporation from it, water will be pulled up the tube continuously. As water evaporates from the sponge, further water from the tube replaces it, the tube being replenished in turn from the reservoir. The cohesion of the water molecules and their adhesion to the glass walls of the tube prevent the water column breaking under the upward pull due to evaporation from the sponge and the downward pull of gravity. This experiment is found to work in the same way if the sponge is replaced by a twig that fits tightly into the top of the tube. The equivalent radius of the spaces between the structural elements in the cell walls in the sapwood of trees is of the order of $10^{-7}$ m; the surface tension of water in pores of this size can be shown to be capable of supporting a water column at least 100 m high [8]. It is also known that gas-free water has a tensile strength more than sufficient to support a column of this height, that of the tallest trees. That water is under tension in a tree that is transpiring is revealed by precise measurement of the trunk's diameter; a continuous trace reveals fluctuations in phase with transpirational activity, higher activity being accompanied by decreased diameter. Although not entirely free from theoretical objections, the cohesion-tension theory is the only one that can account for the large amounts of water that are rapidly translocated in an actively transpiring tree.

At the lower end of the tree is the root. This has three functions to perform: to anchor and feed the tree and to supply it with water. To some extent the first two are in conflict. For the most effective anchorage deep rooting is called for. On the other hand, it is the surface layers of the soil that are richest in those elements that are essential to the growth of the tree and which it cannot obtain in any significant amount through its leaves from the atmosphere. These materials are soluble in water (else the tree could not readily absorb them) so that they are easily leached out of a soil by rain; they are replaced largely by those released from the decaying vegetable matter on or near the surface. Trees differ in their root forms: some genera, which may grow to considerable heights, such as spruce, poplar and beech being shallow rooters while many species of fir, pine and oak retain a tap root for many years. But even a very strong tap root alone would be inadequate to maintain the tree upright and a suitably distributed system of lateral roots permits a distribution of the strain in the stem as it moves to the wind. Lateral and surface roots are not, of course, necessarily the same but in fact tend to be so.

Water is absorbed for the most part not by the roots themselves but by the root hairs with which they are provided, the fineness of which affords a large absorbing area. This water has several roles to play: it is the vehicle by means of which the mineral salts and organic nutrients are translocated to the leaves and the elaborated foods conveyed from them; it is a source of oxygen and hydrogen for the vital processes of the tree; and, if it is the case (which is by no means certain) that one of the functions of transpiration is to moderate the temperature of the leaves, then water is the evaporating medium.

Between the root and the crown is the stem. Its functions are to provide mechanical support and height for the crown of foliage and a means of storing elaborated foods and water. As Fig. 2.1 shows it consists of (working radially inwards) the outer bark, the inner bark or phloem, the cambial layer, where cell division takes place, and the (secondary) xylem — or wood of common speech. Although it is possible to conceive of a solid stem growing by division of the cells at any point across its transverse section, this would call for a constant spatial readjustment of those cells already laid down and further from the axis. A simpler arrangement is for new cells to be added near the circumferential layer. However, a tissue undergoing active cell division is a delicate structure requiring protection against rain, mechanical and thermal shocks and attack by other living organisms. This is the role of the bark. New layers are constantly being added to the outer surface of the xylem, the diameter of which consequently increases progressively. This in turn requires a constant enlargement of the bark. In those species with thin barks, such as some maples, birches and cherries, it results in a peeling of the papery surface layers: in others, such as the sequoias which have bark that may be 30 cm thick, the bark cracks and is filled in with new tissue on the inner surface, the outermost parts often being shed as relatively thick flakes.

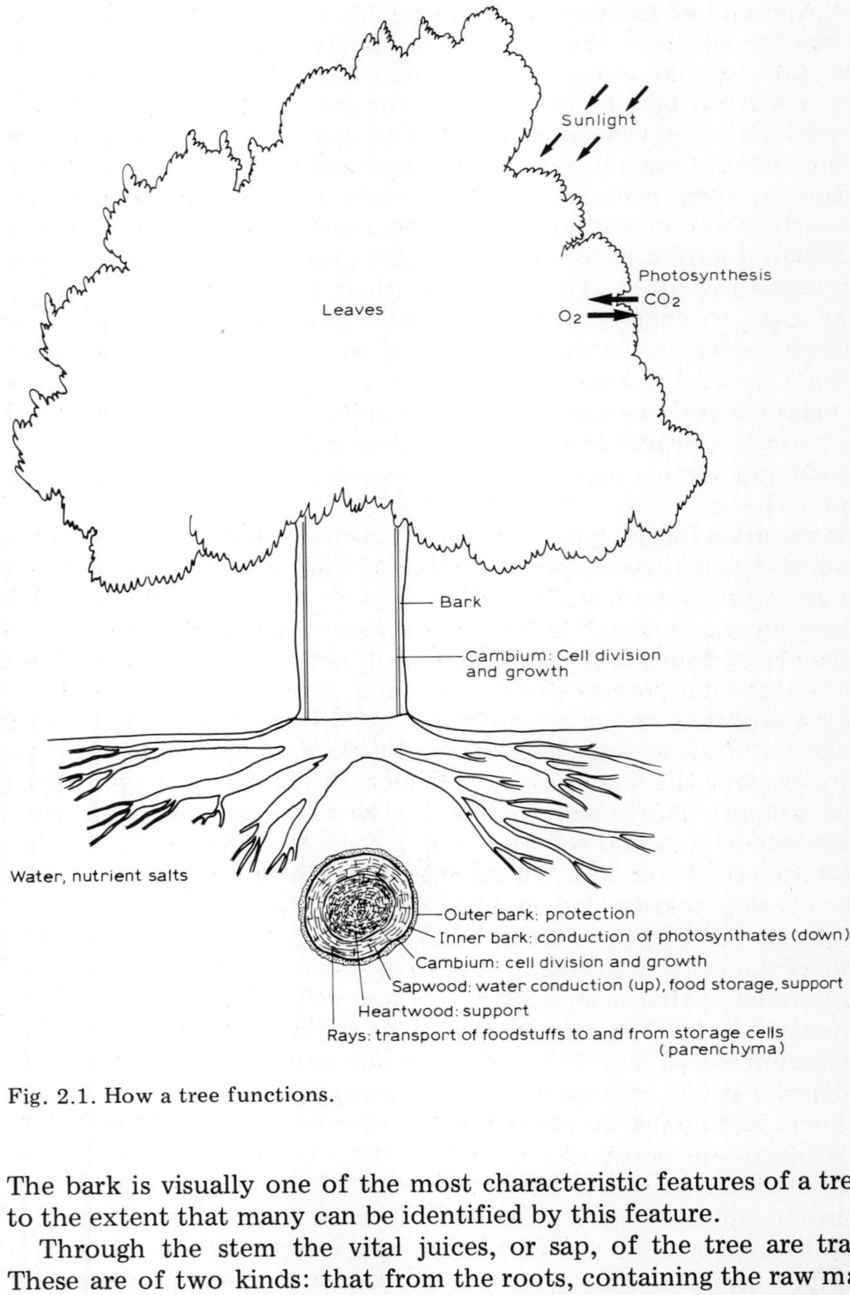


Fig. 2.1. How a tree functions.

The bark is visually one of the most characteristic features of a tree species, to the extent that many can be identified by this feature.

Through the stem the vital juices, or sap, of the tree are transported. These are of two kinds: that from the roots, containing the raw materials of the food, and that from the leaves, containing synthesised foodstuffs, being for the most part conducted up and down, respectively. The former pass up the xylem and the latter, in the form principally of sucrose, chiefly down the phloem, the inner (and living) part of the bark.

The cambium is a thin layer of dividing cells from which the phloem, lying without it, and the xylem, within it, are derived. This layer, only a few cells thick, is the source of all the wood that makes up the stem and it is interesting to reflect that a stratum less than a millimetre thick should continuously be adding to a cylinder of wood, which in the extreme case of the Great Tree of Tule (see p. 94) has at the present stage of its development attained an average diameter of more than ten metres.

## 2.4 WOOD — GROSS STRUCTURE

Within the cambium, then, lies the xylem, or wood, and it is with this part of the tree, as a source of papermaking pulp, that we are from now on concerned.

Xylem tissue is laid down in temperate climates during an annual growing season, coinciding with the warm weather period. The rate at which the cells are formed varies. The flush of growth with which the season opens produces cells that are relatively big but thin-walled. As the season advances, growth proceeds less rapidly and thicker-walled, and usually rather smaller, cells are grown. In some cases, the radial width of the early wood cells measures five or six times as much as that of the outermost late wood cells. The difference in size is particularly marked between the last of the latewood or summerwood cells and the first earlywood (springwood) cells laid down in the succeeding spring. This gives rise to the pattern of annual growth rings that is a striking feature of a transverse section of the stem. In trees of tropical regions, the growth of which is not regulated by alternating hot and cold seasons, such a pattern may not be found. In these regions, growth may be stimulated by the alternation of wet and dry seasons and if, as is sometimes the case, there are two such cycles a year, there will be twice the usual number of rings, so that the term annual growth rings is no longer appropriate. Yet again, the recommencement of normal growth following an interruption, occasioned by, for example, a period of drought or defoliation by insects, will cause more than one ring to be formed in the course of the year, an anomaly referred to as a false annual ring.

The width of growth rings varies between tree species. Some trees, for example the common sycamore (*A. pseudoplatanus*) and many of the eucalypts, grow normally at a fast rate producing wide rings. Others grow more slowly: in the case of box (*B. sempervirens*), for instance, not more than 4 or 5 cm are normally added to the tree's diameter in twenty years. The fine grain to which this gives rise made this wood particularly suitable for the manufacture of mathematical instruments and for wood engraving blocks; it also results in the wood being denser than water so that, alone of British trees, it will not float. The growth rings of the long-living bristlecone pines (*P. aristata*) are usually so close together that they require a microscope to resolve them; one of these trees was recently found to contain more than 1100 rings in a radial segment only 12.7 cm wide [9].

Not only are there variations between species; the width of the growth rings within a tree depends upon many other factors. It varies at different heights in the plant and even in different parts of the circumference of the trunk at a given level. In addition, the rings throughout a tree are affected by the nature of the soil and the growing conditions in general, and particularly by climatic conditions: the narrow rings of dwarfed Sitka spruce trees found growing on Vancouver Island have already been referred to (p. 95). The chronological sequence of annual rings in trees and aged wood is the basis of dendrochronology. Although interest in this began when Leonardo da Vinci wrote on the subject in the 15th century, it was not until the middle of the present century that it began to contribute significantly to knowledge. The same climatic conditions affect all the trees over a large geographical area in the same way. Two cool seasons followed, say, by a warm moist one will be reflected in the relative widths of the annual rings of trees throughout the area. Computer analysis permits the patterns of rings in wood to be rapidly compared and in this way a 7,100 year chronology has been established in east central California by the addition of long dead specimens to the 4,600 year record from living bristlecone pines [9]. The technique confirmed (what had been suspected) that the basic assumption underlying the use of the carbon-14 method of dating objects — that the amount of carbon in the atmosphere is constant — is not valid for periods spanning back more than 3,500 years: wood from the oldest dated trees is now being utilised as a standard for calibration of radiocarbon analysis.

For several years the outer rings of xylem cells continue to participate in the processes of sap conduction and the storage and distribution of food materials. There comes a time, however, when this zone, which is known as sapwood, undergoes marked changes in character. Those cells (parenchyma, described below on p. 122) that were concerned with food storage and transport and which had hitherto remained alive, die and release their contents. The heartwood that results is now completely dead. It ceases to play a part either in the conduction of water or the storage of food and henceforth contributes only to the mechanical strength of the tree. The formation of heartwood involves the hardening of the cell walls and the deposition within them of various compounds: gums, resins and tannins. This often causes it to appear appreciably darker than the sapwood, an effect that is particularly marked in, for example, the common yew (*Taxus baccata*). However, this colour differentiation is not invariable; spruces and poplars, for example, show little, if any, difference in colour between sapwood and heartwood. Generally speaking the more vigorous the growth, the wider the zone of sapwood. Where the roots or branches are more strongly developed on one side of a tree than the other, this results in the boundary between the heartwood and the sapwood not following the pattern of the annual rings.

Two other features of wood that are readily observable are the rays and knots. Rays may be formed at any distance from the axis but necessarily extend from this point to the phloem, down the conducting elements of which elaborated foods are conveyed from the leaves for storage in the ray cells. As the formation of new wood increases the diameter of the stem, neighbouring rays diverge radially. This never proceeds far, however, without a new ray being formed between them; in this way, the frequency of rays at the progressively enlarging perimeter remains more or less constant. The cellular structure of rays and their chemical nature will be discussed below.

Although rays are essential to the life of the tree this is not the case with wood knots. Of no botanical significance they are, however, not without importance technologically. They form where the basal remains of side branches that have fallen, or have otherwise been removed, are overlapped by subsequenctly-formed wood in the main stem or branch. This locally disrupts the parallel arrangement of the fibres.

These are the gross anatomical features of wood. At this stage the difference must be raised between two broad classes of tree. All extant seed-bearing plants are classified either as *gymnosperms*, a group that include the conifers, or as *angiosperms*, which include the trees commonly referred to as broadleaved. The distinction is a phylogenetic one. The older group, the gymnosperms (literally "naked seed": the seed is exposed), represent an early form in the evolutionary history of the higher plants and, as a consequence, their woody tissues are made up in a relatively simple way. By far the greater proportion of their wood is comprised of a type of cell known as the tracheid. In the stem of the gymnosperm tree, the tracheid performs two of the three primary functions of wood cells, namely the mechanical support of the crown and the conduction of sap.

The angiosperms (angeion — "a receptacle": the seed is enclosed) are of much more recent origin and are the highest expression of the evolutionary trend in the vegetable kingdom. In these plants, the functions of support and conduction are performed separately, and more efficiently, by specialised cells. The angiosperms are the dominant plants of the earth's land surface, accounting for two thirds of the world's forest area and being represented by more than a quarter of a million species. A further feature of their evolutionary advance over the gymnosperms is the appearance among this group of plants of an herbaceous habit, one way of coping with freezing conditions: none of the conifers is herbaceous and few are even shrubby in form.

Throughout this work, the terms conifer and broadleaf will be employed for gymnosperm and angiosperm trees, respectively. The terms softwood and hardwood are also used: these do not, of course, accurately reflect the physical properties of the timber, some "hardwoods" being relatively soft and some "softwoods" hard. The use of the terms evergreen and deciduous is always to be avoided as some conifers, notably the larches, shed their needle leaves in winter while some broadleaf trees, particularly those of tropical forests, are evergreen.

The much greater efficiency of the broadleaf trees is thought to be attributable to the larger diameter of the conducting elements. This, however, has a drawback as it makes the trees more vulnerable to the effects of severe winter weather. If the sap becomes frozen, as can happen, the continuity of the water column in the relatively wide conducting elements is likely to be ruptured when the sap thaws, leading to the collapse of the column which may not subsequently be restored. Broadleaf trees are therefore pre-eminently trees of the tropical and temperate zones while conifers predominate in the higher latitudes and at higher altitudes. Conifers tend also to be found to a greater extent in moist than in dry situations, a further consequence doubtless of their relatively poor conducting systems.

The relative tolerance of frost by coniferous and broadleaf trees emerges clearly from consideration of Table 2.5. It will be seen that the coniferous forests are located principally in the large continental masses, with their low winter temperatures, of the northern hemisphere, the U.S.S.R. with 45% and N. America with 36% together disposing of more than 80%. The picture is reversed for broadleaf forests of which Central and S. America, together with Africa, contain more than 60%.

In general economic terms, conifers are more valuable than broadleaf trees as the trunk is longer and straighter, their timber more uniform and more easily worked, and the forests in which they grow are more homogeneous with respect to the species present. In addition, coniferous woods yield a greater proportion of fibrous elements and these are of a kind that is more

TABLE 2.5

Composition of world's forests — based on ref. 3

| Region | Conifers | | Broadleaf | |
|---|---|---|---|---|
| | million ha. | % of total | million ha. | % of total |
| N. America | 440 | 36 | 260 | 10 |
| Central and S. America | 45 | 4 | 836 | 34 |
| Africa | 4 | negligible | 676 | 27 |
| Europe | 80 | 7 | 57 | 2 |
| U.S.S.R. | 553 | 45 | 175 | 7 |
| Asia | 90 | 7 | 400 | 16 |
| Pacific area | 4 | negligible | 84 | 3 |
| Total | 1,216 | ~100% | 2,488 | ~100% |

suitable for most types of papermaking. It is therefore not surprising that, although conifers comprise not more than one third of the total forest area of the world (Table 2.5), three quarters of the wood removed for industrial use is coniferous (Table 2.3). The morphological nature of the wood of conifers will therefore be described here more fully; subsequently attention will be drawn to the respects in which that of broadleaf trees differs significantly.

In 1665 Robert Hooke, having built himself the earliest compound microscope, first figured a biological tissue, a thin slice of cork resembling, in his drawing, moth-fretted lace. To the structural units that he observed, he gave the name *cell*. These were in fact dead; mere empty compartments contained within more or less rigid envelopes. Almost two centuries elapsed before the significance of the protoplasm in living tissues was comprehended and the living cell recognised for what it is: namely, the fundamental unit of life in terms of structure and function; the smallest biological entity capable of growth and reproduction. The large majority of the cells constituting the xylem of a tree are, however, physiologically dead: nevertheless they have a most important part to play in its economy, that of support and to some extent of conduction. The term cell is therefore often extended to refer to the non-living wall together with the space that it encloses, even though this may no longer be filled with protoplasm.

The cellular array of the xylem of a typical softwood comprises, to the extent of 90-95%, cells that are parallel to the longitudinal axis of the stem or branch. They lie up and down the page in Fig. 2.2 and it is convenient to refer to their axial direction as vertical. Transverse to these, but confined to a radial plane, are bundles of horizontal cells forming the rays: these may be seen in section in the tangential plane (C). Thus there are only two directions in which the cells are disposed. (Almost all) the cells aligned vertically are known as *tracheids*; (most of) those radiating horizontally as *ray parenchyma*.

All cells arise from the division of pre-existing cells. There follows a process of enlargement and this is succeeded by a phase in which thickening of the cell wall occurs. Before this is complete the shape of the cells has been determined. In some unicellular organisms, such as some of the algae and the spores of many of the lower plants, the shape is determined solely by surface tension: it therefore tends to be spherical. In tissues, the presence of contiguous cells pressing on one another additionally influences the shape. If, as in most softwood tissues, there are no air spaces between the cells, the tracheid in the early stages of growth approaches in shape the tetrakaidecahedron. This 14-sided polygon is the stable form of the elements of uniform volume present in a foam and subject only to surface tension and the presence of neighbouring cells. It was argued by botanists that, if these were the sole forces determining the shape of plant cells in tissues, they would have such a shape. The true tetrakaidecahedron is, nevertheless, only rarely realised in plant tissues. However, irregularities also occur in soap foams and it is now generally agreed that the shape of all plant cells in tissues, although for various known reasons departing somewhat from it, is based on this geometrical figure.

The ability of the 14-sided polygon to fill space is not affected if it is extended in any direction. This occurs in the case of tracheids, which reveal four to six sides in cross-section and tend to have tapered, chisel-like, sometimes bifurcated ends. Thus they form in the tissue a system of structural units which are elongated, parallel, overlapping and to some extent interlock-

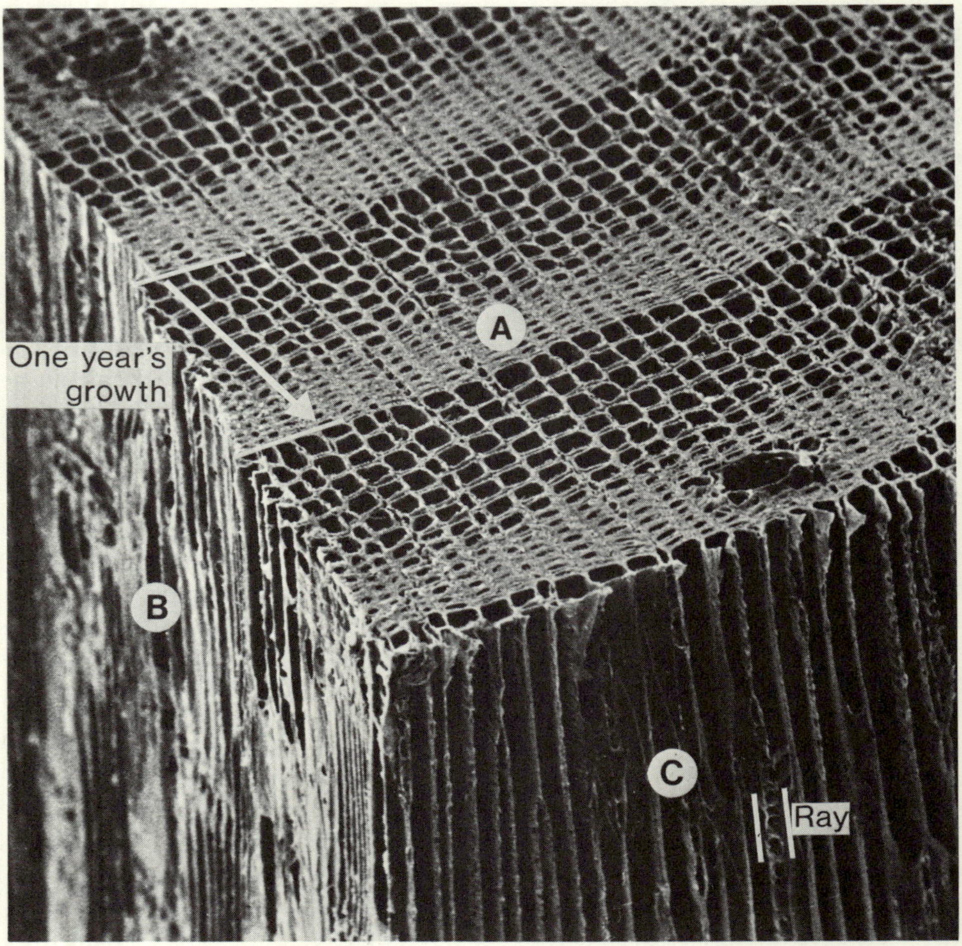

ing. Tracheids, after separation one from another and the removal of their encrustments by pulping, are the (soft-) woodpulp fibres of the papermaker. They measure some 3—5 mm in length and tend, within any one growth ring, to be longer in latewood than in earlywood. Of equal technological significance are the diameter of the lumen, or central canal, and the thickness of the wall that surrounds it.

But before the finer structural details of the units comprising wood are discussed, it will be convenient to consider their chemical constituents, of which the principal are cellulose, hemicellulose and lignin. Associated with

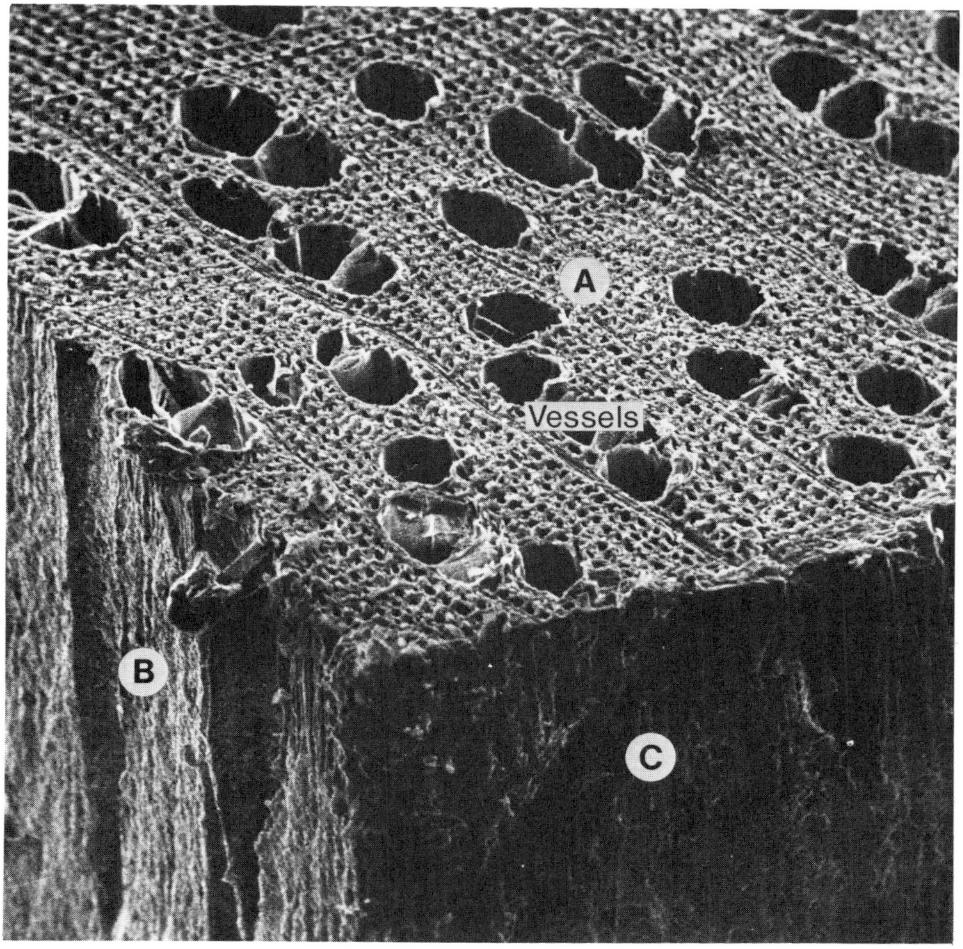


Fig. 2.2. Cellular array of wood elements: (a) softwood; (b) hardwood (M-S Ilvessalo-Pfäffli — reproduced by permission of The Finish Pulp and Paper Research Institute).

these are pectic substances and certain highly complex organic compounds that are deposited in small amounts in the heartwood: but these are readily removed during pulping.

## 2.5 WOOD — CHEMICAL NATURE

A generalised statement of the proportions of the principal constituents of wood is given in Table 2.6 from which it will be seen that the content of cellulose is similar in the normal wood of both coniferous and broadleaved trees; somewhat less than half the total. However, whereas in typical softwoods the lignin content is only marginally greater than that of hemicellulose, in hardwoods the amount of hemicellulose greatly exceeds that of lignin so that softwoods contain appreciably more lignin and less hemicellulose than hardwoods. (The significance of the values for compression wood and tension wood will emerge on p. 127, 128). Of these three major constituents of wood, cellulose is essential for papermaking and the hemicelluloses are mainly beneficial: lignin, on the other hand, is detrimental and its removal from between, and within, the walls of the tracheids is the object of chemical pulping, and, to a large extent, of bleaching.

### *2.5.1 Cellulose*

Cellulose is a clearly definable chemical substances which, in fact, exists in isolation, and in an almost pure form, in the fine hairs attached to the seeds of cotton. Comprising at least a third of the world's vegetable matter, it is widely distributed throughout the plant kingdom as a structural material, the chief exceptions being certain green algae and most fungi (in which xylan and chitin, respectively, provide the structural basis). It is also absent from

TABLE 2.6

Generalised analysis of wood — as a percentage

| | Cellulose | Hemicellulose | Lignin | |
|---|---|---|---|---|
| Conifer — normal wood | 44 | 27 | 29 | (100) |
| | *(41—47)* | *(22—32)* | *(26—32)* | |
| Conifer — compression wood [a] | 30 | 31 | 39 | (100) |
| Broadleaf — normal wood | 44 | 35 | 21 | (100) |
| | *(40—48)* | *(28—42)* | *(17—25)* | |
| Broadleaf — tension wood [a] | 57 | 30 | 13 | (100) |

[a] Values found by Meier [11] for a single species in each case.

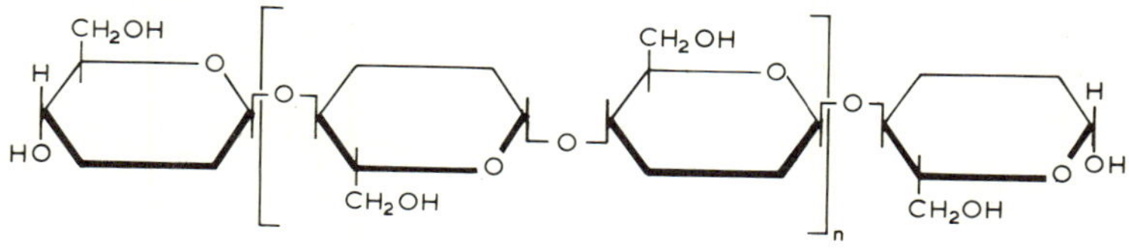


Fig. 2.3. Cellulose molecule.

the cell walls of bacteria, although it is actually synthesised by some species of *Acetobacter*. It appears in some animals, notably in the mantle of certain marine species, and has been found [12] in small quantities in human tissues. The possibility has been advanced of the presence of cellulose in space, as a component of inter-stellar dust[13].

Chemically cellulose is a linear polymer of β-D-glucopyranose units linked by 1,4 glycosidic bonds (Fig. 2.3). The anhydroglucose units are not quite plane but assume the "chair" configuration of Fig. 2.4, from which it will also be noted that, as successive glucose residues rotate through 180°C about the molecular axis, it is the cellobiose unit that is the repeat unit of cellulose. Isolated samples are found to have a molecular weight varying from, perhaps, 50,000 to more than a million, the latter equivalent to a degree of polymerisation of upwards of 7,000 and a length exceeding 0.003 mm. Cellulose is insoluble in water and, generally speaking, relatively inert, though susceptible to oxidation of the primary alcohol groups.

Preston [14] has drawn attention to the significance of the β-1,4 links in cellulose, namely that a chain of sugar residues linked in this way is potentially straight: in cellulose, it is held straight by intra-chain hydrogen bonding along its length (Fig. 2.5). Neither with an α-1,4 link, as in starch, nor a β-1,3 link, as in certain other polysaccharides, is a linear molecule possible. This linearity of the cellulose molecule is of far-reaching consequence in papermaking: only thus is it possible for cellular units to be formed that are endowed with the mechanical properties of plant fibres.

In addition to the glycosidic links, and the intra-chain hydrogen bonding, hydrogen bonds form between the free hydroxyl groups of neighbouring cel-

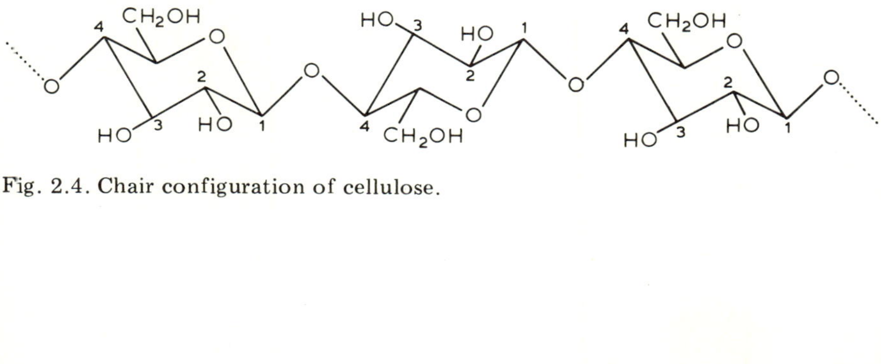


Fig. 2.4. Chair configuration of cellulose.

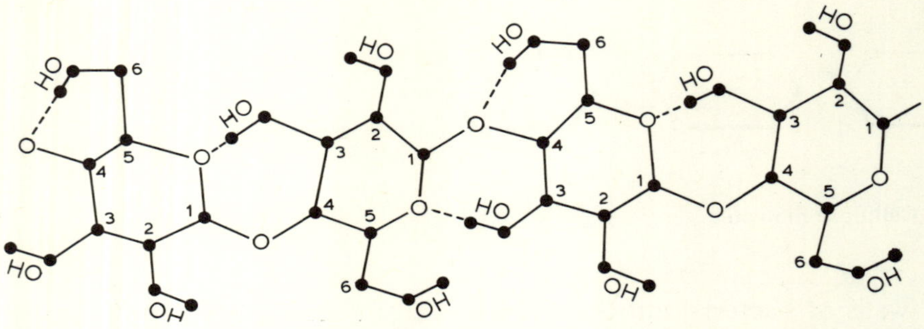


Fig. 2.5. Intra-chain hydrogen bonding of cellulose, according to Hermans [15].

lulose molecules, binding them tightly together. Alone of the three principal constituents of wood, cellulose is (partially) crystalline. The term crystallinity, when applied to macromolecular substances, the individual molecules of which pass through local areas of high order, is not amenable to precise definition. It may, however, be said that a region is crystalline if it is sufficiently large to give a sharp X-ray diffraction diagram and if it behaves thermodynamically as a distinct phase. It is found that typically about two thirds of the cellulose in conifer tracheids fulfils these conditions.

Several crystal habits of cellulose are known but all naturally occuring cellulose assumes the same form, the lattice known as native cellulose or cellulose I, described in 1937 by Meyer and Misch [16]. Many modifications of this have since been proposed but the only one to find wide acceptance is that of Liang and Marchessault [17], illustrated in Fig. 2.6. It is a rhombohedron measuring 0.835 and 0.79 nm in the direction of the *a* and *c* axes, respectively and 1.03 nm, the length of a cellobiose residue, along the *b* axis. The *a* and *c* axes meet an angle of 84°; the *b* axis is orthogonal to the *ac* plane. In the central position between the four corner molecules is a further chain displaced by half a glucose unit which facilitates hydrogen bonding between adjacent molecules. Throughout the period since a lattice structure was first proposed, there has been controversy as to whether adjacent chains along the diagonals of the cell are in fact parallel or, as illustrated, antiparallel; that is, whether all the chains are aligned with their reducing end in the same direction or whether half lie in one direction and the alternate half in the opposite direction. This uncertainty has not yet been resolved.

Cellulose is not uniformly crystalline: extensively distributed throughout the cellulosic material are regions, some 3 by 7 nm in cross-section and at least 50 nm in the axial direction of the chains, that are highly ordered in the crystallographic sense. To these zones, through each of which some one hundred molecules pass for about 100 anhydroglucose units of their length, was originally applied the name *micelle* by extension of a word introduced a cen-

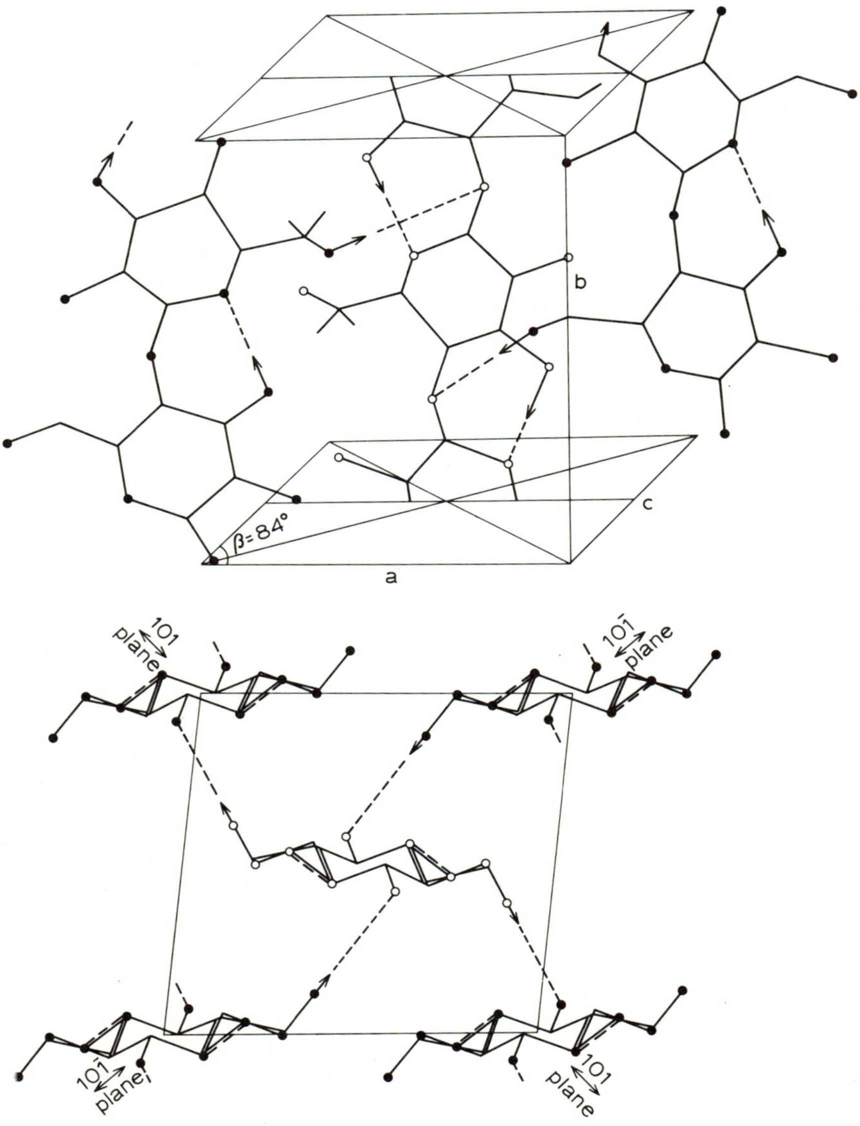


Fig. 2.6. Cellulose unit cell, according to Liang and Marchessault [17]. The arrows depict inter- and intra-molecular hydrogen bonds.

tury ago by Nägeli to describe somewhat similar areas in starch grains: this has tended to yield place, however, to the term *crystallite*. The cellulose molecules, as they continue beyond a crystallite, diverge and twist — to a modest degree, let it be emphasised — and, to some extent, reform in the next crystallite.

Although the crystallites are not individually distinct, the threadlike entity that arises from their linear association is. This is the *microfibril*, the basic structural unit of the plant cell wall. Some 10—30 nm wide, somewhat less in thickness and indefinitely long, it contains 100—2000 cellulose molecules in its cross-section. Around the core of mainly crystalline cellulose in the microfibril, there is believed to be a sheath of paracrystalline polyglucosan molecules, which is itself in turn overlaid by hemicelluloses. Microfibrils are often fasciated into broader fibrils of indeterminate dimensions and the fibrils in turn into sheets or lamellae.

The dimensions of the foregoing structural components are collated in Table 2.7.

Because of the forces between adjacent molecules the crystalline regions are more stable than the amorphous regions, so that it is the latter that are involved in sorption and swelling phenomena and in the early stages of chemical reactions. To the crystalline regions of its microfibrils a tracheid owes its tensile strength. A completely crystalline tracheid, however, would be rigid and it is from the amorphous regions that it derives its flexibility. Bearing in mind that individual molecules pass through several highly ordered and less ordered regions, this combination of strength and flexibility will approach the optimum attainable.

Before passing from the consideration of cellulose, it may be remarked that the precise mechanism and route of synthesis in nature is not fully established. Less obscure are the processes of degradation, so essential for the viability of the ecological system of which it forms so prominent a part. It may also be noted that the industrial use of cellulose is not limited to papermaking; in addition to textile fibres, both natural and regenerated, it finds use in a wide range of products, such as thickening agents, explosives, lacquers, plastics, adhesives and foodstuffs.

TABLE 2.7

The dimensions of the elements that cellulose comprises

| Element | Length | Width |
|---|---|---|
| Unit cell of cellulose I | 1.03 nm | 0.835 nm |
| Cellulose molecule of DP 6000 | 3 $\mu$m | |
| Crystallite | 50 nm | 7 nm |
| Microfibril | indefinite | 10-30 nm |
| Fibril | indefinite | indefinite |

*2.5.2 Hemicellulose*

Closely associated with cellulose in plant cell walls, particularly those of lignified tissues, is a group of chemically ill-defined polysaccharides of low molecular weight to which, almost a century ago, the name *hemicellulose* was given [18]. The term "wood polyoses", also formerly employed for the polysaccharides of arborescent plants, has now largely dropped out of use.

The loose definition of this group stems from the fact that individual molecules contain more than one kind of sugar residue and that these are present in variable proportions, and from the consequent inability to elucidate the polysaccharide structure in other than statistical terms. But although knowledge of the exact chemical structure of these polysaccharides remains incomplete, the intrinsic component residues of pentose and hexose sugars, together with those of some uronic acids, are well established. (The pentose sugars are those with five, the hexoses with six, carbon atoms. The oxidation of the primary hydroxyl to a carboxyl group converts a neutral sugar to a uronic acid.) Only a few are involved in the hemicelluloses of land plants, principally those of D-xylose, D-mannose, D-glucose, D-galactose, L-arabinose, 4-*O*-methyl-D-glucuronic acid, D-galacturonic acid and D-glucuronic acid, the structures of which are shown in Fig. 2.7; less extensively found are L-rhamnose, L-fucose and various *O*-methylated neutral sugars [19]. The carbohydrate composition of wood, based on four softwood and sixteen hardwood species, is shown in Table 2.8.

Wood hemicelluloses are soluble in water or in aqueous alkali, by which they can be isolated from delignified tissues. Although the molecules are

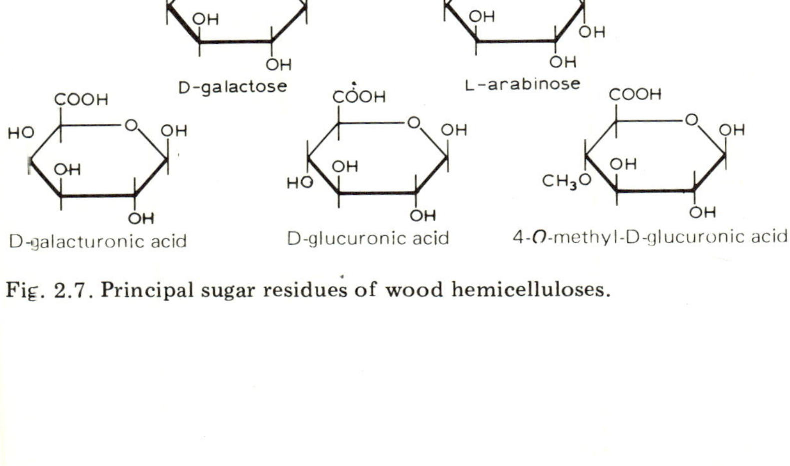


Fig. 2.7. Principal sugar residues of wood hemicelluloses.

TABLE 2.8

The carbohydrate composition of wood [20]

| | Softwoods | | Hardwoods | |
|---|---|---|---|---|
| | Mean (%) | Range (%) | Mean (%) | Range (%) |
| | of 4 species | | of 16 species | |
| Glucosan | 63.6 | *(61.0—65.5)* | 65.2 | *(55.0—74.0)* |
| Mannan | 12.5 | *(7.5—16.0)* | 2.4 | *(0.5—4.0)* |
| Galactan | 10.8 | *(6.0—17.5)* | 2.4 | *(1.0—4.0)* |
| Xylan | 10.5 | *(9.0—13.0)* | 28.6 | *(19.5—39.0)* |
| Arabinan | 2.6 | *(0.5—3.5)* | 1.5 | *(0.5—2.5)* |

essentially linear, the variability of the basic units and the existence of many, albeit short, oligosaccharide side-chains prevents them from aggregating into a crystalline array. The presence of hydroxyl groups endows them with an affinity for water. The same mechanism of hydrogen-bonding between the respective polar functional groups binds hemicellulose readily to cellulose, around the mainly crystalline microfibrils of which they occur as amorphous encrustants. The hemicelluloses are also closely associated with lignin.

No cases are known of a wood hemicellulose having a homopolymeric molecule. Furthermore, substantial differences exist between the hemicelluloses present in softwoods and hardwoods, an indication of which is afforded by Table 2.9. This Table provides only a broad outline, however; species are known, such as those of larch, that have characteristically high contents of hemicelluloses other than those listed.

The normal wood of conifers typically contains some 22 to 32% of hemicellulose. This is chiefly, and indeed in the case of Eastern hemlock (*Tsuga canadensis*) almost entirely, attributable to a series of closely related tripolymers of galactose, glucose and mannose that differ principally in their molec-

TABLE 2.9

The chief hemicelluloses of wood

| Softwoods | Hardwoods |
|---|---|
| *Major* | |
| *O*-acetylgalactoglucomannan | *O*-acetyl-(4,*O*-methylglucurono)xylan |
| *Lesser* | |
| arabino-(4,*O*-methylglucurono)xylan | glucomannan |

ular weight and their content of galactose. Molar ratios of the Gal : Glu : Man of 1 : 1 : 3 for a water-soluble component, and 0.1—0.2 : 1 : 3 for one that may be extracted by alkali, have been reported [21]. (Somewhat different ratios found by other workers may reflect generic or specific differences of the woods examined but changes in the relative sugar ratios seem likely to occur during isolation and purification.) Both, it has been established, have a framework of a linear molecule containing mannose and glucose residues randomly distributed, to which 1,6-linked galactose units form terminal side chains. *O*-acetyl groups are also present, probably attached to $C_2$ or $C_3$ of the mannose residues.

A heteropolymer of a different type present in softwoods in smaller, though substantial, quantities is arabino(4,*O*-methylglucurono)xylan which, in the case of pine at least, contains the uronic acid, arabinose and xylose residues in a molar ratio of 1.9 : 1 : 1.9 [22,23]. The molecule consists of a main chain of $\beta$-1,4-linked D-xylose residues, some of which are substituted at $C_2$ by the methylated uronic acid residues and some at $C_3$ by L-arabinose residues [24]. Uniquely in larch, an arabinoxylan exists additionally as a major constituent of the hemicelluloses [19].

Hardwoods contain some 28—42% of hemicellulose, the upper limit approaching the average content of cellulose. The principal hemicellulose of hardwoods is a glucuronoxylan consisting of a chain of $\beta$-1,4-linked xylose residues, some of which are substituted at $C_2$ by $\alpha$-linked 4-*O*-methyl-D-glucuronic acid or at $C_3$ by acetyl groups [25,23]. In birch, the methylated glucuronic acid residues are present to an average extent of 1 in 10 xylose residues; about 7 in 10 of the xylose units are acetylated [25,26]. According to Timell [19], white birch and silver birch seemed to contain more xylan than any other hardwood studied to that time, namely, 33—35% of the wood, this being seldom exceeded among herbaceous angiosperms, even in the case of graminaceous plants.

Also present in hardwoods, but in smaller quantities, is glucomannan. Unlike the (major) hemicellulose of softwoods this is devoid of galactose residues. The ratio of glucose to mannose units is generally 1 : 2; with birches, in which it is 1 : 1, a notable exception [19].

The degree of polymerisation of the hemicelluloses is very much less than that of cellulose. Thus the number-average DPs of the xylan of ten hardwood species were found [19] to lie within the range 150—200: their weight-average DPs were only slightly higher indicating a low degree of poly-dispersity.

The galactoglucomannan and arabinoglucuronoxylan of softwood and the glucuronoxylan and glucomannan of hardwoods are probably the only hemicellulose occurring in wood in sufficient quantity to be of technological significance to the papermaker. As indicated above, however, other polysaccharides may be present, such as arabinan, arabinogalactan and galactan. Furthermore, it would seem probable that there is a continuous spectrum of uronic acid-containing polysaccharides ("polyuronides") from those with a low proportion of uronic acid residues, as described above, to those containing a very high proportion. The latter are generally classified as pectic substances and are discussed below.

### *2.5.3 Lignin*

Lignin is a complex, systematically-polymerised, highly-aromatic substance with a three-dimensional, highly-branched chain. It is found only in plants more highly evolved than the mosses. It never occurs alone in nature but it is present, as an encrustant, in all woody tissues: indeed, it is the second most common class of organic molecule in living organisms, being about half as abundant as cellulose. However, despite a considerable amount of investigation, it remains ill-defined and, in fact, its constitution may not be precisely definable. There is a marked difference between the lignin from hardwoods and that from softwoods: indeed, its chemical complexity may vary among hardwoods from species to species and possibly within an individual plant at different stages of its growth. Strictly speaking therefore, in the same way that one speaks of hemicelluloses, one should refer to lignins when discussing this type of wood constituent; but the plural form is seldom used.

Part of the difficulty of studying the chemical structure of lignin arises from that of extracting the material without the severe modification that most chemical treatments, such as oxidation, produce. Although lignin is never found in plant cell walls except in close association with polysaccharides, the extent to which this association is physical or chemical, that is, whether it is due to entanglement and secondary forces (van der Waals and hydrogen bonding) or whether to primary chemical bonds uniting some of the components, remains a matter of debate. But whether or not a substantially chemical lignin-carbohydrate complex exists, the presence of some bonds between the lignin and carbohydrate, possibly of a glycosidic nature, seems almost certain.

Lignin can be isolated less drastically by fungal digestion of the polysaccharide fraction, for example by the brownrot fungus *Lentinus lepideus*. Among the unit molecules that are thus obtained the phenylpropanoid units coniferyl alcohol (guaiacyl), present in both hard and softwoods, and sinapyl alcohol (syringyl), which is restricted to hardwoods, are particularly prominent (Fig. 2.8). The phenylpropanoid units are associated, as many as one hundred of them, in the lignin molecule by cross-linkages that are variable and are probably formed randomly. It is this variability of the linkages, and the fact that some may link the lignin molecule with the carbohydrate fraction, that has made the structure of lignin so difficult a field of study.

Lignin then forms a cementing matrix that holds together the cellulose fibrils in the tracheid, and the tracheids one to another, and, because of its physical nature, imparts considerable mechanical strength and permits the growth of tall plant structures. It also provides a measure of protection

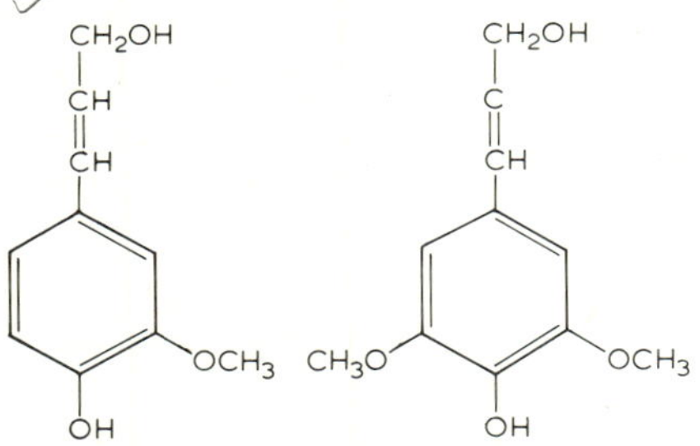

(a) Coniferyl alcohol (Guaiacyl) (b) Sinapyl alcohol (Syringyl)

Fig. 2.8. Phenylpropanoid units found in lignin: (a) coniferyl alcohol (guaiacyl): (b) sinapyl alcohol (syringyl).

against chemical and mechanical damage to the cells. (These, it may be noted in passing, are properties not required of seed hairs and those of cotton, for example, are totally devoid of lignin.) In addition, the hydrophobic properties of lignin confer on the walls of the dead cell a measure of hydrostability that is highly advantageous, and may be essential, to water-conducting tissues in bulk.

In the developing walls of xylem cells lignin is present, if at all, to a comparatively small extent only. There comes a time, however, when the tissue assumes a more rigid form due to the diffuse interposition of lignin between, and within, the walls of adjacent cells. This stage is readily recognisable, for example, in the soft, green tissues of the stem of herbaceous plants, as it is accompanied by a pronounced yellowing of colour and a loss of pliability; that is, the stem becomes "woody". The incidence of lignification accompanying the death of the cells is necessary if they are to continue to function as structural or conducting elements in the tissue, for without lignin they would readily collapse on the withdrawal of the protoplasm. As it has been succinctly expressed: "wood is wood because of lignin".

It has been speculated that it has been the excretion of waste metabolic products by plants — a matter of difficulty in view of their low surface-to-volume ratio and their lack of well-defined organs for the purpose — that has led to the evolutionary origin of lignin, as well as to that of the extractable products of heartwood. It is possible to envisage a complex route of biosynthesis of metabolites that have been excreted into the cell wall; along this path they are acted upon by appropriate enzymes giving rise to lignin and other materials as end-products [27].

With certain stains, such as phloroglucinol, lignin shows characteristic colour reactions that are not exhibited by other fibre constituents [28]. This has been used, together with fluorescence and ultra-violet microscopy, to follow the development of lignification in plant tissues. It is also widely used for the recognition of lignin-rich high yield pulp.

The pulping procedures used to obtain fibrous material for papermaking from wood and other plant tissues are designed to separate the fibres by removing the lignin and pectic substances in the middle lamella. The last few per cent of lignin within the fibre wall can only be removed towards the end of industrial digestion at the expense of considerable degradation and removal of the hemicelluloses and, to some extent, of the cellulose itself. Where, therefore, a good white colour is desired, special bleaching operations, more specific to lignin and the other discolouring substances, are employed. Virtually no lignin is removed by the grinding of wood in the preparation of mechanical pulp, in which case again, if colour is a consideration it must be improved by subsequent bleaching operations.

Vast quantities of lignin derivatives are produced annually in the waste liquors of pulping processes. Much research has been devoted to the chemistry of lignin in the hope that a fuller knowledge of its nature and structure would lead to the development of a major commercial outlet. At present it finds limited use in various ways, in many of them in the form of lignosulphonates from the spent liquors of the sulphite pulping process: as a filler in certain plastic moulding compounds, as a binding agent for dirt road surfaces, as a dispersant and as a source of vanillin, the essential oil of vanilla, which is used in the plating baths of the zinc plating industry.

### *2.5.4 Minor constituents*

Like the hemicelluloses, from which they are usually regarded as distinct, the *pectic substances* are a complex group of polysaccharides for which again it is difficult to frame a precise definition. The former concept of the pectic triad of distinct polysaccharides, consisting of a galacturonic acid (pectic acid), its methyl ester (pectin) and an arabinan, is now known not to have been entirely adequate. Instead there are believed to be several types of polysaccharide, varying considerably in size and structure. On the one hand are highly acidic and relatively linear molecules, such as unesterified galacturonic acid: at the other extreme are polysaccharides that are only mildly acidic, many of the carboxyl groups being methyl-esterified. The pectic substances are widely distributed among plants, occurring in all lignified, as well as many green, tissues. They account for only about 0.5% of the dry weight of mature wood, where they are located in the intercellular regions and the outer layers of the cell walls: from these sites they are completely extracted by conventional pulping processes.

Fats, waxes, resins, mucilages and gums are all organic substances intimately associated with cellulose in its natural form. These, too, are largely removed by cooking but residual resin, which amounts before digestion to upwards of 2.3% of softwoods and 0.7—3% of hardwoods — it is not found in herbaceous plants — sometimes gives trouble to the papermaker by building up as a pitchy deposit on the machine. This is particularly the case with sulphite and high yield pulps.

Mucilages and gums are both groups of complex polysaccharides that are highly branched and contain a relatively large number of different sugar residues. Mucilages are products of normal metabolism; gums, on the other hand, are formed by various organs of the plant in response to injury. They are virtually absent from chemical pulps but are sometimes added in the beater to promote inter-fibre bonding in the paper sheet.

Proteinaceous residues, which derive from the protoplasm that once constituted the cell contents, remain in the lumen when growth ceases. This residual denatured protein, which amounts to 1—2% of wood and cotton and 2—4% of straws, has no significant effect on the pulping and papermaking properties of the fibres.

The mineral content of fibres ranges from about 0.2—1.0% in wood, 0.7—1.6% in cotton and 4—8% in grasses and straws. Rice straw is noteworthy in this respect; its mineral content may be as high as 20%.

## 2.6 WOOD — FINE STRUCTURE

### *2.6.1 Softwood tracheids*

The cells of animals and of the lowest plant forms, such as algae, have only a thin envelope to contain their protoplasm. The case of the higher plants, however, is different: if they are to attain any significant size above ground, the limiting membrane of the cells must have a measure of rigidity to withstand the quite large hydrostatic pressures that then exist. This is conferred by a series of co-axial layers of cellulose microfibrils, embedded in an amorphous matrix of hemicelluloses and, in due course, of lignin. Understanding of the wall structure has been greatly advanced over the last few decades by the electron microscope used both in its original "transition" mode and, more recently, as a scanning instrument. Four superimposed layers, or groups of lamellae, have been distinguished in mature cells. These are known as the primary, and the outer, middle and inner secondary walls, the last being also sometimes referred to as the tertiary wall.

At the time of cell division and during subsequent enlargement a very thin membrane, the *primary wall* (P in Fig. 2.9), alone envelops the protoplast: this is a single, discrete, non-laminated layer. In the developing cell about nine-tenths of it is water: of the solid material little more than one half is cellulose, in the form of microfibrils, the remainder being pectic and hemicellulosic material. The microfibrils are therefore widely spaced and, although to some extent interwoven, have the important ability to move relative to one another as the cell enlarges, which it does chiefly by surface growth at the tips. It is believed that the microfibrils are deposited more or

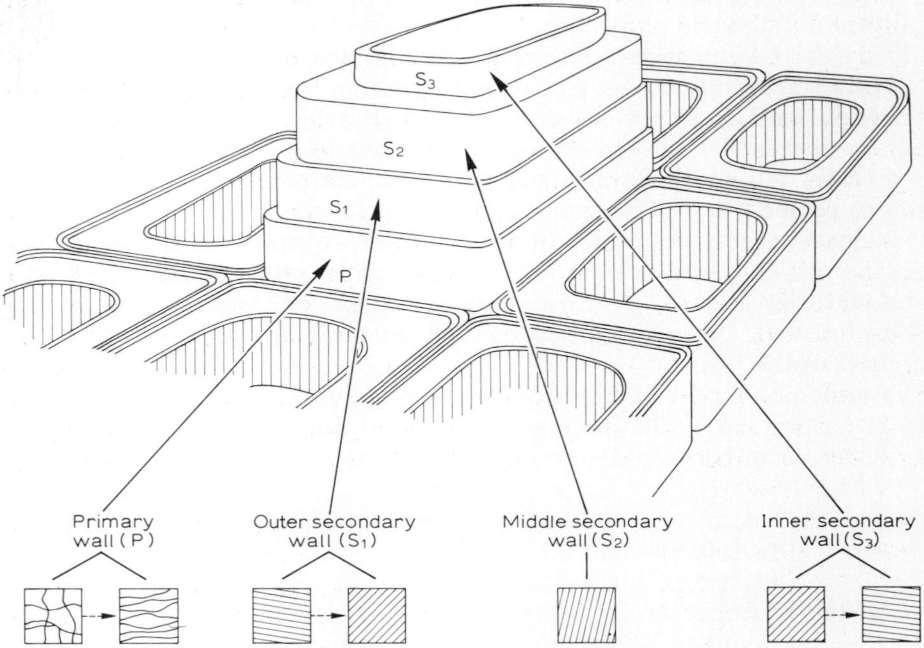


Fig. 2.9. Wall structure of softwood tracheid.

less transversely but that as the cell elongates, and its surface increases, this orientation becomes very much less marked, due to the longitudinal displacement of the microfibrils. Towards the end of this process of cell enlargement, the last microfibrils of the primary wall are laid down with a considerable degree of transverse orientation [29]. The primary walls of two contiguous cells together with the intervening intercellular layer are sometimes referred to collectively as the *middle layer* or compound middle lamella. The widely-separated microfibrils dry down after pulping, which removes the lignin and pectic substances of the middle layer, to a relatively open and very thin membrane, probably only a few tens of nanometres thick.

Within the primary wall, the secondary wall is laid down. Normally this process does not commence until surface growth of the cell has ceased, although it may begin locally while growth is yet proceeding in other parts of the cell [29]. The chemical differences between the primary and secondary walls are not great but, structurally, there are marked dissimilarities. While the thickness of the primary wall before isolation from the tissue is probably only a few hundreds of nanometres, that of the secondary wall of a mature tracheid is of an order or two greater. Whereas in the primary wall the microfibrils form a single layer, are widely spaced, able to move relative to one another, are interlaced and largely unordered, in the secondary wall they form a series of coaxial lamellae, in each of which they lie parallel, or nearly so. They are packed so closely together that to some extent they fasciate laterally, parallel to the outer surface of the tracheid, so that they lose their individuality and become incapable of mutual spatial adjustment during the development of the cell wall. The secondary wall therefore tends to break down, on mechanical disintegration after delignification, into thin membranes and into "microscopic macrofibrils". These latter are generally referred to as fibrils: they are not morphological entities, their width is indeterminate and they can be further reduced towards the individual microfibrils described earlier. The secondary wall defines the shape of the cell and, after lignification, confers rigidity upon it, this providing the main basis of the cell's contribution to the mechanical support of the plant. Three layers of this wall have been distinguished on the basis of the orientation of the (closely packed) microfibrils and, although these layers are no longer believed to be ontogenetically discrete, it will be convenient to describe the structure of the secondary wall in terms of them.

The *outer secondary wall*, $S_1$ in Fig. 2.9, is the first of the three layers to be deposited. In thickness, it is comparable to the primary wall before the extraction from that of the non-cellulsic material. Typically, it consists of four to six lamellae, the outermost of which spiral slowly and in opposite senses about the longitudal axis of the tracheid. The spiral of the inner lamellae becomes progressively steeper in a transition to the orientation of the fibrils of the middle secondary wall [30]. This has been established for a single pine species but it appears to develop, rather than conflict with, earlier findings over a wider range of woods; it would therefore seem reasonable to suppose that it has wider validity. Deposition of each lamella commences midway along the cell and moves progressively towards the tips, so that there is normally less material towards the tapered ends [29].

Deposited on the inner surface of $S_1$ is a series of lamellae, up to 150 in number, each, very broadly speaking, of similar orientation. These collectively are known as the *middle secondary wall* ($S_2$) and, except in very immature cells, constitute the bulk of the cell-wall substance. This wall varies in thickness within wide limits depending, among other factors, upon the position of the tracheid within the tree, the season of the year and the climatic conditions under which the tree grows. Typically, it is less than 1 $\mu$m thick in earlywood tracheids and in latewood cells is up to 5 $\mu$m thick; in the summerwood of some species, for example Douglas fir, $S_2$ is sometimes developed to such a thickness as almost to close the central canal or lumen. The microfibrils of $S_2$ spiral steeply about the axial direction at an angle of some 10 to 20°. A few lamellae towards the outer and inner limits of this wall exhibit a rather less steep helix, transitional between that of the central region of $S_2$ and that of the inner or outer secondary wall which they adjoin. In all lamellae of $S_2$ the sense of the spiral is invariable: indeed, the sense of rotation would seem to be constant throughout a species and possibly to a higher taxonomic level.

The *inner secondary wall* ($S_3$), sometimes referred to as the *tertiary wall*, is the last deposited layer of cellulose and in fact is not always developed. Dunning [30] found twelve lamellae in $S_3$ of longleaf pine and concluded that the true number may be somewhat larger. The microfibrils of the outermost lamellae are almost parallel to those of the adjacent $S_2$, the helix of successive lamellae becoming progressively flatter to an almost transverse orientation. In the innermost four lamellae the sense of the helix is reversed. In some species, notably yew (*Taxus baccata*) and Douglas fir (*Pseudotsuga taxifolia*), prominent spiral thickenings can be observed within $S_3$: these, too, are cellulosic and consist of bundles of microfibrils on the inner face of $S_3$, of which they form a part.

Bordering the lumen of some species are to be found wartlike lumps arranged regularly and, within a species, consistently. To this, when it occurs, the name *warty layer* is applied; an analogous, but somewhat less-developed, layer present in some hardwood fibres is referred to as the *terminal lamella*. The warty layer would appear to be formed by the deposition of protoplasmic material as the cell dies, though why the pattern of warts should be so regular and so specific has yet to be explained.

Seen from a papermaking viewpoint, the wall structure may be briefly summarised as follows. The bulk of the microfibrils form a large number of co-axial lamellae in $S_2$ where they spiral steeply, in the same sense and at virtually the same small angle to the direction of the longitudal axis of the tracheid. Surrounding these, the lamellae of $S_1$ are "cross-gartered" and more nearly transverse, an arrangement that constrains the outward swelling tendency of the $S_2$ layers when they imbibe water. External to $S_1$ is a very thin layer (P) which is of no technological significance once the lignin has been removed from it. Within $S_2$, and bordering the lumen, is a further series of lamellae ($S_3$), which is also believed to be technologically unimportant.

Of particular consequence in papermaking, however, are the length of the tracheids, the overall thickness of the cell wall and the diameter of the lumen. A length of 3—4 mm, which embraces that of many coniferous species, is particularly suitable for many grades of paper. Latewood tracheids are thick-walled and have a narrow lumen: earlywood fibres, on the other hand, are relatively thin-walled with an appreciably larger lumen. Consequently, the latter conform more readily to the compacting forces that act on the fibres during the removal of water on the papermachine to give a denser and more strongly-bonded sheet. A further significant consequence of thin-walled cells is that they yield a larger number of fibrous units per unit weight of pulp; that is, for a given basis weight, unit area of paper will contain more structural elements (of the same length).

The final stage in the differentiation of the tracheid is that of lignification. This process commences in the corners of the primary wall: from here the deposition of lignin spreads outwards into the intercellular layer, where it is associated with the pectic substances, and, to a lesser extent (about one third of the total in softwoods), inwards into the secondary walls where it encrusts the microfibrils of hemicellulose-coated cellulose [29]. This lignin, although permeating the cell was as a continuous phase, therefore also tends to be present in the secondary walls in co-axial layers, although the concentration decreases towards the lumen. Lignification is terminated by the death of the cell. As the thickness of wall substance is greater along a radius of the cell's cross-section that passes through the angled corners, it is not surprising that these corners often retain lignin after cooking; its presence can then be observed microscopically as prominent thick ridges along the length of a fibre. It has been suggested that, apart from changes in the mechanical properties of the cell wall that it brings about, a purpose of lignification is to limit further surface growth of the primary wall by entering into chemical combination with the matrix constituents that otherwise promote it [29]. Be that as it may, it undoubtedly anchors the microfibrils, at the same time conferring on the tissue as a whole a measure of rigidity and providing it with protection against mechanical and chemical damage.

As even a thousand intact tracheids weigh barely one milligram, there are considerable difficulties attendant upon attempts to establish the distribution of the various polysaccharides throughout the cell wall. An approach, adopted by Meier [31,23] to study the tracheids in wood of pine (*Pinus sylvestris*) and spruce (*Picea abies*), involved separating cells into groups of varying age and chemically analysing their constituents at various stages of maturation. More recently, Kibblewhite and Brookes [32] have analysed various fractions of the fines material removed by beating from the outer surface of *Pinus radiata* tracheids from kraft and bisulphite pulps: the source of the material from the cell wall was identified by the orientation of the microfibrils. This method was limited to material from P, $S_1$ and the outer layers of $S_2$ but, with this qualification in mind, the following summarises the results in the four cases, namely pine and spruce tracheids from wood and pine tracheids from the two pulps. The glucomannan content progressively increases radially inwards from P and that of galactan in the reverse direction. In the case of the pentosans, arabinan and xylan, the two pulps were at variance, the kraft pulps being richer in pentosans in $S_1$ than in $S_2$, and richer still in P, whereas the reverse was true for the bisulphite pulp, to an extent that depended upon the yield. No clear pattern emerged for the distribution of pentosans in wood.

One of the functions of the xylem tracheids of conifers, it has been noted, is the conduction of sap. This travels vertically up the trunk through the lumen of the cells but, as these are not open-ended, it is necessary to have some means by which it may pass between adjoining tracheids. This is effected by communicating openings in the cell walls, known as *pits*, that occur in longitudinal rows chiefly on those faces of the tracheid that are radial in the tree. For a pit to be effective it must be matched in the contiguous cell by a complementary pit. As the primary walls of adjoining young cells extend, relatively thin areas (known as primary pit fields) develop. Through the perforations of these sub-microscopic strands of protoplasm pass, providing a connection between the contents of each cell: it is around these strands that the pit pairs later develop in the walls of adjoining tracheids.

The most prominent type of pit found on softwood tracheids is the *bordered pit*. The structural elements of this are illustrated in Fig. 2.10, from which it will be seen that the secondary wall of each cell forms a dome, known as the pit border, the centre of which is the pit aperture. The pit border encloses the pit chamber. The microfibrils of the two primary walls intervening between the two pit chambers become radially orientated and gathered into a system of fine spokes (pit membrane) between which, towards the periphery, aqueous liquids can pass from tracheid to tracheid. The central region of the membrane has a thickening, known as the *torus*, which is more or less impermeable. By moving from side to side, the membrane and torus act jointly as a valve to regulate the flow of sap. In the heartwood of felled softwoods, the torus may be permanently seated across the pit aperture and this impedes the penetration of pulp cooking liquors and timber preservatives. The highly parallelised microfibrils of each lamella of $S_2$ are deflected to one side or the other as they sweep around the pit opening.

*2.6.2 Parenchyma cells*

The cells of the specialised tissue, known as parenchyma, concerned with the storage of reserve food materials differ essentially from tracheids in that they retain their living contents for several years after their development is completed, for food materials can only be stored in, and transported by, living cells. Death of the parenchyma, in fact, is the fundamental change underlying the transformation of sapwood to heartwood.

Parenchyma tissue may be present in the tree stem as vertical and horizon-

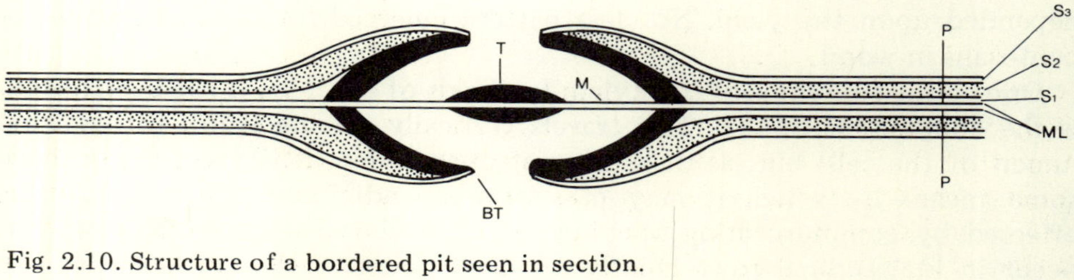


Fig. 2.10. Structure of a bordered pit seen in section.

tal ribbon-shaped strands of cells. Vertical parenchyma is sparse in softwoods; in some species (of e.g. pine) it may be completely absent. The horizontal parenchyma, disposed in a radial plane in the trunk where it forms the familiar rays, is invariably present.

Rays, which comprise 5—10% of the bulk of softwoods, are often only one cell in width in a tangential section but form a broad plate in a radial section. The parenchyma cells are short compared with vertical tracheids, with an aspect ratio of 2 : 1 to 4 : 1, and to a certain extent they pass through the screens and the wire during pulping and papermaking. Within the rays of some species, of e.g. spruce, pine and larch, resin canals are present. These are continuous radial channels lined with *epithelial cells* in which a resinous substance is secreted. When discharged into the canal, this is transported to the site of any injury to the tree caused by wind, animals, insects or fire. The liquid oleoresin, sometimes referred to as balsam — a name in some cases transferred to the tree — oxidises and polymerises as the more volatile constituents evaporate in contact with the air. The resins in the epithelial cells, together with fatty acids also present, are not removed by sulphite digestion. On beating they may be released into the papermaking stock, giving rise to the pitchy deposits on the machine wire referred to earlier.

The structure and chemistry of xylem parenchyma cells have not been studied in a depth comparable to that of tracheids. It would appear that the thin walls of parenchyma cells have a broadly similar arrangement of layers to that of tracheids, the fibril orientation in the outermost and innermost layers of the secondary wall forming a flat helix while those of the middle layer are axially oriented [29]. Attention has recently been drawn [40] to the by no means infrequent occurrence in wood of parenchyma cells with irregular, and in some cases highly-convoluted, ingrowths from the wall into the lumen. The enhanced surface area is believed to assist the diffusion of solutes from cell to cell; for which reason such cells have been termed (xylem parenchyma) *transfer cells.*

In both softwoods and hardwoods, the parenchyma cells have been found to be rich in xylan-containing polysaccharides: the proportions of arabino-4-*O*-methyl-glucuronoxylan present in the vertical tracheids and the parenchyma (and associated ray tracheids) of *Pinus sylvestris* respectively were 17 and 28%; those of 4-*O*-methyl-glucuronoxylan in birch (*B. verrucosa*), 46% and no less than 84% [33].

Bordered pits are characteristic of the vascular system; only rarely do they occur in parenchyma cells. The pits connecting these, known appropriately as *simple pits*, consist merely of an aperture in the secondary walls, covered by a flat membrane of the two primary walls and the intervening middle lamella. These openings vary widely in size and shape between species but in no case does the pit membrane have a torus. Where a tracheid (axial) crosses a ray parenchyma cell (radial), a composite form of these two types of pit occurs; that half of the pair on the living side being simple and the half on the tracheid side being bordered, though this border is often only weakly developed. In these cases the torus is usually wanting or, at most, poorly developed. These *half-bordered pits* are usually sufficiently characteristic in size and shape to permit the identification of species.

### *2.6.3 Hardwood fibres*

Notwithstanding a two-to-one preponderance of hardwood over softwood trees throughout the world, the former provide only between one sixth and one fifth of the world's woodpulp used for papermaking. There are several reasons for this: extensive pure stands of a specific hardwood are rare and the different species of a typical hardwood forest have varying chemical requirements for pulping; the wood in the stem of the hardwood tree is usually a smaller proportion of the entire tree and the want of linearity of the branches makes their handling and debarking difficult; the tissues of hardwoods lack the structural homogeneity of those of softwoods; and the length of hardwood fibres is usually less than that of softwood tracheids and rather below the optimum for those grades of paper where strength properties predominate.

As noted earlier in this chapter, the angiosperms evolved more recently than the gymnosperms. Whereas in softwoods the functions of mechanical support and water conduction are both performed by the tracheid in hardwoods, these are provided separately, and more efficiently, by the fibre and the vessel respectively. In many cases the fibres account for the bulk of the cells in a hardwood, though the proportion varies considerably with the type of tree. Thus horse-chestnut may have as much as 76% of fibres while beech has only 37% [34]. These figures are on a volume basis; referred to the number of cells the values would be higher.

The hardwood fibre is elongated (most commonly 1—2 mm in length), very narrow (say 20 $\mu$m wide) relative to the softwood tracheid, widest at its mid-point but tapering to sharply pointed, and sometimes bifurcated, ends which are spliced into one another. After the completion of lignification they are physiologically dead. In the heavier hardwoods the fibre wall, which has a structure of layers similar to that of softwood tracheids, is so thickened that the lumen is very narrow and the few pits they have are rather small, these two characteristics indicating the insignificant part played by fibres in the conduction of sap. In addition to the fibres, tracheids may be present in the hardwood tissue. These are longer and thinner-walled and have more prominent bordered pits than the fibres. They may be thought of as vestigial cells analogous to softwood tracheids. Usually no clear distinction can be drawn between the fibres and the tracheids of hardwood; the one-cell type merges into the other, the intermediate cells being known as fibre-tracheids. Fortunately, the botanical differentiation between fibres, fibre-tracheids and tracheids can generally be ignored in discussion of the technological properties of a hardwood.

### *2.6.4 Hardwood vessel elements*

In its translocation along the stem and branches of a softwood tree, sap necessarily moves sideways between adjacent tracheids, since these are closed at their ends. In marked contrast, the sap of hardwoods is channelled in nearly straight lines along tubes, known as *vessels.* These are broad (100 $\mu$m is not unusual) and long (tens of centimetres); they may be likened to field drains, being composed of a series of articulated cells (*vessels elements*), the presence of which provides a ready means of distinguishing between hardwood and softwood pulps. More than 50% of the volume of some hardwoods may be contributed by the vessels but the proportion by weight is generally less than 10%. The elements vary considerably in form from a squat drum shape to elongated tubes, according to species, this affording a useful means of identification of a hardwood pulp furnish. A typical length would be rather less than 0.1 mm. Being functionally conducting channels, their lumen has a large cross-sectional area. Upon biological death the end walls, as well as the cytoplasm, break down to give open, or at least widely perforated, ends, so that the large quantities of water transpired by the broad leaves of hardwood trees can pass freely up the stem. Their walls abound in small pits which connect with, and facilitate communication with, neighbouring cells. Some secondary thickening of the vessel elements occurs and their walls are more highly lignified than those of the fibres: but these features are probably necessitated by the need to withstand the pressure of neighbouring tissues and the lumen-collapsing tension of the column of water rather than to increase the general rigidity of the tree stem.

When seen in cross-section the vessels in a hardwood tissue are usually referred to as pores. They are often found to occur in one or other of two general patterns (though many woods are intermediate): in one of these, the vessels may be restricted to the springwood but in any case are distinctly larger than those of the summerwood (*ring-porous*); in the other, they are distributed more evenly and are more uniform in size (*diffuse-porous*). Oak and ash may be cited as typical ring-porous, and beech and birch, illustrated in Fig. 2.2b, as examples of diffuse-porous, hardwoods.

It is the width of the vessels that makes hardwood trees susceptible to frost. A water column of, say, 100 $\mu$m width or more is more readily disrupted, and is much less likely to be re-established, than one passing through the lumina of softwood tracheids and measuring 30 $\mu$m or less. Consequently it is the vessels of the current or most recent season's growth that are principally involved in conduction in broadleaf trees.

As in softwoods, the storage tissue of hardwoods consists of horizontal (ray) and vertical parenchyma cells. These are often much more abundant than in softwoods and exhibit a greater variety of form. During the transformation from sapwood to heartwood the vessels cease to conduct water and pressure differences can arise between the lumina of the vessels and of neighbouring parenchyma cells. As a result, the living contents of the parenchyma cells may be forced throught the many pit openings linking the two types of cell, giving rise to cyst-like growths that partially, or completely, block the lumen of the vessel element. Such a growth is known as a tylosis. The presence of these tyloses inhibits decay of the wood by impeding the spread of fungal hyphae: on the other hand, they hinder the penetration of pulping liquors or of wood preservatives.

In general terms it may be said that, as a source of papermaking pulp, hardwoods are inferior to softwoods owing to the low yield of elongated cells, which more than offsets the advantages acruing from their low lignin content. Some hardwoods have been found to contain up to 30% of ray cells less than 0.2 mm long [35]. Compared to softwood tracheids, hardwood fibres are shorter and thicker-walled. They contain very little glucomannan, which is particularly effective in promoting inter-fibre bonding. Against this may be set the fact that the total hemicellulose content is higher; 28—42% compared with 22—32% in softwoods.

## 2.7 WOOD — REACTION WOOD

Whenever a tree is subjected to forces that tend to deform it, that is to displace the trunk or limbs from the position in which they would grow in the absence of such forces, abnormal wood develops in the cambium in the region of the displacement. This is known as *reaction wood.* It is formed in stems or branches by growth out of their normal alignment as a result of, for example, high prevailing winds, growth on steep slopes, the lodgement of snow or ice on or against them and growth to one side towards the light and away from excessive shade. It is particularly likely to occur in the early growth rings as, at this stage of development, stems and branches are more susceptible to deforming forces.

Although present in both softwoods and hardwoods, reaction wood exhibits a remarkably different form in the two cases. In softwoods, it is found on the lower side of displaced branches and stems and it is known as *compression wood.* In marked contrast, the reaction wood of hardwoods forms on the upper side: this is known as *tension wood.* Reaction wood varies in degree of severity, is not always readily recognisable and is probably present in some measure to a considerably greater extent than is generally realised. Receiving scant attention in the botany textbooks, it is nevertheless of technological significance to papermakers (and even more so to converters of timber) as the tracheids and fibres produced in these zones have physical and chemical properties differing considerably from those of normal wood of the same species.

*2.7.1 Compression wood*

Compression wood is more readily recognised than the tension wood of hardwoods. It usually results in an eccentricity of the stem or branch but even straight stems sometimes contain compression wood as early displacements may be corrected in later life. It can be detected in the transverse section of the trunk or limb by the eccentricity of the growth rings which, in the region where compression wood has developed, are crescent shaped. These regions are brown or dark red in colour, for which reason the term redwood (Rotholz, bois rouge) is sometimes used. In normally grown stems, it tends to be formed immediately below large branches or forks.

Viewed in cross-section under the microscope, compression wood tracheids are seen to be thick-walled and rounded with intercellular spaces between them, in contrast to normal tracheids the angular cross-sections of which fill the tissue space. Furthermore, the transition from earlywood to latewood is less marked. In longitudinal view the tracheids, which are shorter than normal wood tracheids, often exhibit striking fissures that extend from the pit apertures and run parallel to, and between, the spiralling microfibrils. These discontinuous markings afford a ready means of recognising the presence of compression wood. The layered wall structure may be regarded as a modification of the normal structure, $S_2$ being thicker and showing a larger helix angle with respect to the longitudinal axis. $S_3$ is absent or only feebly developed.

Perhaps the most significant characteristic of compression wood is the high degree of lignification of the cell wall. This is reflected in the figures of Table 2. 6 where it will be seen that the ration of lignin to cellulose, 29 : 44 (i.e. 0.66) in typical gymnosperm normal wood, was found to compare with 39 : 30 (1.30) in the compression wood of *Picea abies* — that is, the lignin-cellulose ratio in the compression wood was double that of the normal wood. The zone of $S_2$ immediately within $S_1$, sometimes referred to as $S_2$ (L), is particularly densely lignified [36]. Lignin-rich tissues occurring on the lower side of leaning stems and branches buttress them until such time as the eccentric growth on this side restores them to their normal position. It is interesting to note that wood from the same growth rings on the opposite side of the tree has a lower lignin and a higher cellulose content than the (normal) side wood which again has obvious mechanical advantages [36]. Compression wood is also notable for a high content of galactose units.

*2.7.2 Tension wood*

The tension wood (Zugholz or Weissholz) of hardwoods, although formed in circumstances similar to those of compression wood, is quite different in its location and in the structure and chemistry of its elongated cells. Apart from the obvious symptom of eccentric growth usually evident its presence is not always easily diagnosed, although a longitudinally sawn log, while green, will display a fuzzy or wooly texture. It is thought, however, to be far more common than is generally realised and values as high as 40%, referred to the entire trunk, have been given for the proportion of tension wood in certain poplar species [37]. In contrast to compression wood tracheids, tension wood fibres may be significantly longer that those of normal wood. There are fewer vessels and, where these do occur, the vessel elements are of reduced size.

It has long been known that tension wood fibres are usually characterised by a thick, gelatinous inner wall, rich in cellulose in a highly crystalline state. This layer, often designated S (G), may be formed in addition to the layer $S_1$, $S_2$ and $S_3$, or replaces $S_3$ or replaces both $S_2$ and $S_3$ [36]. The microfibrils are aligned parallel to the longitudinal axis of the fibre and this, together with the high degree of crystallinity of the cellulose, confers on a tissue of (parallel) tension wood fibres considerable resistance to a tensile load applied in the axial direction. In this way, the presence of tension wood on the upper surface of a leaning branch or stem tends to prevent further displacement until it is restored to its normal growing position.

An example of the increased cellulose content of tension wood is provided by the figures in Table 2.6. The ratio of cellulose to lignin, 44 : 21 (i.e. 2.1) in typical angiosperm normal wood, rises to 57 : 13 (4.4) in the tension wood of *Fagus sylvatica* — that is, the cellulose-lignin ratio is doubled in tension wood. (Too much significance must not be attached to these ratios: as explained at the outset of this section, the extent to which reaction wood may be developed covers a continuous and wide spectrum.)

Both types of reaction wood can be satisfactorily pulped by the conventional methods — fortunately! However, the high lignin content of compression wood results in a low yield and the bleach requirements are seriously affected to the extent that sulphite pulps made entirely from compression wood (an unlikely situation in industrial practice) may be practically unbleachable. The high cellulose content of tension wood, on the other hand, results in appreciably higher yields than normal. Compression wood tracheids tend to disintegrate during mechanical pulping but it has been reported [36] that even severely developed tension wood is readily separated into individual fibres by mechanical treatment. The papermaking strength properties of pulps from reaction wood are inferior to those from normal wood. For compression wood, this is particularly the case after sulphite digestion. The deleterious effect of tension wood fibres in a pulp is somewhat greater: the high degree of crystallinity in the gelatinous layer results in a comparatively low uptake of water and hence, as will be explained in Chapter 3, in a slow rate of beating. Furthermore, the resistance of the thick walls to collapse during drying (the same effect is observed with compression wood tracheids) reduces the overall areas of inter-fibre bonding, and the smaller proportion of hemicellulose its density. Tension wood also differs from the normal wood of broadleaf trees in its high galactan content.

## 2.8 OTHER PAPERMAKING FIBRES

The fibrous cells from virtually any vegetable source can, in principle, be used to make paper. In practice, the plants utilised are limited by a number of factors: the seasonal availability of the plant; difficulties and high cost of handling, transporting and storing the often bulky, and always bio-degradable, material; the low yield of elongated cells and the resulting high processing costs and high effluent load; a high silica content with its abrasive effect on the papermachine and, for some purposes, adverse effect on the optical properties of the paper — any of these factors, as well as others, can limit the use of a plant on an industrial scale. Nevertheless, a limited number of species other than trees are in use. The world's productive capacity for papermaking pulp from fibres other than those of wood was, in 1974, less than 7% of the total [39]; nevertheless, these fibres play a highly significant part in the economy of some less highly-developed countries where wood is not readily available. They may conveniently be classified as (a) seed hairs, (b) leaf fibres and (c) bast fibres, according to the part of the plant from which they emanate.

### *2.8.1 Seed hairs*

The seeds of many plants bear hairs that assist their distribution by the agency of the wind. Only a few of the world's seed hair fibres have been found suitable for commercial exploitation and the only one extensively used for papermaking is that of cotton.

Cotton (*Gossypium* spp.) is grown as an annual plant. The seeds, when mature, are covered with a tuft of down — the cotton fibres. After harvesting the longer fibres, known as *lint* fibres, are separated from the seed by the process of ginning. The value of these for textiles, for which purpose they have been used for 5000 years, puts them beyond the economic reach of the papermaker, except in the form of cuttings and rags.

At the time of opening of the seed capsule (boll), the lint fibres are tubular, the superficial angles resulting from the effect of contiguous cells in the case of fibres grown in tissues being absent. They quickly dry out, however, and, due to the large lumen that exists from the torn open basal end to the closed tapered tip of the fibre, readily collapse; a process that is irreversible in water alone. The direction of twist of the flattened tubes is frequently reversed, a direct consequence of similar reversals in the sense of the spiral fibrils that comprise the bulk of the fibre wall. Lint fibres are generally 12—64 mm long and typically 20 $\mu$m wide (Table 2.10), their great length necessitating cutting in the beater. The wall thickness is typically within the range 2.5—6 $\mu$m. A broad lumen results in a large ratio of surface area to weight, this in turn making for greater buoyancy when the seed is air-borne. Seed hairs are, of course, quite devoid of the pits that characterise the tracheids

TABLE 2.10

Typical values of the length, breadth and aspect ratio of some plant cells

| | Elongated cells | | | Short cells | | |
|---|---|---|---|---|---|---|
| | Length (mm) | Breadth ($\mu$m) | Aspect ratio : 1 | Length (mm) | Breadth ($\mu$m) | Aspect ratio : 1 |
| Pine | 2.0—3.0 | 22—50 | 60—90 | | | |
| Spruce | 3.1—3.5 | 19—50 | 70—160 | | | |
| Poplar | 1.5 | 25 | 60 | 650 | 100 | 6.5 |
| Esparto | 1.5 | 11 | 136 | 40—60 | 10—50 | 3—6 |
| Cereal straw | 1.5 | 13 | 115 | 120—600 | 7—60 | 3—16 |
| Manila | 1.8—6.2 | 11—18 | 254 | 90 | 25 | 4 |
| Sisal | 1.3—2.7 | 19—37 | 35—142 | 12—15 | 18—43 | 3—7 |
| Cotton linters | 2—12 | 20 | 100—600 | | | |
| Cotton fibres | 12—50 | 9—23 | 1000—4000 | | | |
| Flax | 10—36 | 11—20 | 1100—1200 | | | |

and, to a lesser extent the fibres, of wood.

The wall structure of cotton fibres has similarities with that of the fibrous elements of wood. The very thin primary wall (0.1—0.2 $\mu$m thick) consists of more or less randomly orientated fibrils, widely spaced before drying and showing a preferred direction that is longitudinal on the outside and transverse on the inner surface. Within this is a thin layer of fibrils, known as the winding: this is analogous to $S_1$ of wood fibres but reverses direction along the length of the fibre. The bulk of the cellulose of mature cells is in the secondary wall ($S_2$ of wood). The fibrils make an angle of about 30° with the direction of the longitudinal axis but again the sense of the helix is frequently reversed. The secondary wall is lamellate, the coaxial layers being laid down as a result of diurnal variations in the rate of growth. Within the secondary wall, a membrane known as the lumen boundary is formed (cf. $S_3$).

Enveloping the cell walls is a very thin outer cuticle of wax and pectic material that serves to protect the cellulose from atmospheric oxidation in the presence of the ultra-violet radiation of strong sunlight. The cuticle is removed by a relatively mild caustic cooking process. Apart from this wax and pectin and that which permeates the primary wall, together with a small amount of mineral matter and of protoplasmic residues in the lumen, the cotton fibre is almost pure cellulose. The complete lack of lignin is to be expected as the function of this appears to be to confer rigidity.

The ginning operation leaves a fuzz of short fibres attached to the seed that can be removed by further processing. These cotton *linters*, as they are called, are not suitable for direct textile use: consequently although less desirable as papermaking fibres, they are more widely used than lint fibres. The world production in 1972 of cotton linters pulp for papermaking was estimated at (a mere) 60,000 air-dry metric tons.

With a length to breadth ratio an order of magnitude less than that of lint fibres, linters are thicker-walled and collapse less readily on drying. Furthermore, although they give pulps of better colour, linters are somewhat brittle compared with lint fibres. Attempts have been made, with some success, to reduce the rigidity of linters by chemical means, for example by low substitution hydroxy-ethylation, for those papers where softness and bulk are not required. A particular shortcoming of linters is the need to bleach minute adhering fragments of the seed integument: if their removal were achievable solely by mechanical means, comparatively mild bleaching would suffice and the inherent strength of the linters would not be reduced; although it would still not be as great as that of lint fibres, which do not require this treatment.

The celluloses obtained from different kinds of hardwood and softwood appear to possess approximately the same properties, although these are influenced to a certain extent by pulping processes. Wood and cotton celluloses, on the other hand, show certain differences. The DP values reported for cotton cellulose are generally substantially higher than those of cellulose derived from wood: it would also appear that the crystallites of cotton cellulose are about a third longer than those of wood [38]. Part of the difference is probably due to degradation during the isolation of cellulose from the hemicellulose and lignin with which it is associated in wood tissues but a real difference may well exist. Because of the absence of non-cellulosic polysaccharides and lignin from cotton, its cellulose molecules exhibit a greater degree of order than those in wood; as a consequence the reduced active surface of cotton cellulose makes it less prone to degradation. It may be said that, in papermaking properties, the particular attributes of paper made from linters are softness, bulk and absorbency; from cotton cuttings and rags papers of high strength -- particularly folding endurance and tearing strength (a consequence of the fibre length) -- good colour and durability may be made, although, with suitable fibre shortening treatment, papers with the characteristics of linters may be produced. The previous history of rags can have a marked effect: repeated flexing of the fibres in a dry state reduces the DP and degrades the structure.

### *2.8.2 Bast fibres*

Angiosperms are divided into two groups: the monocotyledons and the dicotyledons. Although this classification has been in use for more that three centuries, the two types cannot be distinguished by any single characteristic, nor indeed reliably by any set of morphological features. Nevertheless one respect in which they may often be differentiated is in the distribution of the vascular bundles, which in herbaceous dicotyledons usually exist in a cylindrical pattern around a relatively wide pith; whereas, generally speaking, the vascular bundles of monocotyledons are distributed across the stem which has no definable pith. In the inner bark (phloem) of *di*cotyledons, and associated with the vascular bundles to which they lend rigidity and support, are strands, often of the order of a metre in length, of elongated cells known as *bast fibres*. Some of these are used for papermaking, such as flax (*Linum usitatissimum*), hemp (*Cannabis sativa*), jute (*Corchorus* spp.), ramie (*Boehmeria nivea*), *hibiscus*, mulberry (*Broussonetia papyrifera*), nettles (*Urtica* spp.) and baobab (*Adansonia digitata*).

The flax fibres used for papermaking are, for economic reasons, usually in the form of *linen* rags and cuttings, although some use is made of flax tow, the short and tangled strands of fibres removed at the various stages of textile processing. The retting process, by which the flax plant is partially rotted, permits the separation of the fibre strands, some 15—100 cm long; they constitute about 25% by weight in the green stem but not more than 15% of the weight of the living plant is recovered; the figure is, in fact, usually much smaller. The individual fibres, some 10—36 mm in length and with an aspect ratio of the order of $10^3 : 1$, are characterised by the presence of numerous faintly discoloured nodes. Knowledge of the morphology of linen fibres is not so detailed as that of wood fibres but it is known that the bulk of the fibrils from a Z-helix; this is overlain by inner and outer layers spiralling in opposite senses. The fibres are relatively strong, a consequence of their low (2—5%) lignin content (the plant is harvested for its fibre while green, that is, before extensive lignification develops) and the fact that they are separated from the plant tissue without chemical treatment and, provided that retting is carried out with appropriate control, with negligible microbiological degradation. Associated with the fibre strands is a waxy substance: this is responsible for the characteristic lustre of linen and care is taken in the textile processing to retain it. Linen rags for papermaking must therefore be given a gentle cook and a bleach to remove the wax and to improve their colour but, if carefully conducted, this has little effect upon the strength of the individual fibres. These latter on beating break down readily into extensive fibrillar sheets and individual fibrils of considerable length, this permitting the manufacture of strong papers such as bank note paper. For other papers, commercial considerations usually require that it be blended with shorter fibres.

*Cereal straw* pulps are used in many countries for the manufacture of a wide variety of products including both fine paper and board. In particular, they have found favour for the manufacture of corrugating medium. The world's annual production of strawpulp in 1972 amounted to almost exactly 2 million air-dry metric tons, of which about one third was made from rice straw.

The major part of pulp from straw consists of the bast fibres. These average 1.5 mm in length and 13 $\mu$m in width, giving a length-to-breadth ratio of about 115 : 1. The lumen may occupy up to one third of the width of the fibre. A striking characteristic of straw pulp is the large pith cells, the presence of which helps to distinguish this pulp from that of esparto; they vary widely in shape but are mainly thin-walled and collapse to thin sheets when pulped, so that water does not drain readily from the stock even before beating. In addition, hair and epidermal cells are found, these acting as fillers between the readily collapsed fibres when the pulp is formed into paper. Cellulose is present to the extent of 31—40% (dry basis), with 35—50% hemicellulose, which is mainly xylan, and some 15—25% of lignin. Straw has a higher content (4—8%) of mineral matter than other sources of pulp, that of rice straw, 16—20%, being particularly high: this can cause trouble by abrasion of the paper machine, cutting knives, printing plates and other equipment. The papermaking properties of a straw pulp vary according to the pulping method used. Caustic soda alone gives a pulp that forms dense, hard sheets with a rather low opacity and a relatively high rattle. Neutral sulphite cooking yields a pulp, the cells of which collapse less readily, and part with their water more reluctantly, to give paper of higher opacity and greater dimensional stability.

Other graminaceous species are used of which bamboo and bagasse, the fibrous residue of sugarcane milling, are the most notable. In 1972, the estimated world production of bamboo pulp was 1.36 million air-dry metric tons, that of bagasse 0.915 million tons. Broadly speaking, the chemical and morphological features of these two types of fibre vary so widely as to have discouraged detailed investigation and to preclude useful generalisation. Not only do variations arise from climatic and other environmental differences but, in the case of bamboo, there are those that stem from the large number of botanical species — more than a thousand.

### *2.8.3 Leaf fibres*

The so-called *leaf fibres* are derived from the leaves and stems of *monocotyledons* where they provide support and rigidity to the leaves. While many of this group are utilised for cordage, only a few are exploited for papermaking, of which the most notable is esparto grass: others less extensively used include sisal (*Agave sisalana*), Manila hemp (*Musa textilis*) and New Zealand flax (*Phormium tenax*).

*Esparto* grass (*Stipa tenacissima*) is a reed-like grass native to the dry regions of North Africa and the south of Spain. As it is the leaves of the plant that provide the fibres, and not the stem as is the case with other graminaceous plants used for papermaking, esparto pulp is free from the nodes that, from the papermaking viewpoint, are such an objectionable feature of other grasses, although it may be contaminated with the hardened leaf bases, sometimes erroneously referred to as roots. Esparto grass pulps readily, even at atmospheric pressure. A low pressure caustic digestion is widely used, yielding 38—45% of papermaking fibre. Albardin (*Lygeum spartum*), an apparently similar but botanically distinct grass, is sometimes described, and used, as esparto.

Esparto fibres are typically 1.5 mm long with a diameter of 10 $\mu$m. They are thick-walled, so that the lumen is nearly closed, and usually taper to a fine point, although some blunt-ended fibres are to be found. There is a proportion of small non-fibrous cells of two types, namely the serrated and the so-called comma, tooth or tear-shaped hair cells. These are usually lost through the screens during pulp washing, together with an appreciable number of larger vessels that do not wash out so easily. The comma cells, which are characteristic of esparto pulps and provide a ready means of identification, derive from the fine hairs on the leaves of the grass. The wall structure of the fibres has not been definitively studied. About half of the cell wall material is cellulose; this is accompanied by 26—32% hemicellulose, chiefly xylan, and some 16—19% of lignin.

The thick-walled fibres do not collapse but retain a springy sinuous form after drying: this, together with their fineness, results in a large number of bonded areas of small extent (relative to wood pulp), which imparts to paper made from esparto bulk, opacity and dimensional stability, and resilience on the printing press. These properties, together with an even formation to which the non-fibrous cells contribute, make the paper particularly suitable for high quality colour printing.

### *2.8.4 Secondary fibre*

In addition to the fibre pulped direct from the foregoing vegetable sources, and the textile and cordage fibres usually available to the papermaker only as cuttings, used rags, tow or old rope, there remains one other source of importance, namely secondary fibre obtained by repulping paper and board. This is of considerable economic significance to those countries lacking their own sources of virgin pulp. Repulped fibre has long been used in such countries for the manufacture of board and corrugating medium but with steeply rising pulp prices, increased concern for the environment, and the development of de-inking processes to which these provided the incentive, the use of secondary fibre has been extended to many other grades. Thus, in 1974, the OECD countries recycled 24.0% of the paper they used [39]. It may be noted that, at this level and assuming the content of fibre in paper to be constant, of every 100 tons of pulp fibre used almost six tons will be incorporated into three different papers or boards. The figures would be higher if full allowance were made for waste recovered from paper conversion operations.

In view of the likely variability of such fibre, its properties can only be described in the most general terms. Except when the paper to be recycled is limited to a special grade, such as rag paper, the repulped fibre is chiefly of wood. The mechanical effect of handling fibres in the dry state (as paper) or in the wet state (during reprocessing) is progressively to reduce their length and the integrity of the fibre wall, so that paper made entirely from secondary fibre lacks the strength properties of that from virgin pulp. This is further contributed to by the significant loss of swelling capacity after drying, and on resuspension in water, that fibres reveal. This loss of strength is relatively low for weaker pulps such as groundwood but is particularly marked for unbleached kraft pulps, the fibres of which tend to produce an undue proportion of fines when beaten beyond a certain stage. Repulping, by increasing the amount of fine material in suspension, inevitably increases the resistance to drainage on the papermachine wire. Furthermore, the absolute amount of colloidal material of a polysaccharide nature passing through the wire is greater for recycled pulps, so that the biochemical oxygen demand of the effluent is higher. On the other hand, reprocessed fibres are useful as fillers and for closing the surface of printing papers and judicious blending of virgin and secondary fibres can often hold the physical properties of a paper at an acceptable level, if not actually enhance them, at the same time substantially reducing that most important non-physical property of any product — its cost.

## 2.9 NON — FIBROUS RAW MATERIAL

Of the non-fibrous raw materials of paper by far the most important is water; indeed more than important, it is essential. The properties and role of this in papermaking have been dealt with in detail in Chapter 1 and will not be entered into here. A large number of other, diverse materials may be added to the pulp stock. These are conveniently fed batchwise into the beaters, where these are in use, or continuously if only refiners are used: in some cases they are added to the surface of the wet sheet by means of a size press.

### *2.9.1 Internal sizes*

Unsized paper (waterleaf), unless made from well-beaten stock, is highly absorbent. To render it less permeable to aqueous liquids a number of materials are available, the most commonly employed of which is rosin, used with aluminium sulphate. Rosin is a natural resin exuded by several species of pine tree when the outer tissues are cut or damaged: it may therefore be collected in a manner similar to that of latex from the rubber tree, Alternatively, it may be steam-distilled from chips of the wood, those from the stumps providing a particularly rich source. The chemically complex rosin, available to papermakers in a wide range of quality grades, is combined with alkali to form a soap, in which form it is suitable for adding to pulp stock;

from this it may be precipitated onto the fibres by the addition of aluminium sulphate — the papermaker's "alum". A size of this kind, by covering the fibre surface with fine water-resistant particles, greatly reduces the tendency for writing ink to "feather" or disperse too freely over the surface.

If it is desired to take this process considerably further and prevent water from disrupting the inter-fibre bonds, wet strength resins are used, of which the most common are urea or melamine with, in both cases, formaldehyde. From these, synthetic resins are partly polymerised by the chemical manufacturer, the process being carried further during the drying or storage of the paper. The polymerised resin is unaffected by water and holds the fibres together, even though unprotected bonds between cellulose molecules are disrupted.

### *2.9.2 Adhesives*

In addition to their use in coatings and other conversion operations, adhesives are sometimes added to the pulp stock to augment the strength, reduce the air permeability, decrease the tendency for the surface to fluff and enhance the ink hold-out properties of the paper sheet. Various gums and mucilages have been used but by far the most common material is starch. These substances produce their effect as a result of their readiness to form hydrogen bonds with cellulose and of their film-forming propensity, arising from their two-dimensional polymeric molecules. When the surface-modifying properties of such materials are desired, it is usual to apply them at a size press, if this facility is available.

### *2.9.3 Fillers and loadings*

Fillers are incorporated into a paper sheet for a variety of purposes: to provide a closer, smoother surface which is more receptive to printing ink, to increase the opacity, to reduce the rattle and soften the feel, to enhance the dimensional stability and to lower the furnish cost. Particularly widely used for this purpose is China clay (kaolin), approximately one half of the world's production of which is made use of in either filling or coating paper. Other fillers employed include calcium carbonate, calcium sulphate, magnesium silicate and kieselguhr.

Although the terms filler and loading are often used synonymously, a distinction is sometimes maintained, the word loading being reserved for those heavy minerals that are embodied in a sheet chiefly to increase its opacity. Of these, titanium dioxide is pre-eminent. The opacifying effect of loadings (and fillers) arises from the reduction of bonding between cellulosic surfaces that they bring about, the enhanced area of interface increasing the proportion of light that is scattered. In addition, the opacity is augmented by the greater absorption and scattering of light by the mineral particles themselves.

The reduction of the degree of interfibre bonding is accompanied, however, by a loss of paper strength and rigidity: at the same time, engine sizing may be more difficult to attain. Furthermore, mention has already been made of the abrasive nature of some mineral particles.

### *2.9.4 Other materials*

Many other substances may be added to the pulp stock: waxes to confer transparency, water-soluble dyes or insoluble pigments to give colour, optical bleaching agents to improve the whiteness of the paper sheet. Alumunium silicate can be precipitated from sodium silicate by the addition of alum to increase the rattle of the paper; sodium silicate itself is added to impart rigidity to corrugating medium. Alum and other positively-charged colloidal substances assist the retention of cellulose fibres, fines and many fillers that are negatively charged, reducing material losses through the wire and from the system, and the river pollution problems to which these can otherwise give rise.

In general, it may be remarked that the stock conditions under which non-fibrous materials are added are often critical and that some additives interact disadvantageously. Furthermore, additives may produce more than one significant effect on the physical properties of the sheet: it has already been noted that starch and other polysaccharides enhance not only the body strength of a paper sheet but also its surface strength and its printing characteristics; more broadly, they impart a degree of sizing. Again, the voluminous precipitate of aluminium silicate not only affects the handle but increases the retention of fillers and loadings and of sizing agents; some wet strength resins have a similar effect.

ACKNOWLEDGEMENTS

The author is grateful to PIRA and to Pergamon Press Ltd. for permission to draw, in this and Chapter 3, on two of his earlier texts, Fundamentals of the Beating Process (published by BPBIRA, Kenley in 1957) and Chapter 2 of Stuff Preparation for Paper and Paperboard Making (edited by F.M. Bolam, published by Pergamon Press, Oxford in 1965), respectively.

## 2.10 REFERENCES

1 F.A.O. Yearbook of Forest Products, 1974.
2 F.A.O. Yearbook, 1969/70.
3 F.A.O. World Forest Inventory 1963, Rome, 1966.
4 Sir James Frazer, The Golden Bough, Macmillam, London, 1963, Abridged Edition, Chs. ix and x.

5 D.L. Clarke, J. Roy. Hort. Soc., 93 (1968) 478.
6 J.M. Coulter and C.J. Chamberlain, Morphology of Gymnosperms, Univ. of Chicago Press, Chicago, 1910.
7 H.H. Dixon and J. Joly, Phil. Trans. Roy. Soc., London, Ser. B, 186 (1895) 563.
8 L. Leyton, Fluid Behaviour in Biological Systems, Clarendon Press, Oxford, 1975, p. 184.
9 C.W. Ferguson, Science, 159 (1968) 839.
10 M.H. Zimmerman (Ed.), The Formation of Wood in Forest Trees, Academic Press, New York, 1964.
11 H. Meier, l.c. [10], p. 144.
12 A.J. Cruise and J.W. Jeffery, Nature, 183 (1959) 677.
13 F. Hoyle and N.C. Wickramasinghe, Nature, 268 (1977) 610.
14 R.D. Preston, l.c. [10], p. 171.
15 P.H. Hermans, Physics and Chemistry of Cellulose Fibres, Elsevier, New York, 1949.
16 K.H. Meyer and L. Misch, Chem. Ber., 70B (1937) 266.
17 C.Y. Liang and R.H. Marchessault, J. Polym. Sci., 37 (1959) 385.
18 E. Schulze, Chem. Ber., 24 (1891) 2277.
19 T.E. Timell, Advan. Carbohyd. Chem., 19 (1964) 247.
20 C. Gustafsson, J. Sundman, S. Pettersson and T. Lindh, Pap. Puu, 33 (1951) 300.
21 T.E. Timell, Advan. Carbohyd. Chem., 20 (1965) 410.
22 H. Meier, Acta Chem. Scand., 12 (1958) 1911.
23 H. Meier, J. Polym. Sci., 51 (1961) 11.
24 P.J. Garegg and B. Lindberg, Acta Chem. Scand., 14 (1960) 871.
25 C.P.J. Glaudemans and T.E. Timell, J. Am. Chem. Soc., 80 (1958) 1209.
26 H. Bouveng, B. Lindberg and P.J. Garegg, Acta Chem. Scand., 14 (1960) 742.
27 A.C. Neish, l.c. [10], p. 234.
28 G. Jayme and M. Harders-Steinhäuser, Mikroskopie der Holz- und Papierfasern, in H. Freund (Ed.), Handbuch der Mikroskopie in der Technik, Umschau Verlag, Frankfurt, 1970, Vol. 5, Pt. 2, p. 52.
29 A.B. Wardrop, l.c. [10], p. 87.
30 C.E. Dunning, Wood Sci., 1 (1968) 65.
31 Meier, H. and K.C.B. Wilkie, Holzforschung, 13 (1959) 177.
32 R.P. Kibblewhite and D. Brookes, Wood Sci. Technol., 10 (1976) 39.
33 P. Perilä, J. Polym. Sci., 51 (1961) 19.
34 E. Hägglund, Chemistry of Wood, Academic Press, New York, 1951, p. 8.
35 G. Haywood, Tappi, 33 (1950) 370.
36 H.E. Dadswell, A.B. Wardrop and A.J. Watson, in F. Bolam (Ed.), Fundamentals of Papermaking Fibres, Transactions of the Fundamental Research Symposium held at Cambridge, 1957, Tech. Sect. Brit. Paper Board Makers' Assoc., Kenley, 1958, p. 187.
37 G. Jayme, M. Harders-Steinhauser and W. Mohrberg, Papier, 5 (1951) 411.
38 E.H. Immergut and B.G. Ranby, Ind. Eng. Chem., 48 (1956) 1183.
39 O.E.C.D., Prospects and Policies for Waste Paper Recycling in the Pulp and Paper Industry, 1976.
40 J.S. Pate and B.E.S. Gunning, Ann. Rev. Plant Physiol., 23 (1972) 173.

*Chapter 3*

# THE PREPARATION OF PULP FIBRES FOR PAPERMAKING

H.W. Emerton

## 3.1 INTRODUCTION

Chapter 2 of this work opened with a concise, technical statement of the meaning to be attached to the term "paper". A more particular, and less all-embracing, definition may be quoted from a leading English dictionary, as follows: "Paper: substance used for writing, printing, drawing, wrapping up parcels, etc. made of interlaced fibres of rags, straw, wood, etc." [1]. It may be noted that the term "etc.", to which, it would be supposed, lexicographers do not lightly have recourse, is here used twice. The latter occasion reflects the many and various sources of fibre from which paper *can* be made, though as we have seen by far the greatest proportion of the annual production is made from the fibres of softwood trees.

The first use of "etc." is necessitated by the wide range of uses to which paper is put and the variety of properties that can be built into it. The number of different types of paper has been assessed as many hundreds. An engineering scientist, consciously considering paper as an artefact for the first time, expressed his astonishment at "the infinite variety of it uses" arising from "the remarkable diversity of its physical properties. What other man-made material" he demands "is so varied? It can be rough or smooth, greaseproof or waterproof, absorbent or not, soft as cotton wool or stiff as board, heat resistant, even fireproof, chemically resistant, opaque or transparent, white or coloured, shiny or dull, strong or weak, light or heavy, . . . pulpable, cellular, stiff or flexible, laminated, creped, loaded, waxed, sanded, embossed, hinged, corrugated, easily folded and pierced, coarse or fine, even made hairy. Add to this that a large proportion of all paper is, and an even greater proportion could be, recycled and it is clear that paper is indeed a remarkable material" [2].

How, it may be asked by one coming new to the field, is this very considerable diversity of product achieved? First and foremost, by subsequent processes such as coating and impregnation or, as noted above, by lamination, corrugation and creping and a number of other treatments. But when these are left out of consideration, there yet remain many scores of different paper kinds. To some extent, the choice of fibre can play a part: for example, the use of cotton fibres, which are almost pure cellulose, where papers of high chemical purity, such as filter papers, are required or of linen fibres for bank notes where their length contributes particularly to the resistance of the paper to failure from repeated folding. Yet again, the properties of the product can be significantly influenced, even if to only a limited extent, by the manner in which the paper machine is operated, as, for example, in the manufacture of sack kraft when it is desired to minimise the anisotropy of the sheet, or of map papers when dimensional stability is important.

But between the separation of the fibres from the wood, or other plant tissue, and their reconstitution as a sheet on the paper machine, it is possible to treat them mechanically in the presence of water in a way that can have a most profound effect upon the properties of the paper made from them.

In the earliest days of papermaking, the fibre would have been separated by some form of mechanical action such as grinding between stones or pounding in a mortar and it must quickly have been found that this was more readily effected with water present. It would then have been noticed that continuing the grinding or pounding beyond the stage at which the plant tissue was substantially broken up resulted in paper with different, and for some purposes more desirable, characteristics. For this process of mechanical treatment in the presence of water, which has come down to us through the wind- and water-operated stamping mills of Holland to the batch beaters still in use today, the term "beating" is employed. For the use of engines designed to permit continuous operation the word "refining" is applied. This curiously-used term would appear to have been carried over from the milling of flour and the extraction of seed oil, in which industries the refiners first used in papermaking were well established [3]. Refining at consistencies used in beaters produces the same basic effects and, to avoid the tedium of frequent repetition, only the term beating will be used here; this is to be understood to cover the effect of all machines used to modify the properties of fibres in suspension at low consistencies, up to say 5%, before thay pass to the paper machine. With the increasing pressure over recent years to make fuller use of valuable resources, and to reduce the impact on the environment of waste products, various forms of high yield pulp have been developed. In these, the fibres are usually incompletely separated: the aggregates are then reduced to their individual units by treatment in refiners at high (>10%) consistency. This operation concomitantly produces many of the effects of conventional beating [3] and is not otherwise treated here.

A sheet of paper made from unbeaten fibres is characterised by a lack of both tensile strength and resistance to burst, by its bulkiness, by high porosity, absorbency and opacity; by its open irregular surface; and, in many cases, by a wild uneven formation. A sheet made from well-beaten stock, on the other hand, exhibits greater mechanical strength; a higher density; a measure of greaseproofness, which can become quite high; translucency, and in extreme cases transparency; a smoother surface, and a more regular formation. The development of these properties from the unbeaten to the beaten state is broadly illustrated in Fig. 3.1, from which it will be noted that the changes are not, in general, proportional to the duration of the treatment. Nor are they found to be proportional to the amount of energy expended on the stock. It is however possible to interpret them in terms of the increase in density of the sheets made from progressively-beaten pulp and the greater degree of cohesion of their fibres. The explanation of how this is brought about is the main concern of this chapter.

The early theories of the beating process, current in the first quarter of the present century, sought to explain the effects produced solely in terms of the assumed formation of a hydrate of cellulose on the surface of the fibres: this was believed to have gelatinous properties and acted as an adhesive binding the fibres together. This chemical theory fell into disfavour when careful measurements showed that, regardless of whether unbeaten cellulose is itself hydrated, any *changes* that beating might produce in the true water of hydration were small. It was followed by a purely physical explanation in which the strength of paper was supposed to be due to the mechanical entanglement or felting of fibrils raised as a velvet-like pile on the surface of the fibres. According to this theory the increased "wetness", or reluctance of the stock to part with water, is due to retention of the water in the fibril pile.

It was soon realised, however, that this theory was insufficient to explain all the observed effects of beating; indeed, it was found to be in actual conflict with some of them. But it was not long before the seed of the current

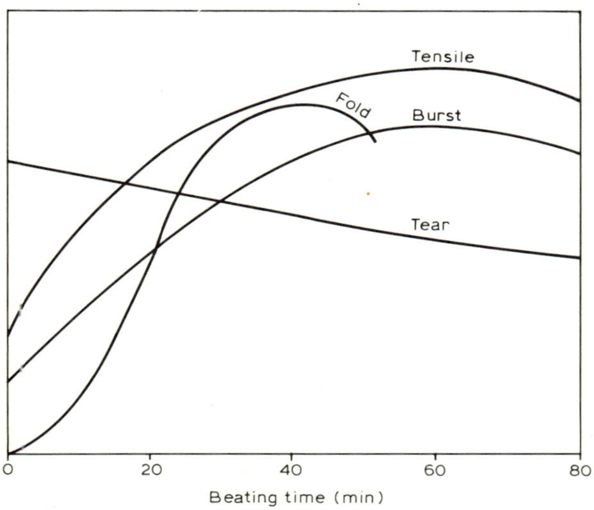


Fig. 3.1. Development of paper properties on beating.

theory, founded particularly upon the work of Campbell in the 1930s [4—6], began to germinate [7—11]. Its growth was concurrent with the general expansion of the study of polymers shortly after the last war, and the development of electron (and other forms of) microscopy, which revealed cellulose to be not a unique substance but a member of a class of materials in the technology of which swelling and plastification effects are of considerable importance.

It is now accepted that, when fibres are immersed in water, hydrogen bonds, chiefly between the lamellae of the wall, are broken. The resulting loosening of the structure is immediately and greatly enhanced by the flexing to which fibres are subjected during beating and refining. Co-axially delaminated and internally lubricated by water, the fibres conform more readily under the compacting effect of surface tension forces which draw them together as the water is removed. As the last layers of water molecules are driven off, hydrogen bonds are re-formed, not only within the fibre walls but, to a significantly greater extent than is the case with papers made from unbeaten pulp, between adjacent fibres.

This is believed to be the major mechanism: its effects are reinforced by those arising from secondary features such as the "draping" effect of teased out, but still attached outer layers of the fibre wall; the adhesive filler properties of detached polysaccharide material of colloidal dimensions; and the fibre packing that arises from the spread of the fibre-length distribution downwards from that of a narrow range of intact fibres, as some of these are cut transversely into two or more shorter lengths.

This is a brief statement of the now generally accepted explanation of the effect of beating. Figure 3.2 is an attempt to portray the several facets of the overall process and their somewhat complex interrelationships. We shall now follow through the various paths of this diagram, examining each group of particularly-closely-linked features in greater detail. With several concomitant effects to be described, requiring frequent anticipation or retrospection, some degree of overlap and repetition will be inevitable.

Throughout this chapter the term "strength", applied to paper without qualification, refers to its tensile or its bursting strength. Although no simple relationship exists between these two properties, increases introduced in the one are almost invariably accompanied by increases in the other, so that for qualitative comparisons of sheet strength it does not greatly matter which is used. On the other hand, gains in tensile or burst strength are brought about, for the most part, at the expense of the resistance to tear so that the word "strength", used alone, should be taken to exclude this property.

Furthermore, the botanical distinction between "tracheid" and "fibre" will generally not be maintained. In what follows, the term "fibre" is to be taken to include the elongated cells of both hardwoods and softwoods, except when it is essential to differentiate between them.

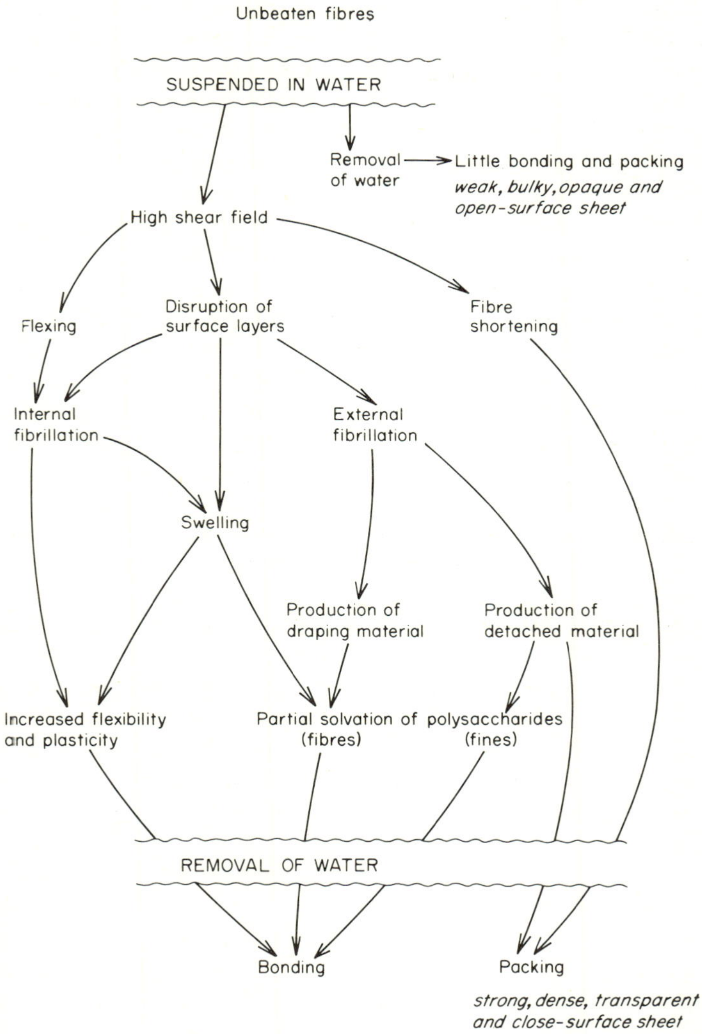


Fig. 3.2. Scheme of the beating process.

## 3.2 THE ROLE OF WATER

Despite its abundance, water has many, and in some cases, unique properties. Several of these are, in the most meaningful sense of the word, vital to the existence of plants and to their growth and some are of considerable consequence in papermaking. We may leave aside its solvent action, which makes it a highly suitable vehicle for the translocation of soil nutrients; its high specific heat, which permits living tissues to gain and lose considerable amounts of heat with only a small change in temperature; its high heat of vaporisation, as a result of which evaporation produces a considerable cooling effect; its lower density as a solid than as a liquid, so that ice forms on a lake from the surface downwards thereby preventing bottom-rooted plants from freezing; its low compressibility, which enables it to play a part in the support of tall plants; and its transparency to visible radiation, which allows sunlight to reach the sites of photosynthesis (chloroplasts) in the cells of the leaves.

Of more importance to the papermaker are the affinity between water and polysaccharides and the remarkable power of cohesion of its molecules. Both stem from the structure of the water molecule. The atoms comprising it have a V-configuration with the oxygen atom at the angle and a hydrogen atom at the end of each arm (Fig. 3.3). The oxygen atom, which is strongly electronegative, attracts electrons away from the hydrogen atoms; this leaves it with a net negative charge and both hydrogen atoms with a net positive charge. This results in electrostatic attraction between the hydrogen atoms of one molecule and the oxygen atoms of a neighbouring molecule — an example of the so-called hydrogen bond.

Hydrogen bonds are present in water to an unusually large extent, giving rise to a high degree of order: indeed, liquid water may locally become almost crystalline. From the heat of fusion, it may be calculated that some 85% of the hydrogen bonds present in ice remain unruptured in liquid water at 0° C and even at 25° more than 80% of the original number are still intact [12]. The strength of the hydrogen bond between water molecules is 4.8 kcal $mole^{-1}$ (which is small compared with that of primary valence bonds; 30

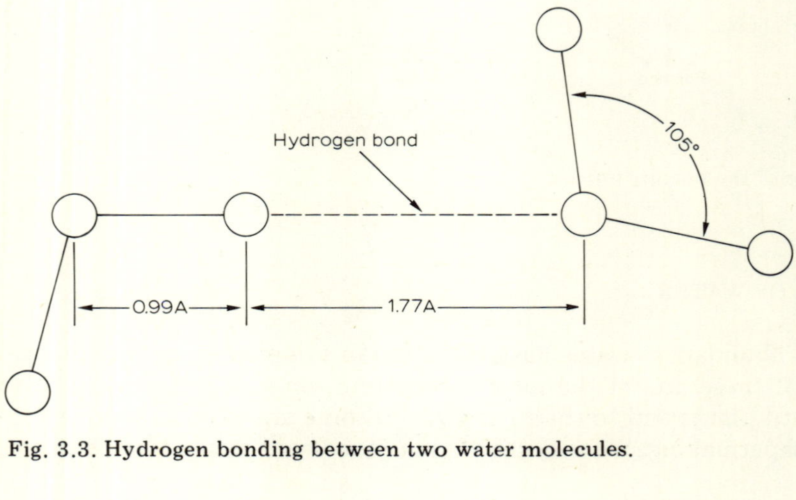


Fig. 3.3. Hydrogen bonding between two water molecules.

kcal/mole in the case of cellulose) so that it is readily broken, particularly as the temperature rises. The locally highly-ordered regions of liquid water differ, in fact, from a true crystal in that the hydrogen bonds are constantly broken and reformed and, as a result of thermal movement, the molecules involved are constantly interchanged. It is this intermolecular attraction acting through many hydrogen bonds which, although individually relatively weak collectively require considerable energy to overcome them, that gives rise to the high heats of fusion and vaporisation, the high specific heat, the high surface tension and the fact that water is a liquid, and not a gas, at normal temperatures.

The same mechanism of hydrogen bonding results in the attraction of water molecules to the hydroxyl, or other polar functional groups, of poly-

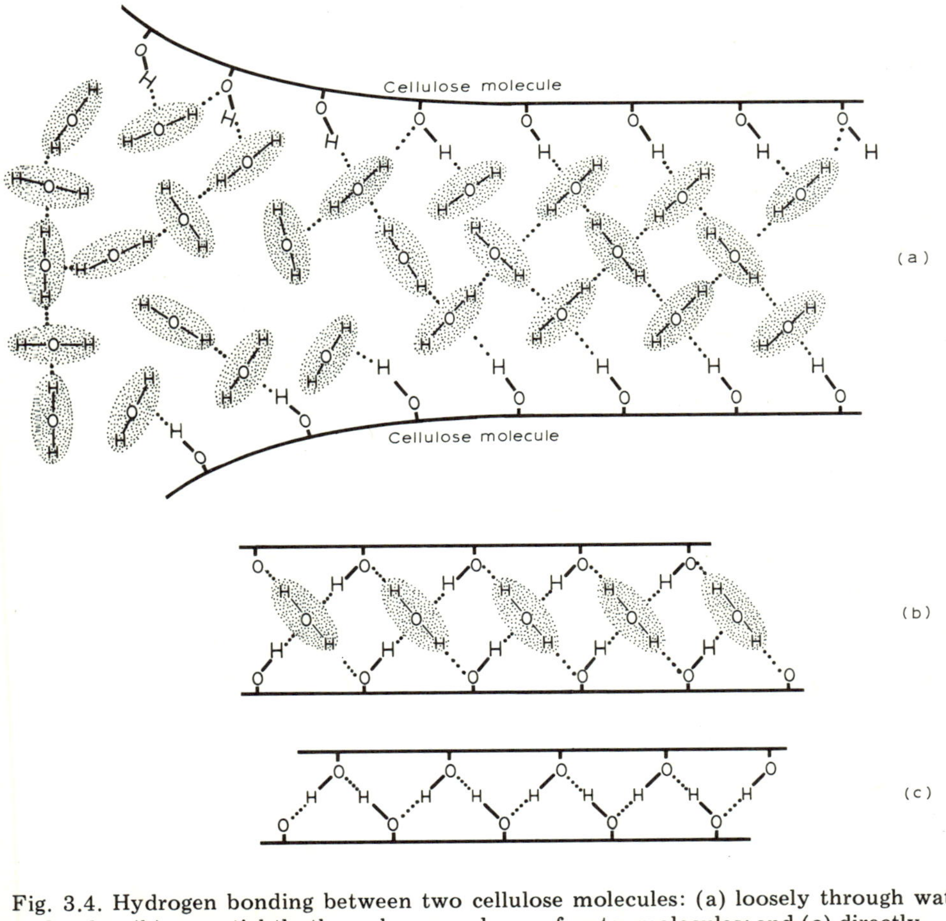


Fig. 3.4. Hydrogen bonding between two cellulose molecules: (a) loosely through water molecules; (b) more tightly through a monolayer of water molecules; and (c) directly.

saccharides and of these latter groups one to another. Figure 3.4(a) (in which for convenience the V-configuration of the water molecules is shown in linear form) is a representation of two cellulose molecules hydrogen-bonded to water molecules and, through these, in effect to one another. Considerable thermal effort is required to drive off these residual water molecules since their bonds with cellulose, even when ruptured, readily reform elsewhere. If carried far enough, however, the drying process reaches a stage at which the two cellulose molecules are bound through a monolayer of water molecules, as in Fig. 3.4(b), and ultimately directly one to another (Fig. 3.4(c) as in the crystalline zones of the cellulose microfibril.

The reverse process may readily be visualised. Water is attracted to the hydrophilic cellulose or hemicellulose on the exposed surfaces of a fibre.

## 3.3 THE ROLE OF THE SHEAR FIELD

In a beater, virtually the entire beating effect occurs in the region of the passing bars. The opposing metal (or sometimes basalt) surfaces, approaching each other through the aqueous suspension at some 10—15 m/sec and at considerably higher speeds in the case of high-speed refiners, create a zone of considerable turbulence, the intensity of which is highly dependent upon the clearance between the passing bars. When this is wide, so that the stock is merely being circulated round the beater trough, or when the cone or disc refiner is backed right off, or even when stock is passed through a centrifugal pump, there is a noticeable, but insignificantly slow, development of the beating effect. But when the plates are brought to within a fraction of a millimetre, the steep shear gradients in the suspension, large enough in some cases, it is believed, to give rise to cavitation, result in considerable longitudinal flexing of the fibres, bruising, disruption, teasing out, and to some extent detachment of their surface layers, and in some cases to actual cutting of the fibres. A high shear gradient alone is able to produce these effects; actual bar-to-fibre contact is probably not essential even under production conditions of beating or even for cutting the fibres. In the presence of water, it may well be unusual for a bar to make direct contact with a fibre (if, indeed, the expression is meaningful when applied at the molecular level), for the same reason that it is difficult to grasp a piece of soap in water, namely, that it is prevented by the lubricating effect of layers of water molecules held by hydrogen bonds to the fibre or soap.

## 3.4 DISRUPTION OF THE OUTER LAYERS OF THE FIBRE

Water molecules sorbed onto the surfaces of closely adjacent and facing cellulose molecules push them apart, as shown diagrammatically in Fig. 3.4.

This is necessarily so; if the elements of the structure cannot move apart, only a limited amount of water can be imbibed. From the figure it is apparent that the swelling of the fibre wall when its cellulose takes up water will be transverse to the direction of the majority of the fibrils. It was explained in Chapter 2 that most of the cellulose is located in the middle secondary wall in co-axial layers, the fibrils of which spiral steeply about the longitudinal axis of the fibre. It therefore follows that the entry of water, or any other swelling agent, into the fibre wall must be accompanied by dimensional changes that are mainly radial, that is, in cross-section the wall will swell either centrifugally or centripetally.

In the living tissue, the developing cell wall is prevented from swelling inwards by the protoplasm contained within it. It seems likely that its indefinite expansion outwards is prevented chiefly by the constricting effect of the fibrils of the outer secondary wall which, it will be recalled, are disposed in several lamellae the outermost of which spiral much less steeply and in alternate senses. (Such a cross-gartered pattern of nearly transverse microfibrils has been reported in elongated cells other than wood, for example those of bamboo, flax and cotton.)

Pulp fibres after drying are often able to swell into the empty lumen on taking up water again and the necessity to disrupt the outer secondary wall early in the beating process, proposed several years ago [13], would now appear to have been to some extent overemphasised. Nevertheless for fibres with thick walls and a narrow lumen, such as many hardwood fibres and latewood tracheids of softwoods, the uptake of water may be limited by an intact $S_1$. That $S_1$ does in certain, rather extreme, circumstances restrict swelling can readily be observed microscopically. When cellulose fibres are immersed in swelling agents, such as cuprammonium hydroxide or phosphoric acid, of suitable strength, they assume the form of a string of beads or balloons. Nodes between the beads are formed by $S_1$ which parts under the strong radial turgor pressure generated in $S_2$ and gathers intermittently along the fibre. This gathering of $S_1$ occurs, in softwood fibres at least, at more or less regular intervals: it is irreversible; when the fibres are deflated, the gatherings remain as rucked-up sleeves. Some evidence has been adduced which suggests that a limited amount of balloon swelling may occur during sulphite digestion but not in the course of alkaline pulping: this would certainly contribute to the markedly different response to beating exhibited by pulps prepared by these two processes [11].

It will not have escaped notice that only the outer secondary wall has been mentioned as a potential constraint on the uptake of water. Because of the thinness of the primary wall, and the relatively wide separation of its fibrils, this wall contains comparatively little cellulose and it is improbable that, once it has been delignified, it contributes significantly as a constraint on centrifugal swelling.

But regardless of any effect that the outer layers may have during beating on the uptake of water by the bulk of the polysaccharide material in the fibre wall, and the consequent increase in its conformability during drying, their disruption leads to their partial or full removal, the significance of which is considered in 3.7 and 3.8.

### 3.5 SWELLING, FLEXING AND INTERNAL FIBRILLATION

Because of their interacting relationships, it is desirable to consider together the sorption of water by fibres which is enhanced by the dissolution of interfibrillar bonds resulting from repeated hydrodynamic flexing.

A rigid gel, such as silica gel, is capable of absorbing considerable quantities of water without significant change in its dimensions. An elastic gel such as cellulose when placed in water or some other suitable liquid also takes up large quantities but in this case the uptake, which may be considered an intermediate stage in the process of solvation, is accompanied by an appreciable increase in volume. This phenomenon is properly termed *imbibition* but the dimensional changes that elastic gels undergo leads to the use of the word *swelling*. In this work, both words are used to describe the uptake of water by the walls of natural cellulose fibres. Water is imbibed; at the same time the fibre wall swells. But it is to be appreciated that, when the word swelling is used in connection with the effect of beating on papermaking fibres, the actual volume change is of no immediate consequence; the essential feature is the uptake of water. It will emerge in later chapters however that, when the reverse process of water-removal and drying is carried out, dimensional changes of the fibres can have considerable significance for the properties of the paper made from them.

Water molecules can enter any pores that may already be present in a cellulosic fibre wall and which are not sealed off. It has been established, however, that when delignified fibres are dried their wall material is densely compacted. The total volume of voids is less than 0.2 $cm^3$ $g^{-1}$, at which level they can have little influence on the relationship between water uptake and the resulting change in the dimensions of the cell wall [14].

The process of swelling may be visualised as follows. Water molecules penetrate any cracks that they are able to prise open by severing intermolecular bonds. Water is sorbed by exposed surface polysaccharides because they are subject to unbalanced electrical forces perpendicular to the plane and water molecules are highly polar. Only polar liquids are able to swell cellulose and the extent of swelling is proportional to the polarity of the liquid. As adsorption decreases the surface free energy, and as the entropy is decreased, the process is exothermic. The evolution of heat — the heat of adsorption or heat of wetting — is significant in amount (of the order of 10 cal $g^{-1}$ for the sorption of water by cellulose) and provides a source of energy for the breaking of lateral bonds between the polysaccharide molecules, provided that the density of such bonds is relatively low. Where adjacent cellulose molecules are disposed in significantly different orientations, as between the microfibrils of the outermost, and of the innermost, layers of the cell wall intermolecular bonding on drying will be less dense and water will enter more readily than between those that are aligned parallel to one another. Water will also enter more readily between molecules of hemicellulose with a significant proportion of side units. But where there is a high density of lateral bonding, as between the molecules of highly parallelised cellulose, or between cellulose and unbranched hemicellulose chains, as in $S_2$, the simultaneous opening up of identical bonds is thermodynamically improbable. The imbibition of water by cellulose is therefore restricted; in the study of gels generally this is termed *limited swelling.*

To promote further swelling in this situation, additional energy for the breaking of bonds between microfibrils must be supplied. One way of providing this is by flexing the fibres in suspension in a shear field. The mechanical effect of this is concentrated on those bonds already under stress from the swelling pressure of the water, many of which will be close to rupture. As these fail, more water is imbibed and further bonds in turn come under load from the turgor pressure. The replacement during beating of polysaccharide-polysaccharide bonds in the fibre wall by polysaccharide-water bonds is termed fibrillation, this word being used both as a substantive to describe the effect produced or as a verb for the action that produces it. Where the severing occurs in a way that does not itself destroy the integrity of the cell wall, the phenomenon is known as *internal fibrillation.*

The hydrophilic (101) plane of the cellulose lattice is disposed tangentially in the cell wall, that is parallel to the circumference, and it would be reasonable to suppose that when the polysaccharide structure is re-immersed and flexed, water molecules will re-enter between these planes, which would ultimately become separated. Evidence has been advanced [15], however, that delamination is only preferentially in the tangential plane, that is, that it is accompanied by the cleavage of a (smaller) proportion of the radially orientated bonds, and that successive lamellae cohere locally. On this model, therefore, the intermediate stages of swelling lead to a honeycomb structure, with lenticular cells the long axis of which is parallel to that of the fibre. Even on this basis, however, there will come a stage, if beating is carried far enough, when the points of local cohesion are broken and coaxial delamination is substantially completed.

Frequent reference has been made to the dimensional changes of cellulose fibres that accompany the imbibition of water and it is desirable to put this in perspective. When a fibre is taken from a dry to a water-saturated state, the increase in its length is negligible, about 1%; indeed the very strong balloon swelling produced by some reagents can cause a considerable longitudinal contraction [16,17]. In the transverse directions, native fibres typically swell by 20--25%. The converse process of shrinkage on drying produces dimensional changes of comparable magnitude. In one investigation, fibres were allowed to dry in close proximity to (though not in contact with) a photographic emulsion and illuminated by a point source of X-rays that impinged normally on the film. It was found that the average shrinkage of eleven unbeaten spruce sulphite fibres was 18% in width and 60% in thickness: the average for thirteen similar fibres after beating to 85° Schopper Riegler was 29 and 75%, respectively [18]. (Width and thickness refer to the form of the collapsed wet fibre and not to the original shape of the wood tracheid.) Thus the transverse shrinkage, on drying, of the fibres, which had been taken from a dried pulp, and presumably the swelling that preceded it, increased by between one half and one quarter on beating to a high degree of wetness. In another study it was found [19] that, after beating to the same degree of wetness (45° SR), the water uptake, as measured by the solute exclusion method, of low-yield sulphite pulp fibres had increased by 22% and that of sulphate fibres by 10%. Fibres that had been dried after pulping imbibed less water than never-dried fibres but could be brought to the same level of imbibition by beating.

There is a notable exception to these magnitudes of swelling, namely cotton, which is said to swell to only half this extent [20,21]. This may be attributable to the absence of amorphous hemicelluloses and also perhaps to a greater degree of irreversible internal bonding when the cotton hairs first dry out.

The degree of swelling is affected by several properties of the pulp themselves and of the water in which they are immersed. It is influenced, in some cases markedly, by the nature and quantity of the hemicelluloses, the molecules of which are largely disordered and hence have a relatively large proportion of hydroxyl groups available for hydrogen bonding. Variations between pulps prepared by different processes are well established: at any given yield sulphite pulps have been found to have a larger pore size than kraft pulps, which is consistent with their higher degree of swelling [14]. Not only does lignin itself not swell in water but, to some extent, it masks the swelling that would otherwise be contributed by the polysaccharides, which it overlies and to which it may be partly bound. As swelling is an exothermic reaction, it proceeds faster at low than at high temperatures. All these factors, by affecting the extent of swelling, have a significant, in some cases a marked, effect upon a pulp's response to beating.

Reference was earlier made to constraints on swelling. Should these exist, swelling is prevented. However, the turgor pressure that arises when a swelling agent is added to a gel is very considerable. Thus a porous pot containing dry gelatin may be broken by placing it in water. Or again, the wetting of a dry wedge of wood was used by quarrymen over many years, and, it may well be, still is in unindustrialised countries, as a means of splitting rock.

An insight into the process of progressively opening up the slit-like interstices to permit the entry of water was earlier gained by studying the exposed surface area revealed by nitrogen adsorption after the suspending water had been removed by solvent-exchange drying (in which the water-swollen fibres are immersed in succession in solvents of progressively reduced polarity, the final solvent in the series being evaporated in a stream of dry gas). From such measurement, it was found [22] that, typically, dry native fibres have a specific surface area of the order of 1 $m^2 g^{-1}$. It was pointed out [23] that, on the simple model of the wall breaking down into 100 separated co-axial layers of equal thickness, the specific surface area would be expected to increase 100-fold to 100 $m^2 g^{-1}$; values of that magnitude had already been reported both for never-dried fibres and for fibres that had been dried and subsequently beaten extensively.

However, it is known that solvent-exchange drying does not retain the swollen structure completely and more recent work has been based on the solute exclusion technique [24]. This measures the proportion of macromolecules, non-reactive to cell-wall materials and of known hydrodynamic size and known concentration in an aqueous solution, not diffusing into the water-filled pores of the fibre, as calculated by the extent to which the solution is diluted. By using solute molecules of different size it is possible to derive the maximum and the median pore sizes and also the internal surface area. This method has been applied to pulp fibres by Stone et al. [19] and Stone and Scallan [25,26] whose work has confirmed the postulate (discussed in ref. 11) that swelling accompanies beating and, in particular, that it increases more rapidly in the early stages of beating when strength development is most pronounced but external fibrillation still a minor effect. This work also confirmed that the loss of potential paper strength that results from drying a pulp before it is used for papermaking can be attributed to a closing of the cell-wall structure, as proposed earlier.

## 3.6 INCREASED FLEXIBILITY AND PLASTICITY

It is a matter of common experience that wood is more flexible before than after drying. In the former state, the fibre walls contain considerable quantities of water: 0.31 to 0.62 g/g has been cited for a number of softwoods and hardwoods [14]. As this water, located substantially between and on the tangentially parallel layers of hemicellulose-coated cellulose, dries out the fibre wall is compacted, thus permitting extensive bonding to build up between the microfibrils.

During pulping, aqueous liquids again enter and open up the structure to some extent but this is chiefly by filling spaces previously occupied by the lignin-hemicellulose gel, the hydrogen bonding between polysaccharides being relatively undisturbed [14]. Should the pulp then be dried, the fibres are further compacted and the density of hydrogen bonding further increased. It is this internal hydrogen bonding that makes dry fibres rigid.

The rigidity to bending of a red or sheet increases as the fourth power of the transverse dimension; thus bonding two lamellae of approximately equal thickness increases the rigidity sixteen-fold.

Conversely, when internally fibrillated in water, fibres undergo a large increase in flexibility from the breakdown of the structure into thinner layers. At the same time, however, the sorbed water produces a state of turgidity and consequent rigidity which in some measure counters this. However, the rigidity arising from turgor pressure rapidly declines as the water is removed (the cause of the wilting of non-lignified plants when their water supply falls) and the fibres become more flexible — until internal bonds are again formed. Thus the effect of beating is progressively to increase the stiffness of the fibres in the water-swollen state but at the same time to increase their potential for flexing as the water dries out, up to the stage at which the lamellae are brought into molecular contact permitting hydrogen bonds to be formed between them.

The progressive increase in fibre-wall flexibility during the early stages of water removal may reach a level at which the fibre wall collapses into the lumen, so that the cross-section is transformed from its approximately rectangular form (almost circular when fully swollen) to that of a ribbon, as in a substantially emptied canvas hose. This event is irreversible (unless the fibre is immersed in a strong swelling agent) and the stage at which it is reached depends, in addition to the extent of beating, principally on the thickness of the fibre wall and on the method and degree of delignification. Thus the state of, or potential for, collapse is reached at an earlier stage by the thin-walled springwood tracheids than by the thicker-walled summerwood tracheids. High yield pulps, in water suspension and before beating, have few collapsed fibres; even after forming into paper and drying, a significant proportion may retain, more or less, their original cross-section. A low yield pulp from the same wood, on the other hand, reveals fewer uncollapsed fibres in suspension and far fewer in the formed and dried paper sheet. Fibre collapse is discussed further, later in this chapter, in the context of the removal of water.

The process of enhancing the elastic flexibility has been likened to the lubrication of a rusty leaf spring. Before oiling, the rust bonds together the leaves, analogous to the lamellae of the fibre wall, so that the entire structure is rigid. The introduction of quite thin layers of cleaning oil between the leaves overcomes some of the bonds and noticeably reduces the rigidity but subsequent mechanical working of the spring is needed to overcome the majority of the rust bonds and recover the full elastic potential of the spring.

However, the bonding that occurs in a drying sheet between fibres that are elastically deformed results in dried-in stresses, which predispose the fibres to spring back when the stresses are relieved, with possible distortion of the sheet (Fig. 3.5(a)). To be preferred, if strength properties are required, is plastic deformation. It is to be supposed that, if beating is carried to the

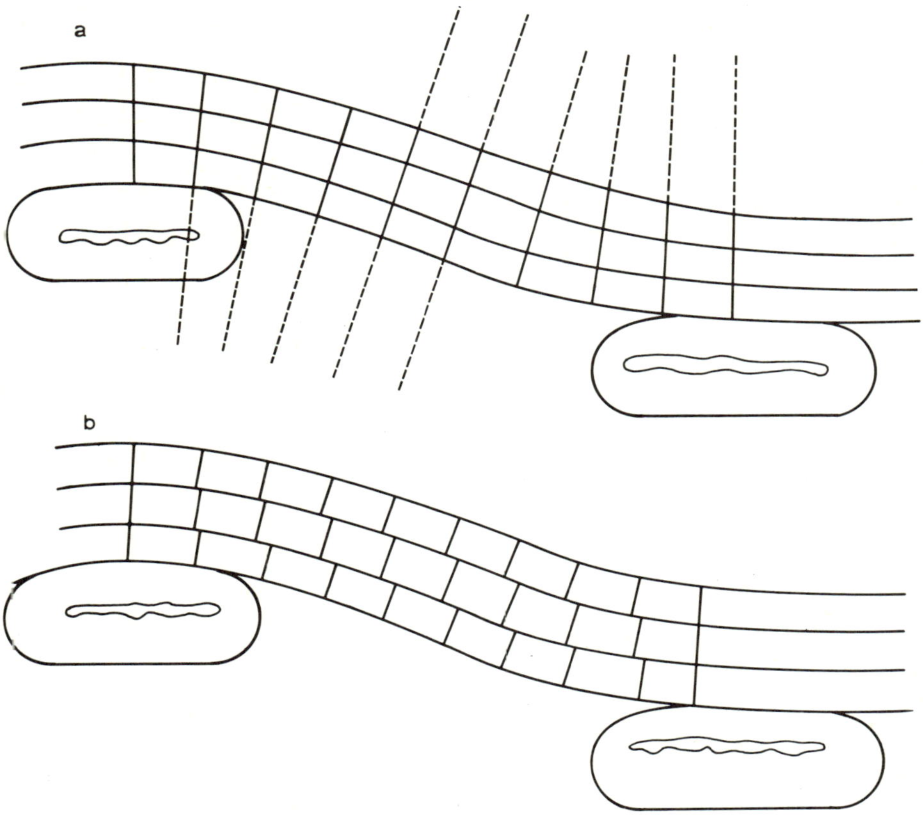


Fig. 3.5. Deformation of a lamellate cell wall: (a) elastic; (b) plastic.

stage where extensive co-axial delamination is effected, the lubrication of the separated wall layers by the imbibed water molecules will enable them to glide during drying into a configuration that minimises the stress (Fig. 3.5(b)). Thus the two effects, namely the reduction of the relatively solid wall into layers that are much more elastically flexible and the development of the potential for plastic deformation, are complementary.

## 3.7. EXTERNAL FIBRILLATION

The term internal fibrillation was defined earlier as the severing of bonds between lamellae, or between fibrils of an individual lamella, within the fibre wall. When primary valency bonds are broken along their length, fibrils lying on the outer surface of the fibre are teased out by the shear field as a superficial fuzz. This process, and the detached fibrillar material to which it can lead, are both known as *external fibrillation.* It is to be noted that the effects of internal and external fibrillation are produced concomitantly.

As beating proceeds the surface layers are locally disrupted, teased out and removed progressively, external fibrillation of the primary wall being followed by that of the outer secondary wall and then by that of the successive lamellae of the middle secondary wall. For most kinds of paper this process is not carried very far and individual fibres, though showing obvious breakdown of their outer layers when viewed under the microscope, essentially retain their identity. For certain types, however, notably greaseproofs and glassines, the process is taken to an extreme, the structural elements being unravelled and dispersed into a gel-like suspension in which individual fibres may no longer be recognisable.

The material detached, completely or in part, during the normal beating of wood pulp fibres may be in the form of membranes [27] or of discrete fibrils, with ribbons of indeterminate width as an intermediate form. The evidence would seem to suggest that the membranes derive from the primary and outer secondary walls, although the middle secondary might also be expected to break down to some extent in this form if the beating treatment proceeds sufficiently gently to enable it to delaminate; more severe treatment tends to sunder it into strips of microfibrils. Because in any layer of $S_2$ the microfibrils have a high degree of parallelism those unidimensional fibres that are teased from it tend to be long. Also to be noted is the tendency of surface tension during water removal to gather in the edges of a strip or ribbon of fibrils so that it becomes more rope-like [27].

Cotton fibres show little, if any, tendency to fibrillate as membranes. Because there is no lignin, and negligible quantities of non-cellulose polysaccharides, encrusting the layers of these fibres, the density of the internal hydrogen bonding that results from their initial drying out is particularly high. A proportion of this cannot be disrupted even on soaking or boiling the fibres in water [28]. Hence the fibrillated material might be expected to have the form of bundles rather than of sheets of microfibrils. The fibres of tension wood, which also have a low lignin content, likewise fibrillate in this way.

It is not generally realised how rapidly the external surface area (a measure of external fibrillation) of wood fibres increases on beating. It was found in one experiment, for example, that the external surface of softwood sulphite fibres, as measured by a permeability method, doubled when the pulp was beaten from 14° Schopper-Riegler (SR) to about 28° SR, quadrupled by the time that its freeness was 42° SR and increased more than forthy-fold on beating to 72° SR [11].

External fibrillation is the most immediately apparent effect of beating on fibres: consequently, it was for long thought to be the most important factor contributing to the development of paper strength. As a result of the increase in external surface, the area available for contact and bonding with adjacent fibres is increased. It has already been mentioned that disruption of the outer secondary wall, which in a sense is to be considered external fibrillation, may assist the imbibition of water and the accompanying internal fibrillation. This apart, however, it is now generally accepted that the early marked increase in paper strength, before extensive external fibrillation of the fibres is apparent, can only be satisfactorily explained in terms of the immediate plasticisation and resulting enhanced conformability of the fibres. Although external fibrillation so often accompanies internal fibrillation, it is this increased conformability which enables larger areas of fibres to come within bonding distance, rather than extensive bonding by means of fibrils and fibril sheets teased out on their surface, that is the important factor in the preparation of stocks for strong papers.

Whether external fibrillation is actually detrimental to paper strength is another matter. It would seem likely that, in some cases, it would weaken the surface layers of the fibre so that they tear away more readily when the sheet is stressed; indeed fibrillar material ruptured in this way may be observed with the microscope. In other cases, particularly where the fibres have been digested by an acid process, the cellulose of the surface layers may be preferentially degraded. In the course of external fibrillation, such weakened material may then be more easily removed during beating, thereby exposing more strongly-anchored layers. Furthermore, the possibility exists that the removal of outer layers exposes surfaces that are more reactive; that is, surfaces potentially capable of a greater density of hydrogen bonding. When surface layers remain undegraded by the pulping process, their external fibrillation will make some contribution to the strength of dense papers by the draping of thin sheets and of fibrils over adjacent fibres and their bonding to them. As discussed later in this chapter, fibrillar material that is actually broken off and passes into suspension also has a significant part to play in the compaction of the paper sheet and the development of its properties.

The foregoing consideration of the effects of external fibrillation has been limited to papers where strength is a major consideration. Its role may be significantly different in grades where other properties are paramount. Where, for example, high absorption or permeability are the need, it will be detrimental as a consequence of its tendency to block and seal off inter-fibre spaces: or where dimensional stability is required, the increased sheet density to which it leads will be disadvantageous.

## 3.8 PARTIAL SOLVATION

When a material of low molecular weight is placed in a solvent, for example common salt in water, the outermost molecules are quickly surrounded by water molecules; that is they are dissolved in the water. When the outermost layers have been dissolved, they move away by diffusion; the underlying molecules are then exposed to the solvent and the process continues, provided there is sufficient solvent, until all the solids have passed into solution.

The solvation or, as it is often called, the dispersion of a linear polymer such as cellulose is somewhat more complex. As with materials of low molecular weight, the macromolecules of cellulose must surround themselves with solvent and move away to expose underlying molecules if dispersion is to be complete. The liquid, strongly attracted to and sorbed by the chain molecules, penetrates so far as the cohesive forces of the structure permit but unless it can overcome these forces, and in particular the forces binding the molecules into a crystal lattice, the liquid causes only limited swelling and total dispersion of the gel is not achieved. When such a state of affairs exists, it is appropriate to speak of the system as a saturated solution of the liquid in the gel. Certain solvents, such as cuprammonium hydroxide and phosphoric acid, are capable of dispersing plant polysaccharides completely: non-polar liquids not at all. Intermediate between these is water, which is capable of partially dissolving the polysaccharide molecules (and of producing limited swelling).

When partially dispersed, polysaccharide molecules, by binding water molecules to their unsaturated hydroxyl groups, are pre-disposed to form bonds, on drying, with neighbouring polysaccharide molecules. It is the tacky condition that results from this that underlies the "wet" or slimy feel that develops in a stock as it is beaten. The addition of gums and mucilages to a furnish, from which dense paper is to be made, is essentially an enhancement of this process.

As has been noted, the object of beating for strength is to increase the bonded area between fibres by making them more flexible so that they deform, preferably plastically, under the surface tension forces that act during drying. Deformation alone, however, merely provides the opportunity for bonding to occur: the extent of the bonding will depend upon the state of the surfaces brought together. Happily it is an accompanying, and inseparable, consequence of the process of increasing the conformability of pulp fibres that the cellulose and hemicelluloses of their surface layers are partially (at least) dissolved. Furthermore, it may be the case that fully dissolved polysaccharide molecules of low molecular weight are, as a result of internal fibrillation and mechanical flexing, worked from the interior of the fibre wall to the surface. Certainly some material of this kind is leached out into suspension from where, together with colloidal particles that have been mechanically detached from the surface, it may be trapped in inter-fibre interstices and re-bonded onto elements of the forming paper sheet.

## 3.9 REMOVAL OF WATER

In the preparation of paper from a pulp stock, water contributes in several ways. In addition to providing a medium through which shear forces may be transmitted to the fibres during beating, it acts as a vehicle of suspension in which the fibres, having been well dispersed, can be brought together to give a sheet having the desired formation; it supplies a mechanism of consolidation of the fibres as the wet web is dried out; and, finally, it plays a major role in the bonding together of the fibres and of parts of the wall of individual fibres. It is with the latter two of these that this and the following sections are concerned.

Free water — that which exists over and above what is required to saturate the structure — is removed, as the name implies, with relative ease; for example by gravity and suction. Capillary water — that which exists in bulk (on a molecular scale) — is held by its surface tension in the interstices between fibres in a paper sheet and within the lumen and any pores of the fibre wall; greater effort, such as pressing, is required to remove it. The most tenaciously held water — that which is molecularly bound by hydrogen bonds — requires still greater effort, usually thermal, to drive it off.

The relative extents to which the processes of thickening and filtration, respectively, are responsible for the structure of the paper sheet cannot be confidently asserted. It appears, however, from practical experimentation as well as theoretical consideration, that after a preliminary thickening on the wire most of the formation results from filtration through the initially deposited mat [29]. As drainage through the wire proceeds the upper surface of the water ceases to be continuous, the fibres progressively emerge through it and the total length of fibre-water contact increases. The force of surface tension being proportional to this length the fibres are brought increasingly closer to one another. This brings about a decrease in the thickness of the forming sheet and an increase in its tensile strength: even when the solids content of the web is no more than 8%, the tension in the menisci between the fibres is sufficient to hold them together [30]. This mechanism is seen at work when a paint brush is withdrawn from water: when immersed, the hairs may stand apart but on removal they are drawn together.

Later, during drying, evaporation occurs and the air-water interfaces retreat into smaller spaces between the fibres. The effect of surface tension is now determined less and less by the total length of fibre-water contact, which is contracting, and more and more by the internal tension of the water. For any given value of the ambient relative humidity there is a maximum diameter of a capillary in which water can remain: from capillaries larger than this it evaporates. If the capillary is of uniform size, it will be completely emptied when the humidity falls to this critical value. If, on the other hand, it is tapering, the meniscus will retreat towards the smaller end: at the same time, the curvature of the water meniscus increases. Because the internal tension is inversely proportional to the radius of curvature the force drawing together the capillary walls rises rapidly as drying proceeds. For example, the tension, additional to that behind a level surface, increases more than 300-fold as the equivalent radius of the meniscus reduces from 1 $\mu$m to 690 nm, a radius which is still considerably greater than the range of hydrogen bonds — of the order of 10 nm. (The two radii are those above which water at 25°C evaporates when the ambient relative humidities are 99.9 and 80%, respectively.) Externally-fibrillated material, by reducing a relatively large space between adjacent fibres and creating new menisci of smaller radius, assists this process of compaction.

The same force acts within the walls of individual fibres causing compaction of the cell-wall lamellae onto each other. It also operates within the lumen where it may be sufficiently large to bring about the collapse of the fibre. A striking feature of micrographs of transverse sections of paper sheets made from fully digested and moderately or well-beaten pulps is the low proportion of fibres with a lumen of appreciable cross-sectional area: the impression may therefore be gained that almost all the fibres have collapsed, e.g. Fig. 4.6. In fact in the case of latewood tracheids, and the thicker-walled fibres of a hardwood pulp, the lumen was comparatively small in the wood and may largely have retained its shape. Earlywood tracheids, and the thinner-walled fibres, possess a large lumen and their relatively thin walls have a low elastic modulus, so that they readily collapse into thin ribbons, two opposing faces being brought into close contact over large areas.

There are two particular consequences of fibre collapse: the flexural rigidity of the fibre as a whole is reduced and larger areas are made available for inter-fibre bonding. Furthermore, as gravity, suction by the table rolls and the effect of the presses all act across the thickness of the forming web, the collapse proceeds mainly in this $z$ direction; that is, the collapsed ribbon-like fibres lie parallel to the paper surface. This may be accounted a fortunate circumstance for the papermaker, as it results in the maximum cohesion of fibres in a plane perpendicular to the thinnest dimension of the sheet.

One other effect of the removal of water from individual fibres may be alluded to. Most of the fibrils in the fibre wall are aligned in a direction close, but not parallel, to that of the longitudinal axis: on immersion in water the swelling pressure that develops gives rise to torsional stresses that tend to align the fibrils of the outer secondary wall more nearly parallel to the axis. This effect is reversed when the fibres are dried: if the fibres are free, and especially if they are thick-walled (latewood) fibres, they twist considerably as the wall shrinks onto itself [31,32]. In practice, however, much of the twisting is inhibited in the forming sheet by mechanical entanglement but microscopic study of the fibres lying in the surface of paper reveals many examples of the effect [33]. Twisting may also be constrained by the formation of some bonds between fibres before the torsional stresses have fully built up.

As drying empties the capillaries, the water molecules remaining become progressively more highly orientated by the attractive forces between them and the polysaccharide molecules, acting through a declining number of layers of water molecules, and hence less mobile. In this way neighbouring fibres, and adjacent polysaccharide molecules within the fibres, are hydrogen-bonded at first weakly through water molecules and ultimately with the maximum strength if all the water is eliminated. Individually, hydrogen bonds are weak: it is to their frequent occurrence along the fibrils and fibres that the reasonably strong adhesion that develops between fibres must be attributed.

There is a particularly important consequence of this progressive effect of bonding building up from a low level before all the water is evaporated. Briefly stated it is that, in a sheet free to shrink during drying, a bonded area between two fibres crossing, say, more or less at right angles, results in a longitudinal contraction of one fibre as a consequence of the stronger transverse shrinkage of the other [34].

Support for the foregoing model of the compaction of the network, and bonding of fibres, a model first proposed by Campbell in 1933 [4—6], has been provided by several experiments. Surface tension has been prevented from operating at the air-liquid interface by freezing the water and subsequently removing it by sublimation [35]; by passing the pulp through a series of solvents of progressively lower polarity, the last of which is a non-polar liquid of low surface tension (solvent-exchange drying) [36]; or by introducing surface-active agents that depress the surface tension [37]. In all cases, a reduction in the strength of the paper indicated that the fibres and fibrils had been brought insufficiently close together for extensive bonding to be effected. It has further been demonstrated that when sheets dried by solvent exchange are exposed to a humid atmosphere (85% r.h.) their strength, after subsequent reconditioning at 65% r.h., increases over long period of humidifying until it attains a value comparable with that of a normally formed and dried sheet [36]. It seems reasonable to suppose that at such high humidities water condenses in the capillaries and that, while subsequently being brought to a state of equilibrium with an atmosphere of 65% r.h., the normal processes of compaction and bond formation take place. If the hydroxyl groups of the amorphous polysaccharides are to a considerable extent substituted by other groups, not only is less water imbibed but inter-fibre bonding when the fibres are dried from water is greatly reduced. When such substituted fibres are suspended in appropriate solvents, such as methanol for methylated and acetone for acetylated fibres, the substituted polysaccharide molecules are partially dissolved, that is, the fibre wall imbibes the liquid more freely and swells; under these conditions, paper sheets of a reasonable strength can be made. A high lignin content, both by stiffening the fibres and by reducing the density of hydrogen bonding, inhibits the development of paper strength during formation and consolidation of the web.

But just as the extent of bonding can be limited by reducing the efficacy of surface tension forces, so it can be promoted by bringing the fibres closer together by applying external pressure to the wet sheet. This happens at the wet presses of the paper machine and may occur to an embarrassing extent at the suction couch, where too intense a suction may locally compact the sheet to an extent that variations in the optical transmission and scattering are produced; the so-called couch mark.

The strength of inter- and intra-fibre bonds was investigated by Nordman et al. [38] who measured the energy required to rupture a certain number of them. The energy irreversibly lost in taking a paper strip through a loading-deloading cycle was calculated from the stress-strain curve and the area of the fibres which, during this cycle, was converted from the bonded to the unbonded state was measured in terms of the increased scattering coefficient. The value of the bonding strength per unit bonded area was found to be independent of the degree of beating, the fibre orientation, wet pressing and basis weight, so that these variables can only affect the tensile strength by varying the number of bonds formed. The bonding strength was, however, dependent upon the type of pulp, a rough correlation existing with the pentosan content. This was interpreted as due to a variation in the number of bonds per unit bonded area, depending upon the nature of the hemicelluloses present.

So far, bonding has been referred to only in terms of the more or less entire fibres: the part played by partially- and completely-detached fibrillar elements must also be considered. Teased-out fibrils were earlier thought to be the source from which paper derives its strength properties but it has long been recognised that, for most papers, their role is secondary. Because of its thinness — of the order of $10^4$ nm — externally-fibrillated material is very flexible and readily drapes itself over neighbouring surfaces, into intimate contact with which it is drawn. When a membrane or ribbon is bonded over appreciable areas of each of two fibres, the intervening portion, if freely suspended, may be gathered onto itself by surface tension of the retreating water: a structure such as this, broadly anchored at the ends and bonded to itself between these regions, may be supposed to provide a somewhat stronger link.

Attention has already been directed to the part played by externally-fibrillated material and fines in the promotion of surface tension forces. Mineral fillers and loadings may be expected to produce the same result but, in this case, do not usually participate significantly in the overall bonding together of the structure. Indeed, they may hinder it by keeping apart polysaccharide surfaces that would otherwise have been brought together and united. Another effect that may prevent the full bonding potential of the exposed polysaccharides from being realised is that which arises from the strong shrinkage stresses in the drying network: there is evidence that bonds formed at an early stage of water removal and drying are sometimes subsequently broken when these forces cannot otherwise be dissipated by shrinkage of the sheet as a whole [39].

## 3.11 FIBRE SHORTENING

So far the description of the processes of beating, sheet formation and sheet consolidation has assumed that the fibres remain essentially entire even though their surface layers may be unravelled, teased out and, to some extent, broken off. In practice, a proportion of the fibres are cut across so that their length is reduced. Indeed, in certain cases, this is essential if a reasonably even formation is to be achieved. This is particularly the case with textile fibres the length of which is typically an order of magnitude greater than that of wood fibres. For example, cotton fibres without shortening tend to form threads which can be caught up on parts of the machine on their approach to the wire. These then grow by entangling other fibres that are passing until they attain such a size that they break away and are swept on to the wire, where they can create substantial problems. Furthermore, the longer the fibres the greater the tendency for the stock to flocculate so that such fibres, if not shortened to some extent, would be ill-dispersed and would produce a sheet with a "wild", highly-flocculated structure. The resistance to oil and fat penetration of greaseproof and glassine grades requires as homogeneously dense a sheet as possible: this is achieved not solely by extensive external fibrillation but also by reducing the length of sufficient of the fibres to ensure that the interstices between the fibres are filled. Some measure of fibre shortening is often required to give the uniform closed surface needed for printing papers. On the other hand, for some grades made from wood pulp, such as kraft paper, where strength properties are paramount, shortening of the fibres is deliberately minimised as far as possible.

The relative proportions of fibrillation (to promote water uptake or "wetting") and shortening can be varied by adjustment of the beater working distances and by choice of beating equipment. The expression "Paper is made in the beater", if now less frequently heard than hitherto, and largely requiring the term "refiner" as the last word, remains not without significance.

Not only does the operation of the beater affect the degree of cutting: the fibres themselves respond in different ways. Generalising, it may be said that the less flexible the fibres the more likely they are to be cut. Hence, for a given set of beating conditions, relatively brittle high-yield pulps will be shortened more readily than soft, well-cooked pulps; thick-walled latewood fibres more than earlywood; and cotton fibres least of all.

Cutting is more likely to occur at an early stage of beating, before the flexural rigidity of a fibre is reduced by imbibition, as a result of flexing in the shear field. It is general papermaking practice to fibrillate a stock before lowering the beater roll for any cutting that may be necessary. There would seem to be sound theoretical grounds for this: the longer the fibre, the greater in general will be the length of the shear gradient through which it passes and the more efficient will be the flexion action.

## 3.12 PACKING EFFECT

In any stock flowing onto the paper-machine wire there will be a distribution of fibre lengths and this will be reflected in the structure of the web made from it. Particularly in the case of bulky sheets, the distribution will influence the way that the fibres pack together, the smaller structural elements settling to some extent into what would otherwise be interstices between the larger ones.

## 3.13 CONCLUSION

By way of conclusion, let us briefly consider the extent to which the two major raw materials of the papermaker are in accord with his needs. Although paper could be made from fibres having groups other than hydroxyl groups, this would require, as a suspending and forming medium, a liquid to match the functional groups (for example, methanol where these are methyl groups): this is not seriously to be considered as a practical or commercial proposition. The chief drawbacks to natural cellulose fibres are that they grow most readily in tropical parts of the world and that associated with them in trees are appreciable quantities of lignin ("Wood is wood because of lignin"). Of the former, it may be said that the diversity of species in tropical forests is disadvantageous; and the lower yield of hardwood fibres and their shorter length make them less attractive than those from softwoods: of the latter, that unwanted lignin in the wood adds to the cost of handling from the forest to the pulp mill, as well as requiring costly chemicals and energy to remove it in capital intensive industrial plant. Is also presents the pulp mill with a major waste-disposal problem.

This having been said, it is to be recognised that natural cellulose fibres go a very long way towards meeting the papermaker's requirement. Their layered fibrillar structure confers great strength with flexibility, particularly when wet. Their hydrophilic nature enables them to imbibe water freely *but to a limited extent* — complete dispersion and subsequent regeneration in sheet form, as for example from a xanthate solution, produces a material with very different properties; it permits hydrogen bonds to be disrupted with modest mechanical effort and to be readily reformed so that, in effect, from existing entirely within individual fibres they are re-disturbed throughout the fibre assembly; and it allows a large proportion of paper to be recycled after it has served its purpose. The water used in making paper may also largely be recovered for further use and both water and fibre are comparatively cheap. All these factors combine to make the papermaker's chief raw materials as near ideal as might reasonably be hoped for. It has been remarked before, but will bear repetition, that it is a most fortunate circumstance that these two major and essential constituents of paper, natural cellulose fibres and water, are the world's two most renewable resources.

## 3.14 REFERENCES

1 H.W. Fowler and F.G. Fowler (Eds.), Concise Oxford Dictionary, Clarendon Press, Oxford, 5th edn., 1964.
2 C. Cherry, in F. Bolam (Ed.), Consolidation of the Paper Web, Transactions of the Fundamental Research Symposium held at Cambridge, 1965, Tech. Sect. Brit. Paper Board Makers' Assoc., London, 1966, p. 2.
3 D. Atack, in Fibre-Water Interactions in Paper Making, Transactions of the Fundamental Research Symposium held at Oxford, 1977, Fund. Res. Comm., Tech. Div. Brit. Paper Board Inc. Fed. London, 1978, p. 261.
4 W.B. Campbell, Pap. Trade J., 95 (1932) Pt. 8, 29.
5 W.B. Campbell, Bull. For. Ser., Dept. Int. Can., 84 (1933).
6 W.B. Campbell, Ind. Eng. Chem., 26 (1934) 218.
7 B. Steenberg, Sven. Papperstidn., 50 (1947) 155.
8 H.F. Lewis, Tappi, 33 (1950) 418.
9 W. Gallay, Tappi, 33 (1950) 425.
10 L.G. Cottrall, Tappi, 33 (1950) 471.
11 H.W. Emerton, Fundamentals of the Beating Process, B.P. and B.I.R.A., Kenley, 1957.
12 P.S. Nobel, Plant Cell Physiology, W.H. Freeman, San Francisco, 1970, p. 36.
13 H.W. Emerton, in F. Bolam (Ed.), Fundamentals of Papermaking Fibres, Transactions of the Fundamental Research Symposium held at Cambridge, 1957, Tech. Sect. Brit, Paper. Board Makers' Assoc., Kenley, 1958, p. 35.
14 A.M. Scallan, l. c. [3], p. 9.
15 A.M. Scallan, Wood Scie., 6 (1974) 266.
16 H. Bucher, Bull. Tech. Ind. Pap., 4—5 (1955) 95.
17 H.W. Emerton, Tappi, 40 (1957) 542.
18 P.A. Tydeman, D.R. Wembridge and D.H. Page, l. c. [2], p. 119.
19 J.E. Stone, A.M. Scallan and B. Abrahamson, Sven. Papperstidn, 71 (1968) 687.
20 F.F. Morehead, Textile Res. J., 17 (1947) 96.
21 A.T. Moore, L.W. Scott, I.V. de Gruy and M.L. Rollins, Textile Res. J., 20 (1950) 620.
22 F.H. Forziatti, R.M. Brownell and C.M. Hunt, J. Res. Nat. Bur. Std., 50 (1953) 139.
23 H.W. Emerton, Pulp Pap. Mag. Can., 56 (1955) 65.
24 L.G. Aggebrant and O. Samuelson, J. Appl. Polym. Sci., 8 (1964) 2801.
25 J.E. Stone and A.M. Scallan, Tappi, 50 (1967) 496.

26 J.E. Stone and A.M. Scallan, Pulp Pap. Mag. Can., 69 (1968) 288.
27 H.W. Emerton, Fundamentals of the Beating Process, B.P. and B.I.R.A., Kenley, 1957, Figures 98—113.
28. H.E. Dadswell and A.B. Wardrop, Proc. 8th Int. Conf. Botany, Paris, 1954.
29 B. Radvan, C. Dodson and C.G. Skold, l. c. [2], p. 189.
30 L.M. Lyne and W. Gallay, Tappi, 37 (1954) 698.
31 H.W. Emerton, Proc. Tech. Sec. Brit. Paper Board Makers' Assoc., 36 (1955) 595.
32 H.W. Emerton, Br. J. Appl., Phys., 11 (1960) 94.
33 H.W. Emerton, D.H. Page and W.H. Hale, in F. Bolam (Ed.), Formation and Structure of paper, Transactions of the Fundamental Research Symposium held at Oxford, 1961, Tech. Sect. Brit. Paper Board Makers' Assoc., London, 1962, p. 53.
34 D.H. Page and P.A. Tydeman, (a) l. c. [33], p. 397. (b) l. c. [2] p. 371.
35 J.A. Van den Akker, Tappi, 35 (1952) 13.
36 G. Broughton and J.P. Wang, Tappi, 38 (1955) 412.
37 H.P. Dixson, Pap. Trade J., 111 (1940) Pt. 3, 29.
38 L. Nordman, C. Gustafsson and G. Olafsson, Pap. Puu, 34 (1952) 47 and 36 (1954) 315.
39 G. Jayme and G. Hunger, l. c. [33], p. 171.

*Chapter 4*

# FORMING THE WEB OF PAPER

B. Radvan

## 4.1 INTRODUCTION

Perhaps the most unique feature of the papermaking process is that operation, in which the dilute suspension of fibres in water drains through a woven fabric, to leave behind a coherent web. The papermaker uses the terms "drainage" and "formation" to denote this operation, and to distinguish two different aspects of it: firstly, a process of material separation or filtration, in the chemical engineering sense; and secondly, a process of building up a structure of fibres — the web.

In its most typical form, the operation takes place on the "wire table" of a Fourdrinier papermaking machine. Figure 4.1 is a diagrammatic representation of the common elements of such a machine: the stock issuing from the slice is a suspension of fibres in water, typically containing 1—10 g of dry solids per litre in a layer some 6—18 mm deep and up to several metres wide. It is deposited on, and drains through, an endless band of a woven fabric, the "wire" (originally made from wire, nowadays more often from polyester filament). At very low speeds — say one-third to two m $sec^{-1}$, the force of gravity predominates in causing the drainage; at higher speeds, it becomes negligible compared with the pumping action of the drainage elements — "table rolls" and/or "foils". A visible change occurs in the appearance of the stock on the wire when its concentration reaches some 20 g $l^{-1}$: its surface ceases to appear mobile, and soon afterwards loses its liquid sheen, and becomes matt. The drainage elements are no longer effective; the web is formed, and its consolidation begins, assisted by the action of the "vacuum boxes". Some slight re-arrangement of the fibres may still be achieved by the pressure of a light, permeable "dandy roll". The web will finally leave the

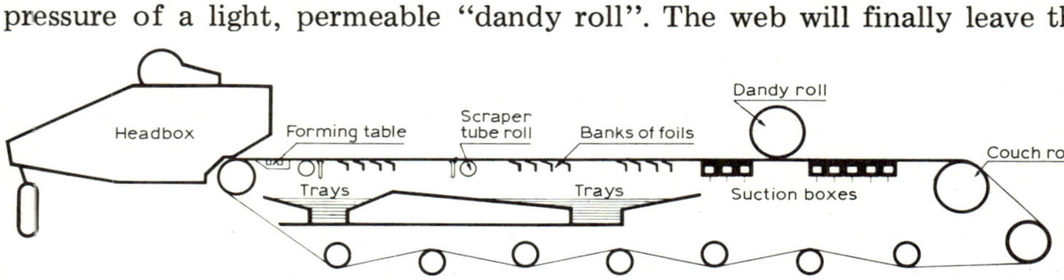


Fig. 4.1. The elements of a Fourdrinier wire part.

"couch" with a solids content of some 100—200 g $kg^{-1}$, for subsequent pressing, drying and finishing.

A chemical engineer may view this operation as an example of the unit operation of filtration, in which a suspension of solids in water is separated into a cake, and a filtrate. A useful description of this operation would then be based on the rate equation:

Rate of the Process = Driving Force ÷ Resistance

Investigation of the operation consists in determining the values of the above three quantities. Many such investigations of the paper-web-forming process have been carried out, and some of their results are summarised in Section 4.2.

But, the operation of forming the web also presents another aspect: that of building its structure, in addition to merely separating it from the suspension. The arrangement of the fibres, which comprise the web, profoundly affects all subsequent operations as well as the properties of the finished product, and itself results from the phenomena which take place during the drainage. This complex interaction of two simultaneous processes is perhaps the fundamental problem of papermaking [1]; some of the most penetrating research work in the last few decades has been directed towards understanding it. As yet the understanding is only qualitative and descriptive. Section 4.3 discusses this aspect.

The operation of forming the web can also take place on paper and board machines involving drainage systems different from those of the Fourdrinier machine. In recent times, machines involving drainage through two wire meshes, one on each side of the forming web, have been developed and have already gained a significant degree of commercial acceptance in practical papermaking. But the single-wire Fourdrinier machine remains the dominant type, and the newer designs are still undergoing rapid development. Consequently, the discussion in this chapter will be limited to consideration of the classical Fourdrinier machine.

## 4.2 FORMING THE WEB AS A PROCESS OF FILTRATION

The accepted chemical engineering practice for describing continuous processes such as filtration, is to represent the rate of the process as being the ratio of a "driving force" — in this case, a pressure drop — to the "resistance" — in this case, the frictional drag of the mat. The latter is often replaced by its inverse, the "permeability". The three factors will now be considered in turn.

The measurement of the rate of filtration presents an immediate difficulty, and illustrates an obvious difference from most other examples of the process of filtration; it is largely discontinuous. The appearance of the underside of the wire shows large discharges from the downstream sides of the drainage elements which only extend for distances which are short compared with the separations between the elements. Some filtration probably occurs between the elements (with a thin film of the filtrate clinging to the underside of the fabric, until scraped off) but this is usually neglected, and the discharge from each element is considered to be due to its action alone. Thus, whether one measures the discharge from a single element or collects the combined discharge from several of them, the true area over which it occurs is unknown. An ingenious method has been proposed [2] for measuring the depth of stock at any point by the reflection of ultrasonic waves off its surface. On occasions, this has made it possible to measure the instantaneous rate of discharge over a selected area.

It is precisely because the area over which the measured discharge takes place is not known, that the application of laboratory measurements to the design, and to predicting the performance of Fourdrinier-forming sections, is uncertain. In contrast, the rate of drainage over known areas exposed to known vacuum — i.e. over wet-forming boxes or on vacuum-forming machines other than Fourdrinier — may be predicted with some accuracy [3, 4].

If no satisfactory means yet exists for determining the rate of drainage, the understanding and determination of the second factor in the rate equation, i.e. the driving force, has been one of the success stories of engineering science in the paper industry in the recent decades. The view was expressed in 1937 [5] that drainage takes place under the action of the forces of gravity. Several investigators demonstrated during the 1950s, both theoretically [6—9] and experimentally [10] that the high speed separation of the surfaces of the table rolls from the underside of the wire develops a vacuum, and thus exerts a pumping action, which accounts for the major part of the drainage. The nature of this force is inertial; it results from the curvature of the streamlines in the flooded nip. Burkhard and Wrist [10] succeeded in measuring the development of the vacuum in the few millimetres distance downstream of the line of contact between the roll and the wire surface, and demonstrated that it reaches the predicted value — equal to the dynamic head of the stock flowing on the wire — before the film of water in the nip becomes unstable, and breaks in a complex manner (see Fig. 4.2).

Alternately, consider the flow in the nip between a table roll and the underside of the wire. Water at the point of stagnation is a rest; therefore its pressure is atmospheric, except for the negligible capillary depression. Immediately after draining through the mat, it is moving forward with the velocity, $v$, of the wire; its pressure must be accordingly less by an amount equal to the velocity head. Thus, the maximum pressure drop across the forming mat reaches the value given by the familiar formula

$$P = dv^2/2g$$

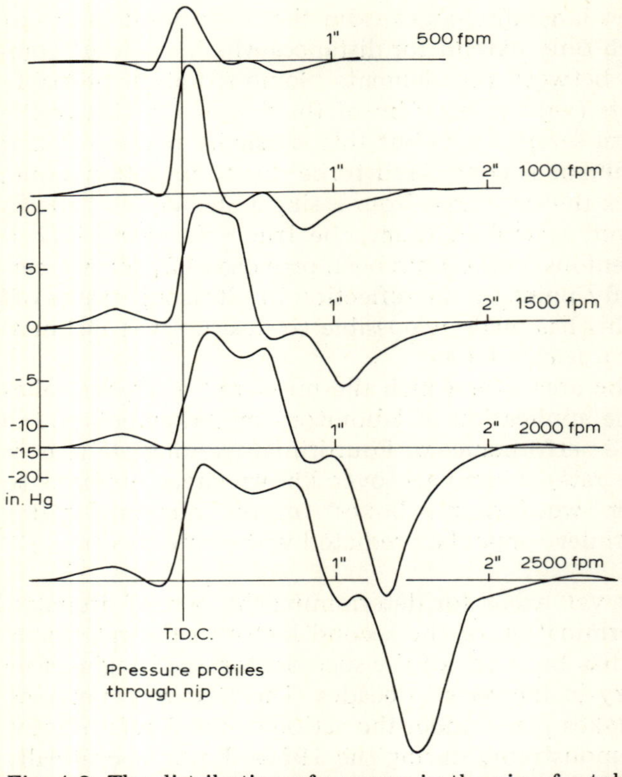


Fig. 4.2. The distribution of pressure in the nip of a tube roll, after Burkhard and Wrist [10].

where $P$ = pressure drop, $v$ = velocity of the wire, $g$ = gravitational accelaration, $d$ = density of water (in consistent units).

More detailed calculations of the distribution of pressure in the nip (rather than of just its peak value) have been published in a number of references [6—9] but, in addition to hydrodynamic considerations, they also involve assumptions about the progress of drainage through the forming mat, and the resulting growth of the resistance to it. As discussed in the next section, this introduces another order of complexity. It now appears accepted that such calculations have not reached a degree of reliability sufficient for detailed design.

Furthermore, as is often the case, the progress of technology has overtaken the need for detailed understanding of the original process — the recent developments being mainly concerned with other methods of forming. The table rolls have now been largely replaced by the gentler-acting hydrofoils, and wet-forming boxes [11]; and indeed, the Fourdrinier-forming table is being replaced by other drainage systems, such as two-wire machines. But that very progress has been stimulated by, or even resulted from, advances in the understanding of the Fourdrinier-table-forming action.

In contrast to the success in discovering the nature of the driving forces of drainage, the understanding and prediction of the third factor in the rate equation, viz. of the resistance of the web to flow, has proved a more complex and difficult problem.

The distinction between "slow-draining" and "fast-draining" stocks must have been apparent from the very beginnings of the craft of papermaking, and many attempts have been made to devise empirical methods of measuring this property. The two most successful, i.e. the Schopper-Riegler Drainage test, and the Canadian Standard Freeness test, are now established as conventional means of process control. But, with the advent of fast newsprint and other paper machines, these tests were found of less use in predicting output capacities of equipment and drainage behaviour of different stocks. Studies of the permeability of the forming web of paper were undertaken in earnest during the 1950s, in parallel with, and no doubt stimulated by, the successful discovery and control of the forces of drainage, which took place at that time. Indeed, a distinction between the two lines of research can hardly be observed; for instance, the clearer understanding of the forces of drainage would be used to determine the resistance, and vice versa.

Again, the theory of the unit operation of filtration provides the starting point. The underlying assumption — so generally accepted that it is often left unsaid — is that the web of paper forms a discrete mat, which grows in thickness as the remaining suspension filters through it. (In contrast to another possible mechanism, by which the concentration would increase uniformly throughout the depth of the suspension during the formation.) Although so widely accepted, this assumption is supported by very little experimental evidence; Linberg [12] and Radvan et al. [13] succeeded in obtaining a film record of the mat separating from the suspension under laboratory conditions. Indirect support for the assumption is provided by the layered structure of the resulting web (see p. 200). Very convincing evidence that a discrete mat forms even at high speeds comes from the recent introduction of "stratified sheet forming" [14]: a specially designed headbox delivers a jet of stock on to the wire, consisting of two or three layers of different composition, which drain in succession, and form a correspondingly layered web.

Given the existence of a discrete mat, its resistance may be examined according to the theories of filtration. A series of papers by Ingmanson and co-workers at the Institute of Paper Chemistry, Appleton [15—17] based their treatment on the Kozeny-Carman equation;

$$\frac{dQ}{dt} = \frac{1}{K} \frac{(1-C)^3}{S^2C^2} \frac{1}{\mu} \Delta P$$

where $dQ/dt$ = rate of drainage, per unit cross-sectional area of the web, $\Delta P$ = pressure gradient across the web, $C$ = volume fraction of the web occupied by solids, $S$ = specific surface area of the solids, per unit volume, $\mu$ = viscosity of the liquid.

The Kozeny-Carmen equation is, of course, only a more detailed form of the familiar rate equation of the unit operation of filtration. It proposes that the operation may be represented by a particular model: (a) the flow is laminar; (b) the web does not compress appreciably during the period of the measurement; and (c) its resistance is equivalent to that of a bunch of capillaries of a suitable size and number. This last, seemingly preposterous approximation, is in fact made valid through redefining $C$ as the volume not accessible to flow; and $S$ as an empirical value, which often may differ from those obtained by other means by an order of magnitude; and finally, by selecting a value of the Kozeny constant "$k$" to represent the departure of the shape of the pores (or of the particles), from the assumed. Typical values of the specific surface, and of the specific volumes, of a range of woodpulps, are given in Table 4.1.

The other two assumptions are less easily accommodated in describing the behaviour of a forming web of paper. Firstly, the flow may not be laminar at drainage rates corresponding to the operation of medium-speed and fast machines [17,18]. The resistance of the web is then more closely represented by a sum of two terms: the laminar term, proportional to the rate of drainage and the turbulent term, proportional to its square. It is estimated that the latter may account for the greater part of the whole in the extreme cases of very fast drainage and very porous mats.

Secondly, the mat is not incompressible. As in the case of most porous, or expanded, solids its compression may be represented by the empirical equation:

$$d = d_0 + MP^N$$

where $d$ = density of the web, when compressed with a superficial pressure intensity $P$; $d_0$ = its density in the uncompressed state; $M$ = an empirical constant, which is intended to represent the effect of the flexibility of the solids in their unexpanded state; $N$ = an empirical constant, which represents the increase in the compression resistance of the mat with increased compaction.

This equation is found to apply over several orders of magnitude of $P$, for a wide range of expanded solids, although the reasons for it are not understood; that is to say, no simple model has been generally accepted which would enable numerical values of the constants $M$ and $N$ to be predicted without resort to measurement — nor explain why their values remain constant over such a wide range of conditions. Figure 4.3 shows a correlation of the two quantities for a range of woodpulps.

However, much of the insight into the filtration behaviour of paper webs,

TABLE 4.1

The effect of beating on the specific surface (after Robertson and Mason, Pulp Paper Mag. Can., 50 (13) December 1949)

| Sample | Beating time (min) | Freeness (Canadian Standard) (cc) | Specific volume (cc/g) | Specific surface (sq.cm/g) |
|---|---|---|---|---|
| Sulphite 1 | 0 | 665 | 3.08 | 13,300 |
| (dried) | 10 | 589 | 3.92 | 18,300 |
| | 20 | 426 | 4.28 | 26,600 |
| | 30 | 285 | 4.22 | 39,600 |
| Sulphite 2 | 0 | 682 | 3.13 | 13,000 |
| (dried) | 10 | 515 | 4.03 | 21,400 |
| | 20 | 403 | 4.30 | 29,300 |
| | 25 | 275 | 4.60 | 39,500 |
| | 30 | 178 | 4.60 | 49,400 |
| Sulphite 3 | 0 | 633 | 2.80 | 11,400 |
| (dried) | 10 | 564 | 4.13 | 21,700 |
| | 20 | 390 | 4.30 | 29,300 |
| | 25 | 305 | 4.16 | 36,700 |
| | 30 | 181 | 4.45 | 48,200 |
| Sulphite 4 | 0 | 669 | 4.40 | 10,200 |
| (wet) | 7 | 559 | 4.90 | 14,700 |
| | 11 | 498 | 4.86 | 19,100 |
| | 15 | 416 | 5.25 | 23,400 |
| | 18 | 320 | 5.12 | 32,000 |
| | 21 | 242 | 5.15 | 38,500 |
| Sulphite 5 | 0 | 661 | 4.50 | 10,600 |
| (wet) | 7 | 560 | 4.90 | 14,700 |
| | 11 | 500 | 4.88 | 19,000 |
| | 15 | 421 | 5.04 | 22,200 |
| | 21 | 270 | 5.10 | 35,800 |
| Kraft 1 | 0 | 718 | 4.27 | 9,100 |
| (wet) | 10 | 708 | 4.60 | 11,500 |
| | 33 | 598 | 4.60 | 28,300 |
| | 44 | 454 | 4.70 | 37,300 |
| | 48 | 384 | 4.91 | 43,000 |
| Kraft 2 | 0 | 738 | 3.66 | 9,500 |
| (wet) | 10 | 703 | 4.57 | 9,700 |
| | 41 | 541 | 4.70 | 31,600 |
| | 54 | 429 | 4.70 | 41,600 |
| | 66 | 292 | 4.67 | 58,400 |

comes from the study of model webs, formed from suspensions of synthetic fibres. Jones [22] found the expected effect that increased fibre flexibility increases both $M$ and $N$; for instance, $M$ is approximately inversely proportional to the 0.24 power of the modulus of the fibre material, whilst both $M$

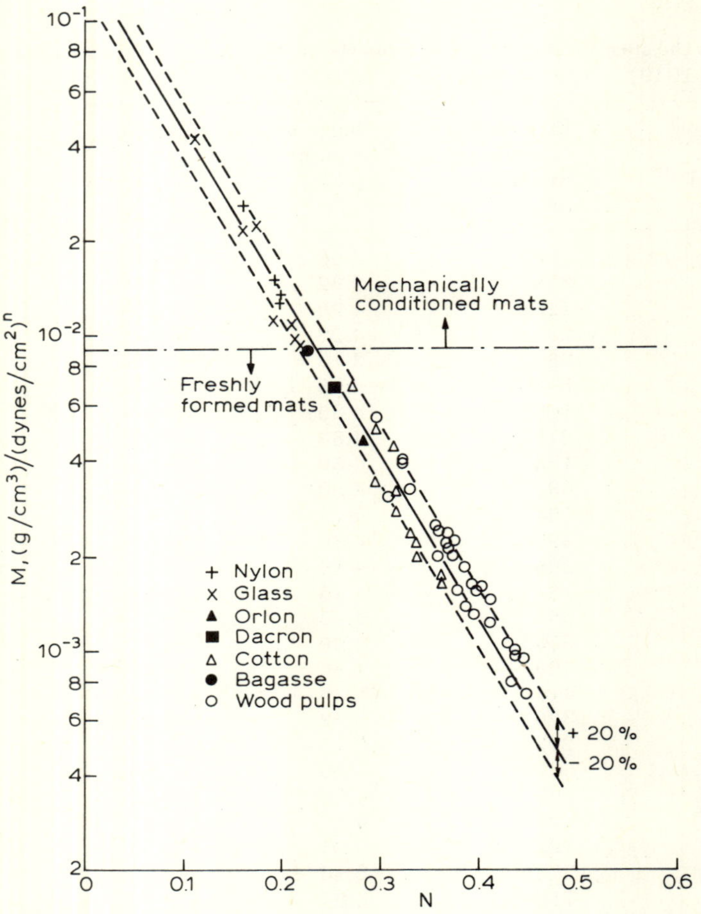


Fig. 4.3. The correlation between the coefficients in the compressibility equation for woodpulp, after S.T. Han [104].

and $N$ increase with increasing fibre length. Qualitatively, these conclusions apply to compressive behaviour of wet mats of woodpulp fibres, but the data are scarce, and capable of several interpretations. Further complications arise from the limitation that the compressibility equation only describes the state of the mat at equilibrium, which may take several minutes to be reached. Under dynamic conditions, the effect of the time-dependence of compressibility is difficult to distinguish from that of turbulence.

Several attempts have been made to include the compression equation in the Kozeny-Carman (so as to predict the change in the void volume with pressure), but it is recognised that the results are not valid for predicting drainage behaviour of paper stocks (see e.g. ref. 3).

In a recent examination of the effects of fibre morphology on drainage resistance, Arnevik and Helle [23] also conclude that only empirical equations are sufficiently reliable, and use the Wahlstrom and O'Blenes equation [24] to correlate their results:

$$t = \frac{G}{S} W^{\alpha}(\Delta P)^{n}$$

where $t$ = time to form a web of basis weight $W$, under a constant pressure drop $\Delta P$; $S$ = concentration of the suspension; $G$, $\alpha$, $N$ = empirical constants.

The empirical constants are determined in laboratory experiments, using a constant vacuum applied to the underside of the wire, on which the experimental web is formed. The value of the "$\alpha$" exponent falls usually within the range of 2—4. whilst that of the "$N$" exponent, within $-\frac{1}{3}$ to $-\frac{1}{2}$. Such measurements correlate well with the results obtained on vacuum-forming and some two-wire-forming machines, but are of only indirect use in predicting the more complex Fourdrinier drainage action.

Finally, the effect of the supporting medium — the wire — on the drainage resistance of the forming web, may not be neglected. These fabrics are woven from bronze wire, or more commonly, from polyester strands, to obtain the best compromise between least filtration resistance, minimum tendency to impress and mark the forming web, and maximum resistance to stretch and wear. A great variety of weaving patterns are available [25]; with spacings between the strands approximately equal to strand diameters, the superficial open area is approximately 20%. This is sufficient to render the drainage resistance of such fabrics negligibly small compared with that of even very thin webs of paper.

However, it is the common experience of papermakers, confirmed by laboratory measurements, that the drainage effects of different fabric weaves is much greater than could be accounted for by their own, negligibly small resistance to the flow of water [26,27]. It appears that the action is indirect; the fabric affects the structure of the first layers of fibres deposited on it. Helle [28] and Kufferath [29] demonstrated in laboratory experiments that the topology of the fabric surface is decisive: if most of the depressions between the fabric knuckless extend in the direction of shear, then fibres will accumulate in them, and a dense impermeable initial mat will result. Conversely, a wire weave with knuckles and depressions extending across the direction of shear will give a loose, open, fast-draining web. Much research work is now proceeding in this field, stimulated by the increasingly severe requirements of the two-wire machines, and aimed at understanding these effects, and hence producing improved fabric structures.

Earlier work by Estridge [30] had attempted to describe the retention of single fibres on wire grids from considerations of geometrical probability. However, Abrams [31] demonstrated that deviations due to the flocculation of fibres, to the relative movement of the suspension and the fabric, and to the complex geometry of the surface, render this approach unsuitable for describing the phenomena which occur during the forming of a web of paper.

To summarise, the process of forming the web of paper may be represented as an example of the unit operation of solid/liquid separation. It proceeds at a rate which is related — but not quite proportional — to the pressure drop, and inversely related — but not quite inversely proportional — to the basis weight already formed. Both observations demonstrate the complexity of the problem, which renders its description by the Kozeny-Carman equation only a very artificial approximation. Qualitatively, the causes of the deviations are understood — the flow is not laminar, the web is not incompressible, and the compression is not even instantaneous. Work still proceeds aimed at a more quantitative understanding of these effects but in the meantime, empirical correlations of laboratory measurements suffice to interpret the differences between different stocks and to predict their drainage behaviour on some of the newer machines.

Such predictions fail when applied to the more complex phenomena which take place on the Fourdrinier wire table, largely because the driving forces of drainage (although their nature is now understood) operate intermittently over distance/time intervals which are not predetermined by the designer, but themselves result from the forming of the web.

A considerable part of the uncertainty also stems from the fact that the drainage behaviour of the forming web depends on its structure, as illustrated for instance, by the unexpectedly large influence of the forming fabrics. The build-up of the structure forms the subject of the second part of this chapter.

## 4.3 FORMING THE WEB AS A STRUCTURE BUILDING PROCESS

The structure of paper differs from those of other common sheet materials in a unique way:

(i) unlike plastics films and metal foils, paper is fibrous;

(ii) unlike woven cloth, it is composed of single short fibres, arranged largely at random instead of in a regular array;

(iii) unlike woven cloth, or woollen felt, or leather, it is laminar, in that each fibre is disposed largely in plane of the sheet, and occupies a particular level within its thickness.

However, paper resembles the other materials in that its structure is anisotropic in its plane, with most of the component fibres oriented along the "grain", or the "machine" direction. All these structural features are discussed under the three headings listed on p. 174; in each case, a very brief description of the principles of measurement or assessment of the feature is followed by a discussion of its origins in the process of forming the web. Much of this information is qualitative and descriptive, reflecting both the complexity of the problems and our imperfect understanding of the underlying physical phenomena.

### *4.3.1 Gross departures from uniformity in-plane*

A major structural feature of paper webs is their lack of uniformity in weight per unit area, the so-called basis weight. It is convenient to attribute a size, or a scale of magnitude to these variations: a small scale over distances comparable to the fibre length, or to floc size: and a larger scale which is the result of imperfect spreading of the flow of stock approaching the slice, or of gross disturbances of the flow on the forming table.

The minimising of these gross variations ranks among the major tasks of the papermaker, in which he has only recently been assisted by advances in instrument and computer technology, and by the application of the statistical methods of the analysis of variance, and of time series. The most important class of measuring instruments are the various forms of radioactive gauges, (invariably using beta radiation), either mounted in a fixed position on the paper machine, or traversing its width at right angles. Alternately, large samples of the web in the form of narrow strips may be removed off the machine and tested by running through a separate instrument. A further refinement is to use a combination of two gauges in tandem, recording the dry and wet basis weights. A basic limitation of these instruments arises from uncertainty over the exact area of the web being measured at any instant during the traverse, so that the recording is of a running average value. Secondly, the response times exceed some 100 milliseconds, largely because of the limitation in the rate at which single beta particles are emitted per unit surface of even the strongest radioactive sources. Trade literature abounds with the descriptions of such instruments.

With both these limitations, the radioactive beta gauges offer the advantage that the quantity which they measure, i.e. the extinction of beta rays in their passage through paper, depends mainly on its basis weight, and much less on its structure and composition.

For this reason, earlier attempts to use optical rather than radioactive sources have been abandoned. The former are much cheaper and act faster, although they give readings subject to greater uncertainty in interpretation.

Whichever, type of measurement is used, a mass of data is soon accumulated (sometimes threatening to rival in quantity the output of the paper machine itself!) usually in the form of machine direction, or cross-direction profiles. These are most useful — singly — in monitoring the steady running of the machine, but have to be reduced to a more manageable set of figures, in order to be of use in understanding the pattern of the variability, and tracing its sources.

Cuffey [32] described an elegant method of presenting a visual display of the continuously changing distribution of the basis weight of the web based on an optical traversing gauge. Each traverse was displayed as a single scan on the screen of an oscilloscope, and then repeated in successive rows slightly displaced in an X-Y diagonal direction. Several scans displayed simultaneously produced a continuously changing "map" of the distribution of basis weight (see Fig. 4.4).

The use of the statistical method of the analysis of variance became established during the 1950s (e.g. see ref. 33). It is a basic tool of the industrial statistician, and consists in partitioning the observed variance into orthogonal components: in this case, those due to the consistent differences in the basis weight across the width — the *C/D* component; those due to consistent changes of the *C/D* profile from time to time — the *M/D* component; and those not accounted for — the residual. The three components add vectorially.

The assumption is that the *C/D* component, i.e. the "stable" part of the profile, lies within the control of the operator adjusting the slice, but that the residual component represents an "intrinsic" lack of stability of the machine, and limits the range within which the profile may be maintained. Since most machines tend to be operated near that limit, the average range of the *C/D* profile may itself be taken as an indication of the intrinsic stability of the machine, and as a reflection of its capacity to produce an even web [34].

A serious disadvantage of the analysis of variance method is the arbitrary definition of the two "stable" components. The computed value of each component will depend on the spacing between the points of measurement; say, the spacing between the slice-adjusting screws; or the interval between the scans of a traversing gauge. Hence in turn, the value of the residual component — ostensibly a measure of the intrinsic stability — will include some of the variability between the omitted points of measurement, and may be changed almost at will.

Also, the profile of the finished paper may be a complex and unknown transform of its control, i.e. of the profile of the slice. Any local change in the depth of the stock at the slice will propagate sideways with the velocity of a shallow gravity wave; with stock depth of say, 10 mm, this will amount to approximately one-third m $sec^{-1}$ — sufficient for a disturbance to travel an appreciable distance across the web, before it becomes set in the formed web. The final profile of the web then becomes a complex result of the criss-crossing of such waves, and of any lengthwise movement of the position on the wire table, at which the web becomes set [35]. Attempts have been reported to stabilise the profile by means of computer control of the slice

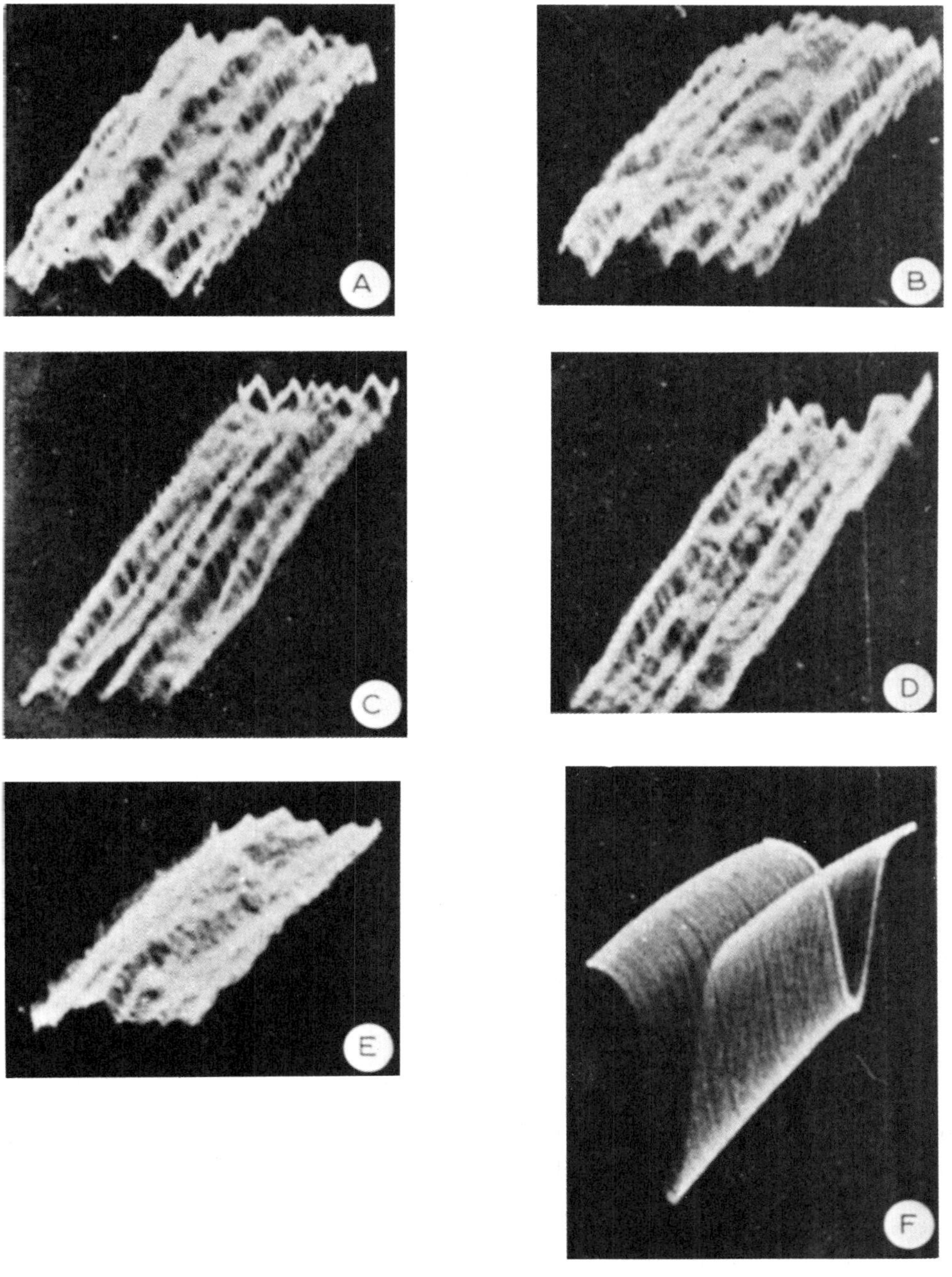


Fig. 4.4. A visual representation of the distribution of basis weight, after W.H. Cuffey [32].

opening — automatic "back tending" [36–38].

The arbitrary nature of the *MD* component of the analysis of variance is even more apparent. Clearly, basis weight will vary to a greater extent say, from day to day than from hour to hour, or in general, the variability is greater within larger time intervals. One may extend the technique of the analysis of variance and partition the total *MD* component into sub-components generated by the variability between successive seconds; or minutes; or hours, etc. This approach leads ultimately to Time Series techniques by which the *MD* variability is represented as a "power spectrum", i.e. the partitioning of the variance between different frequency ranges, or as an "autocorrellogram", i.e. the plot of the correlation coefficients between successive readings, against the length of the interval between the readings. The techniques are described in textbooks of statistics, and their applications to papermaking, in several recent articles (e.g. see ref. 37). Published examples of basis weight *MD* spectra are, as yet, rare.

Identifying the existence, and measuring the relative importance of a periodic disturbance provides the incentive for, and a guide to, locating its source.

A particularly elegant analogue method for tracing the sources of periodic, or quasi-periodic disturbances, is described by Parker and Epton [39]: it relies on very fine tuning of the frequency and phase relationship between the signal generated by the disturbance, i.e. the output of say, a beta gauge, and the signal of the suspected cause of the disturbance — say, the output of an accelerometer attached to a vibrating piece of machinery.

All such measurements and analyses demonstrate the existence of gross disturbances in the process of web-forming only indirectly, from their effects on the uniformity of the finished web. Direct measurements are scarce. Mason et al. [40] demonstrated the existence of large scale turbulence in the headbox, and on the wire of a paper machine, using tracer techniques. Samples of stock tagged with radioactive iodine were added at various points within the headbox and were detected sideways over distances of several decimetres within the finished web. A major source of such turbulence are stable vortices which originate from uneven distribution of stock velocities, or from boundary flow separation within the headbox. Van der Meer [41] describes the common causes of these flow disturbances in flow approach pipes, and in headboxes of the older types of design. The resulting eddies and vortices will cause characteristic flow streaks in the slice jet. Such streaks tend to wander sideways, and to stop and start at irregular intervals. Small eddies are thermodynamically less stable, and tend to coalesce into fewer large ones; it is a major objective of headbox design to minimise such deterioration of the uniformity of stock flow through breaking up the larger eddies whilst maintaining a high intensity of small-scale turbulence.

Reiner and Wahren [42] quantified this concept by measuring the turbulence spectra of several types of commercial headboxes, using fast-response impact probes immersed in the jet of water — not stock — issuing from the slice at a representative rate of flow. The spectrum (see Fig. 4.5) then becomes a criterion of the hydrodynamic performance of the headbox, the aim of the designer being to maximise the intensity of the small-scale end of the spectrum — say 10 mm and less — and to eliminate the large-scale features of the spectrum — at say 200 mm or more, (e.g. see ref. 43).

Another common type of a gross departure from uniformity of a paper web results from edge effects in the flow on the wire table.

The jet of stock emerges from the slice unsupported along its sides. Immediately therefore, a shallow gravity wave will spread outwards, and a corresponding depression in the level of the stock will move towards the centre, travelling with a velocity of $\sqrt{gh}$. On slow, and medium-speed machines, this effect will produce the familiar "light edges" extending up to some 300 mm into the web. Supporting the edge of the jet with a solid deckle offers only a partial remedy, because the retarded boundary layer of stock flowing next to the slice cheek plates will emerge from the slice at a lower rate of discharge; the resulting depression in the level will again cause an outward-moving wave which after reflection off the deckle, will eventually produce another common M-shaped profile of the basis weight. Bleeding the retarded layer from just upstream of the slice minimises this fault, but attempts to counteract it by moving the deckle either in, or outwards will only aggravate the formation of the deckle wave through wakes.

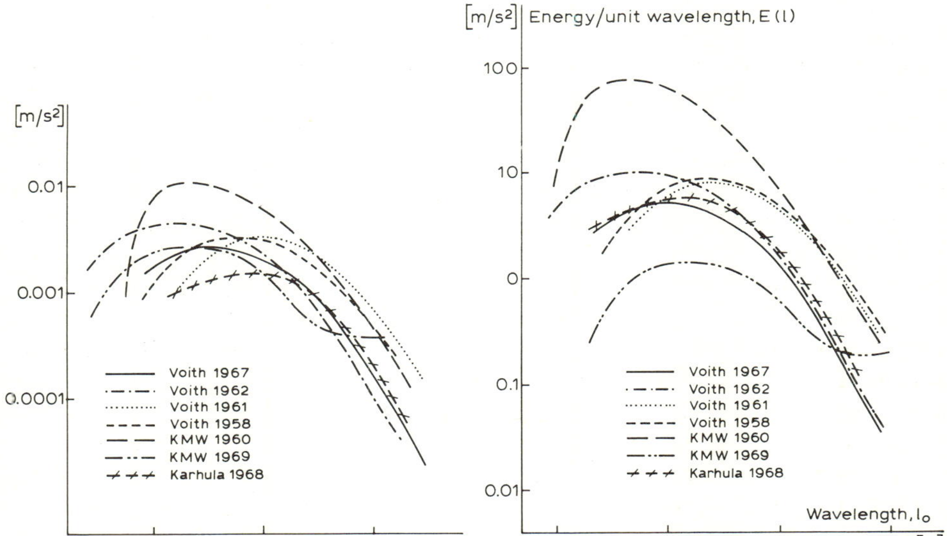


Fig. 4.5. Spectra of turbulence in the slice, after Reiner and Wahren [42].

The problem becomes negligible with increasing machine speeds, since the wave can only travel a short distance before the web is set, and the resulting light edge is trimmed off. Such machines, however, tend to suffer more from another gross disturbance of formation known as "barring". Straight narrow ripples on the surface of the stock extending across the width of the machine may be seen moving down the wire table, usually at a regular spacing corresponding to a frequency of a few Hertz. They are almost invariably caused by pulsations of pressure in stock lines, or by vibrations of rotating heavy machinery, being transmitted to the supports of the wire table or slice. Parker [44] points out that even slight pulsations in the stock pressure behind the slice will be amplified to appear as large momentary changes in basis weight. He calculates the maximum variation in the depth of the stock on the wire table by assuming that a standing gravity wave develops from a given pulsation at the slice. If a node of the wave occurs at that point on the wire where the web is set, then a correspondingly heavy weight "bar" will appear in the formed web. The increase in the depth of stock, resulting from a given increase in pressure at the slice is calculated from the formula:

$$\Delta h/h = (\Delta H/H)\sqrt{H/2g}$$

where $h$, $H$ are the mean depth on the wire, and the head behind the slice, respectively.

The node will occur at the set point if the latter is a quarter-wavelength downstream from the slice, the wavelength being given by

$$\lambda = f\sqrt{gh}$$

where $\lambda$ = the wavelength, $f$ = the frequency of pulsation, $g$ = gravitational acceleration, $h$ = mean depth of stock on the wire.

The derivation of the first formula is not clear, but it appears to equate the shallow wave velocity $\sqrt{gh}$ with the velocity of the wire. The assumption that a standing wave exists is also open to query, in the absence of a reflecting boundary. Nevertheless, numerical calculations yield frequencies in the range of 10—20 Hz, and amplification factors $\Delta h/h : \Delta H/H$ of the order of 5—10, in agreement with the observations of fast machines.

Moen [45] suggests that fluctuations in the shape of the trajectory of the slice jet may supply another possible mechanism for the amplification of pressure fluctuations into basis weigth fluctuations. He also speculates that further amplification of such waves may occur during the travel on the wire table.

Neither author appears to consider a third possible mechanism, the "Taylor instability" which occurs whenever a liquid/gas interface is accelerated towards the former phase, as happens during drainage (see p. 185).

As regards the amplification of pulsations from the slice, a simpler mechanism may also be suggested as explained in Fig. 4.6. Suppose the slice to oscillate with a frequency $f$; a shallow gravity wave will then recede from it with a velocity

$$U + \sqrt{gh}$$

where $U$ = velocity of the wire, $g$ = gravitational acceleration, $h$ = mean depth on the wire; and resonance will occur whenever $U$ is a multiple of $\sqrt{gh}$. Typically $h$ = 0.01 m; $\sqrt{gh}$ = 0.3 m sec$^{-1}$; resonance will occur every ±3% of the mean wire velocity; or, if $U$ is 15 m sec$^{-1}$, every ±1%, etc. rendering fast machines more liable to barring. Further resonance may also result from vibrations of the elements of the wire table.

Secondly, the velocity of the shallow bar wave must decrease as the depth of the fluid stock remaining on the wire decreases, so that its wavelength will also decrease. Figure 4.6 illustrates the effect: the change in the wave veloc-

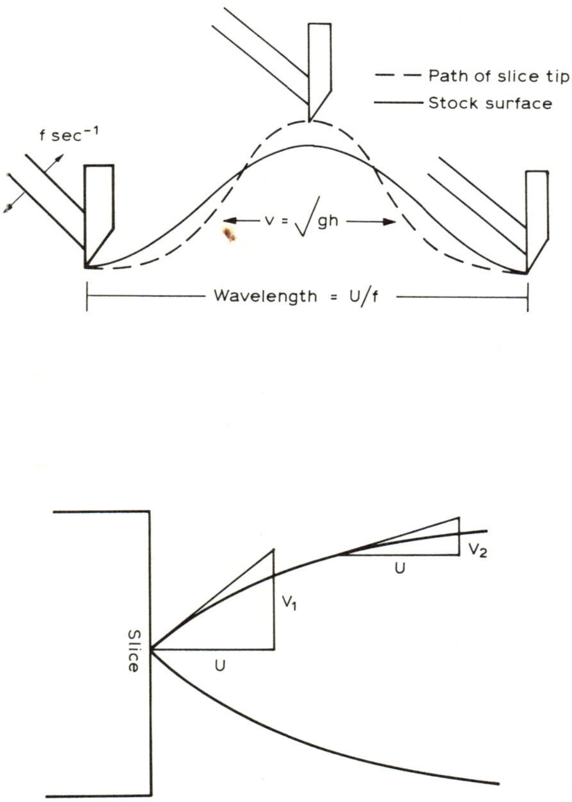


Fig. 4.6. A mechanism for the amplification of waves on the wire table.

ity may be demonstrated by the observation that any Vee type disturbances on the surface of the stock (resulting from the vector addition of the wave, and the wire velocities) become distorted as shown. The resulting reduction in the wavelength of any forward-running waves will cause them to increase in height, and eventually overrun each other, in the manner of surf lapping at a shallow beach. A similar effect is observed, if the slice jet speed much exceeds wire speed as discussed later.

A further amplification of the basis weight bars may also occur through differential retention because the heavy areas of the web act as more efficient filters. Mardon and Cutshall [46] suggest that this effect is large.

To conclude, gross disturbances of the uniformity of the forming web may be regular, such as those caused by edge effects, or quasi-periodic such as barring, or they may be of the more intractable large-scale turbulence type. Much progress has been made in the methods of measurement during the last decade, and equally important, in the mathematical methods of analysing the results. As a consequence, objective methods of assessment have already produced improved headbox designs, whilst better investigational tools offer the promise of easier tracing and rectifying of faults on existing machines. The possibility of computer control of the large-scale variability is still open.

### *4.3.2 Small-scale departures from uniformity of the web*

Even on a most cursory inspection, a web of paper will be found non-uniform on small scale, i.e. over distances comparable with the length of fibres, or of fibre flocs — say up to several centimetres. The visual impression of this type of non-uniformity is termed "formation" or "look-through", and is related to, but not identical with, the distribution of basis weight on that scale — the "mass density distribution", a term used by Corte [47]. The aesthetic effect is obvious, and may be drastic; and the objective effects on the physical properties of the paper can be highly significant, though not immediately apparent. These are thoroughly reviewed [48] by Norman and Wahren, who also discuss and describe the available methods of measurement, both optical and radioactive. In this chapter, we are concerned with the origins of this type of non-uniformity in the process of forming the web.

Perhaps the most common, and easily recognisable defect of formation is "rolling" — the appearance of elongated fibre flocs, arranged roughly in parallel to the cross-direction rather like flotsam on a shallow beach (see Fig. 4.7). It is worth nothing that although the appearance of the fault is unmistakable, and its origins quite specific, no objective means is known of distinguishing it from a general "cloudiness", or lack of uniformity. One may speculate, that perhaps, an areal autocorrellogram could be used to recognise the cross-direction extended shapes of the rolling flocs.

It is now well established, (e.g. see ref. 49), that this fault results most

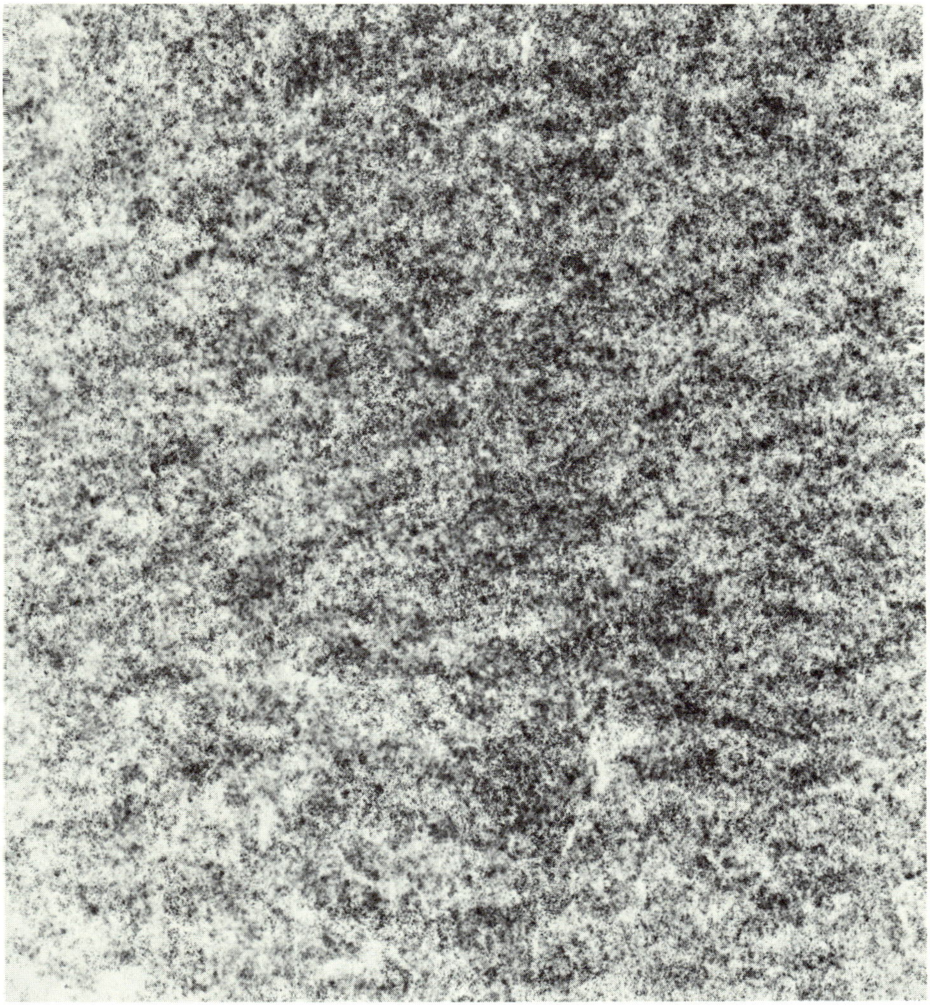

Fig. 4.7. The appearance of "rolling" flocs in paper due to excessive Jet/wire speed ratio.

often from a mis-match between the speed of the jet issuing from the slice, and of the wire fabric at the point of impact, and subsequently during the early stages of formation. Such a mis-match is virtually impossible to avoid because firstly, the jet velocity itself is far from uniform either across the width, or especially across its thickness; and secondly, because the angle of impact on to the wire is never quite zero (see ref. 50 for methods of calculating the angle of the jet). In practice, the papermaker will attempt to adjust the two variables for optimum results. This is likely to be found when the nominal jet speed (i.e. as calculated from the hydrostatic head behind the slice) is some few per cent less than the speed of the wire (see Fig. 4.8). Conversely in the other extreme case of very large excess jet speed, the rolling flocs become very pronounced and may merge into a form of "barring" as discussed above.

The author is not aware of a published rigorous explanation of the mechanism of the "rolling" fault. It is intuitively explained as being due to the lifting action of the shear between the already formed mat, and the stock still flowing above it. Such an effect could be due to the viscoelastic properties of the stock (similar to the well-known Weissenberg effect in which tangential shear produces radial tension), or, more likely, to hydrodynamic lift exerted by the flow over a rough surface, not unlike the mechanism involved in the sculpturing of beach sand, and the formation of sand dunes [51]. This explanation is consistent with the practical observation that very large differences in jet/wire speeds may be tolerated and will not cause "rolling" provided that the web is held down and protected from disruption, by vacuum applied to the underside of the wire.

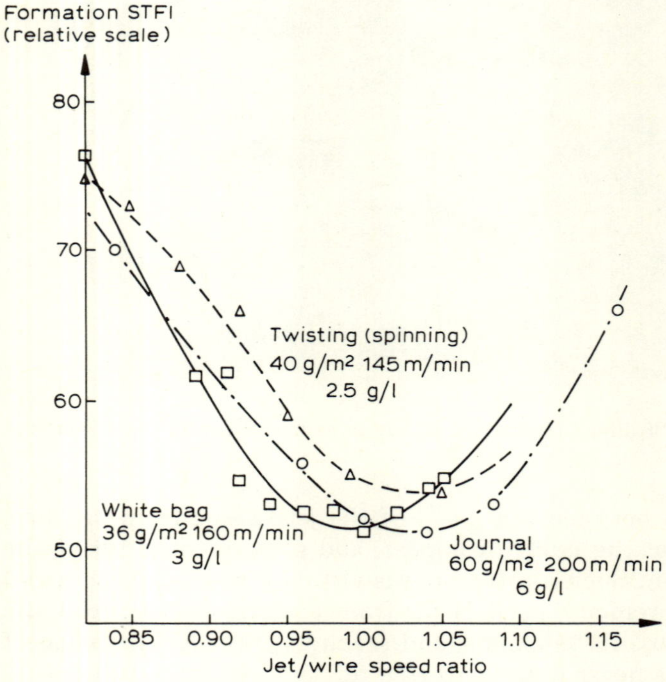


Fig. 4.8. The effect of the jet/wire speed ratio on formation, after Schroder and Svenson [49].

Another type of disruption of a partly-formed web occurs on fast-running paper machines, as the result of the violent phenomena which take place during fast discontinuous drainage. Many spectacular photographs have been published of the break-up of stock surface on the wire table, and the subsequent growth of the resulting spouts and ripples (e.g. see ref. 52). Several likely explanations have been advanced. Burkhard and Wrist [10] demonstrated by direct measurement that a film of water carried round a table will cause a pulse of positive pressure on entering the nip (see Fig. 4.2) and may be expected to disrupt a partly formed web. A special investigation sponsored by TAPPI confirmed the theoretical predictions of Taylor that the main cause of free-surface instability is the rapid acceleration downwards which it suffers on passage over drainage elements, and which may be aggravated by sudden release of the wire fabric from its partial wrap-round a table roll [52—54]. The resulting paper presents a typical "broken-up" appearance; again, no objective method is known of identifying and measuring this fault specially. The minimising of this fault is one of the main reasons for the widespread replacement of tube rolls by the gentler-acting hydrofoils; and its avoidance is a major objective in the development of two-wire forming machines, in which it is completely suppressed in the absence of a free surface [55].

Very occasionally, a fault may also occur which although similar in appearance, results from very different causes during the early stages of consolidation, rather that of forming. The dandy roll, whether plain or watermarking, acts as a very gentle press, forcing the water expressed from the top layers of the very loose web to flow in the direction opposite to that of the moving web. Should the drag of this flow exceed the yield strength of the web, then the latter will rupture in a distinctive pattern. The same phenomenon occurs more frequently, and indeed constitutes the major limitation on the operation of wet presses, where it is known as "crushing" [56]. Since the effect of the dandy increases with increasing moisture content of the web, whilst the resistance of the latter to crushing decrease at the same time, the efficient operation of the dandy takes place under conditions of just incipient crushing.

However, by far the most common and most important cause of uneven distribution of basis weight on the small scale, is the natural tendency of the fibres suspended in the stock to form relatively dense assemblies, or "flocs". These may then persist during the forming process, and remain as recognisable heavy-weight patches within the finished webs [13,57,58]. The word "flocculation" used by papermakers to describe the phenomenon has a slightly different connotation than in the case of the aggregation of suspensions of less extended shapes; its mechanism is probably different. The validity has even been queried of the very concept of flocs as discrete entities, rather than merely random fluctuations in the concentration of the suspension. However, the generally accepted view now is that fibre flocs are objectively definable entities, and that their nature and mechanism of formation is understood if only qualitatively — the latter largely as the outcome of some very revealing research carried out in the last two to three decades.

The technical literature of the 1930s and 1940s contains many references to the observed effects of ionic and colloidal additives on the tendency of paper stocks to flocculate. For instance, Erspamer [59] and Gundersen [60] explain the effect of alum in terms of the adsorption of the positive ion of aluminium on to the negatively-charged surface of fibres. As indicated by slow drainage, the dispersion of fibres is most even at concentrations of alum corresponding to the isokinetic point, whilst larger rates of addition of alum cause the stock to flocculate again. The implication here is that the mechanism of fibre flocculation is similar to that of colloidal particles, i.e. by surface attraction. A number of authors investigated the action of various mucilage-like additives in improving the dispersion of long-fibred stocks; interestingly, Broadbent and Harrison [61] identify high viscosity as an essential property which confers the outstanding dispersing capacity upon solutions of vegetable gums used in the traditional Japanese papermaking craft.

In a series of investigations summarised in ref. 62, Mason and co-workers adopted a different approach, treating flocculation as the result of mechanical entanglement of fibres, itself caused by random collisions in a sheared suspension. Using a modified Couette apparatus, they observed and classified the movements of single particles in a shear field: spheres, rigid cylinders, and flexible fibres. The behaviour of the two former shapes is predictable theoretically, by extension of Jeffery's formula for oblate spheroids, and observations confirm its validity. In particular, extended rigid cylinders rotate in a double conical orbit with a predictable constant velocity. Flexible fibres twist and rotate in many complex ways, but qualitatively, the effects are similar in that each fibre effectively occupies an extended volume in the suspension, which is swept by its rotating orbit. When two particles approach, collisions may occur. Again, these are definite events in the case of spheres, and their calculable frequency of incidence is confirmed by observations. Rigid rods and fibres need not actually make contact in order to affect each others paths, and an interpenetration of the orbits is sufficient. Their rotations are also affected; it is found that such particles in a dilute suspension under shear will gradually assume a preferred orientation in the direction of the shear, as the result of multiple collisions between the particles affecting their rotational orbits and velocities.

Mason postulated that flocculation is the result of a dynamic equilibrium between collisions causing the formation of doublets, triplets, etc., of the fibres, and of a simultaneous breakdown of such aggregates by shear. In particular, if each particle is assumed to occupy a space equal to the volume of its orbit, then a critical concentration may be calculated at which the whole volume of the suspension is occupied, and no free rotation is possible.

It is a simple function of the ratio of length to diameter:

$$C_0 = 3/2r^2$$

where $C_0$ is the critical concentration, volume/volume $r$ = the ratio of length to diameter.

The formula is only an order-of-magnitude approximation in the case of flexible wood fibres, which rotate in complex ways, and whose hydrodynamic volume is not well defined. Even so, the calculated values of the critical concentration range from some $\frac{1}{2}$ percent (for very short fibres) to one-hundredth or one-thousandth of a percent by volume, for fibres of a shape similar to that of wood fibres.

Such concentrations are much below the value encountered in the papermaking process. In a later work [63], an attempt was reported to deduce the mechanism of flocculation of paper stocks from their behaviour during flow in pipes. In particular, the concept was introduced that the flowing suspension consists of a coherent structure of flocculated fibres, which has a measurable strength. The strength was assessed simply and directly, by measuring the maximum length of a floc that will support its own weight in water, and used to predict the Reynold No. of a pipe flow, at which the plug of stock in the centre of the pipe will break up under the action of turbulence. The values of the network strength at 0.8% concentration were found to be of the order of 3 dynes per sq. cm of the nominal area, for a high-yield softwood pulp, to 16 dynes $cm^{-2}$ for a softwood sulphite pulp, in agreement with the expected increasing tendency to flocculate. Significantly, the network strength could be halved by the addition of a deflocculating agent.

The concept of a coherent fibre network, the strength of which determines the propensity to flocculate, proved very productive. In an extended investigation Wahren and co-workers [64,65] proposed and demonstrated another mechanism by which such networks may form. The starting observation is that a fibre may only become a part of a network if it is in contact with, at least, three other fibres (see Fig. 4.9). Secondly, it is not necessary to assume that adhesion, or frictional forces need act at such points of contact, in order to maintain a coherence of the structure. Instead, it is proposed that the fibres possess an appreciable rigidity; when being dispersed in a suspension (e.g. in a pump, or by a stirrer), they bend out of their unstrained shape. In the interval after the dispersion, and before movement of fibres in the suspension is stilled — or, before a network is formed — the fibres will rotate, collide, and unbend at the same time. If the concentration is sufficiently high, the movement will cease, and fibres become locked into their positions, whilst still strained — and it will be these frozen strains that will maintain the coherence of the network, or, if the latter is broken up — of the resulting fragments, i.e. the flocs.

In order to test this model experimentally, two measurable quantities

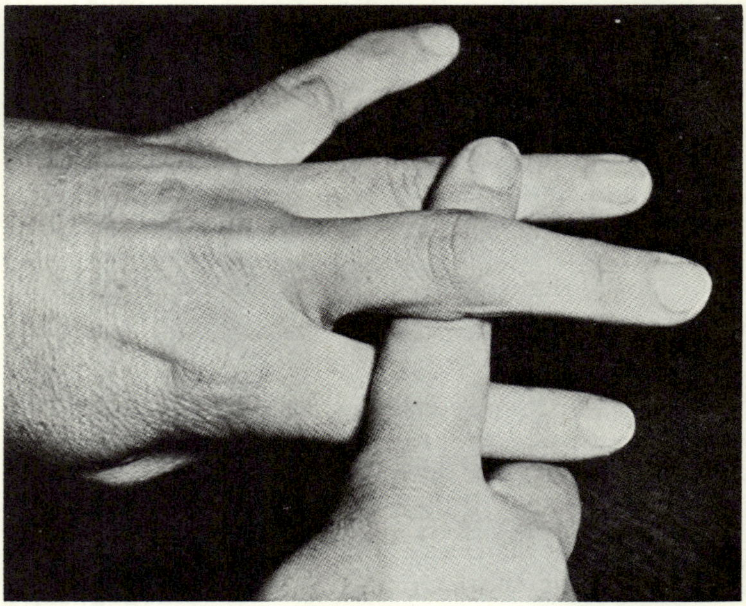

Fig. 4.9. One-plus-three fibres make a structure, after J.D. Parker [102].

were derived by assuming that the network consists of a three-dimensionally random arrangement of fibre segments and of fibre-fibre bonds, the probability of a bond being found on any given segment being proportional to its length. Firstly, a minimum concentration was calculated at which each fibre would be in contact with a given number of other fibres (at least three), i.e. the minimum at which a coherent network could be found. This value was identified with the sedimentation concentration for the given type of fibre. A very complex formula resulted, if allowance was made for a plausible distribution of fibre lengths and length/diameter ratios, but for a suspension of model fibres, say glass or synthetic, these distributions are single-valued, and a very simple formula then results:

$$C_{\text{min.}} = 108\pi/A^2$$

where $C_{\text{min.}}$ = sedimentation concentration, by volume, and $A$ = the length to diameter ratio of the fibres.

This formula resembles that of Mason's (see above), in that it predicts the limiting value to be inversely proportional to the square of the length/diameter ratio, but the value is greater by nearly two orders of magnitude, which brings it into the range of practical papermaking. Comparison with the measured values of sedimentation volumes of model fibre suspensions shows good agreement, if the number of contracts per fibre is taken to be 3—5 rather than just 3.

Secondly, the shear modulus of the network was derived from the same assumptions, and predicted to be proportional to the flexural modulus of the fibres, and to the length diameter ratio. The very complex formula which resulted was tested against measurements of the modulus of model fibre suspensions in a Couette-type apparatus, and found to be in qualitative agreement.

Furthermore, methods of dimensional analysis were used to correlate measured network modulus with the rigidity of model fibres, a good degree of agreement suggesting that most of the significant variables had been included in the model.

Suspensions of woodpulp fibres were found too complex for direct calculation. Even so, the proposed model has made it possible to interpret the observed effects of mixing different stocks and of beating, if only in a qualitative way. For instance, Fig. 4.10 shows the change in the shear strength of suspensions of a mixed groundwood-sulphite stock, with changing proportions of the two components. At low concentrations, the short, rigid groundwood fibres cannot participate in the formation of the network, the strength of which varies linearly with the amount of sulphite. At higher concentrations, sufficient contact density is obtained for even the groundwood fibres to be entangled, when their superior rigidity produces a very strong network. In contrast, the behaviour of a mixture of long kraft fibres and their fines suggests that each component forms an independent structure.

It is clearly an oversimplification to attribute the formation of flocs merely to the action of locked-in fibres straining against each other, and to disregard the contribution of adhesion and frictional forces at points of contact between fibres. Reports appear continuously of the flocculating effects

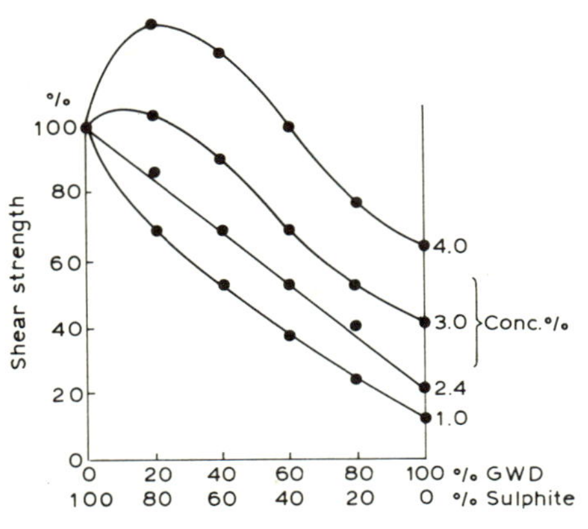


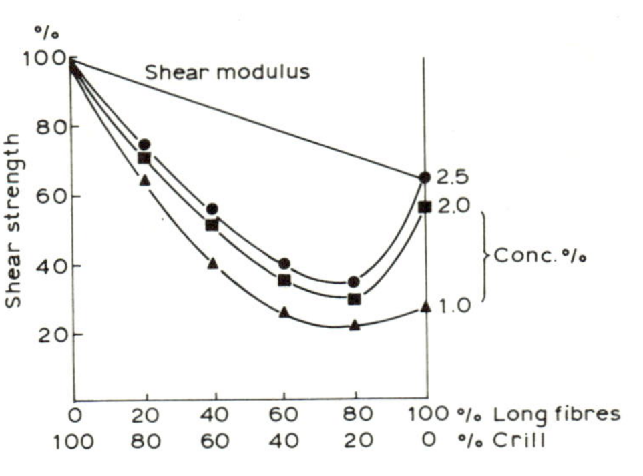


Fig. 4.10. The shear strength of mixed furnishes, after Steenberg et al. [65].

of "retention agents" and of the deflocculating effects of gums and mucilages, but there does not appear to exist a quantitative theory connecting the magnitude of these forces with their flocculating effects, beyond repeated observations that the tendency to flocculate is often most at rates of addition near the isoelectric point. Predictably, cationic additives, e.g. polyacrylamides, modified starches [66,67] are most effective in causing flocculation. It is also well know that small bubbles of dispersed air tend to promote flocculation [68]. Gavelin [69] speculates that the effect is due to a bridging action, with a small bubble acting as a bond between two slightly hydrophobic fibre surfaces; this appears a plausible explanation, especially by analogy with the bubble-stabilising action of some powders [70]. Indeed, a similar phenomenon has been used industrially to accelerate the formation of large aggregates in settling suspensions [71]. The extraordinary deflocculating effectiveness of the vegetable gums used in the ancient Japanese papermaking craft cannot be explained merely by the high viscosity of their solutions; the existence of very bulky adsorbed layers, and of resulting long-range surface forces, has to be postulated [72,73].

However, the mechanical entanglement theory of flocculation, although a gross oversimplification in that it neglects the effects of surface forces, has proved extraordinarily fruitful through the insights which it provided into the working of the forming section of the paper machine.

Flocculation of the stock is seen as inevitable at practical papermaking concentrations; rather than attempting to prevent it, the efforts of the designer are instead aimed at providing conditions for dispersing the single mass of entangled fibres into acceptably small fragments, through the application of intense, small scale turbulence [55,74]. Success may be objectively assessed by measuring rhe spectrum of turbulence at the slice by means of an impact probe [42]. When working in water, or in very dilute fibre suspensions, the dimensions of the probe may be reduced so that the intensity of eddies of only some ten millimetre size may be measured, and that of the smaller ones, inferred from extrapolation. The ultimate development of this approach is probably represented by the design of the High Consistency Forming Headbox [75] which uses, as the slice, a long channel of only some 3 mm depth, to ensure that the scale of turbulence is correspondingly small. As the result, the device can produce webs with an acceptable degree of uniformity from stocks at concentrations of an order of magnitude greater than would be possible using conventional designs.

The traditional papermaking practice of shaking the wire table during this period of formation, in order to improve the uniformity of dispersion, belongs in the same category. Its mechanism has not been analysed theoretically, but it is clear from many studies (e.g. see refs. 76,77) that the intensity of shear resulting from the shake is sufficient to improve the dispersion on slow, and medium-speed machines. The disruptive effect of the table rolls, described above, is also claimed to have a similar effect.

Paradoxically, the opposite approach — attempts to prevent flocculation from developing, rather than to disperse the flocculated mass — has also profited from the insights provided by the mechanical entanglement theory. Since entanglement results from the free rotation of dispersed fibres being restricted by collisions with others, then it should be preventable if the dispersing medium has sufficiently high viscosity to dampen out the rotation before too many collisions take place. Such an effect of high viscosity has been observed elsewhere and indeed has formed the basis of industrial processes (e.g. see refs. 78, 79) which use solutions of thickening agents, such as carboxymethylcellulose, to achieve very even dispersion of very long, easily flocculated fibres. Steenberg et al. [65] illustrate their explanation of the effect of viscosity by means of the graph shown in Fig. 4.11; the shear modulus of a model fibre suspension decreases steeply, once a critical value of viscosity of the suspending solution is reached. The many reported observations of the dispersing action of aqueous foams (e.g. see ref. 80) may also be explained by the high "viscosity" of that medium (more likely the high value of yield stress of the structure of tightly-packed bubbles of air). In addition, the foam medium offers the advantage of a pronounced degree of shear-thinning so that its resistance to the — relatively large — drainage forces is much less than would be that of a Newtonian liquid with the same viscosity at a lower shear. The combination of very effective dispersion with relatively fast drainage forms the basis of the Radfoam forming process [81]. These high-viscosity media make it possible to obtain very uniform dispersion of very long fibres at normal concentrations, or of conventional woodpulp fibres, at concentrations much above normal. In effect, the value of minimum network concentration in the Meyer/Wahren formula quoted

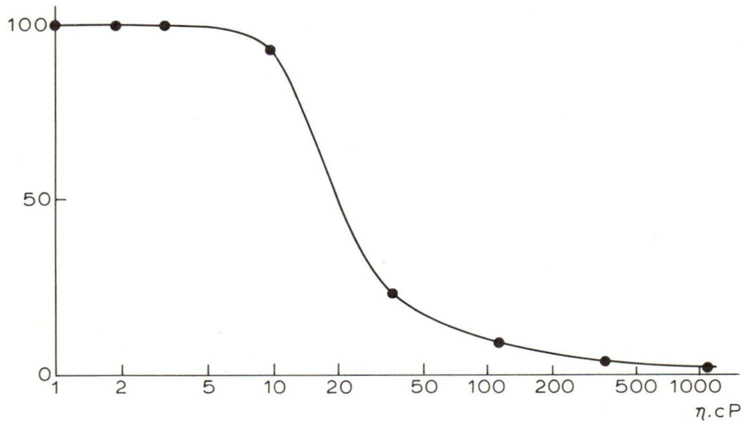


Fig. 4.11. The effect of the viscosity of the suspending fluid on the floc strength of a fibrous suspension, after Steenberg et al. [65].

above, is increased by an order of magnitude. It will be noticed that the derivation of that value makes no allowance for the viscosity of the medium, nor indeed for any mechanism by which the network is formed; it merely specifies the geometrically random arrangement of rigid fibres, in which a number of contact points per fibre must unavoidably be made. With flexible fibres, no such necessity exists, and an arrangement may be visualised at virtually any concentration, in which no fibre touches any other — as appears to be approximately the case in very viscous suspensions.

The more traditional methods of preventing flocculation through using fibre suspensions at very low concentrations are of course, the basis of a number of modified Fourdrinier designs (e.g. see ref. 82), which fall outside the scope of this chapter.

Thus, a consensus on the mechanism of fibre flocculation may be discerned: the phenomenon is dominated by mechanical entanglement, especially of long and rigid fibres. It is not, however, independent of the forces of surface adhesion, or friction between the fibres. The extent of flocculation may therefore be minimised by the addition of surface adsorbing ions or polymers, or by increasing the viscosity of the suspending medium. It may also be counteracted by maintaining the flocculated stock in a state of vigorous small-scale turbulence.

The understanding of the phenomenon, although only qualitative, has led to a number of practical advances in the recent years. A quantitative theory is still missing, which would enable one to predict the extent of flocculation for a given stock, at a given concentration, and under given flow conditions. Such knowledge would probably have to be based on the cross-correlation function between simultaneous measurements of flocculation and turbulence, but no such measurements have yet been reported [83]. Together with the obtaining of cross-correlations against the third distribution — that of mass density — the problem forms the outstanding task for research in paper physics at the present time.

For instance, circumstantial evidence exist, but is contested, that the state of dispersion reached by the stock on the wire may become more even through the very forming action. The most common example of this effect is the frequent observation that the wire side of paper made on slow, and medium-fast machines, is more uniform than is the top side, presumably because of being exposed to the deflocculating action of shear during formation. A more controversial view was first expressed by Wrist [84] that the web being formed is more uniform than would be its projection from the original suspension, because preferential drainage through any incipient low-weight spots will tend to fill them up. Conversely, less stock will be drained through, and fewer fibres be deposited on, the heavy-weight areas. In particular, a web which consisted of fibres deposited separately at random — i.e. with the ultimate in the evenness of dispersion available by any turbulent action — would be unacceptably non-uniform, in the absence of the preferential drainage effect. A computer simulation of the random deposition process was used to support the argument. Corte [47] surveyed the distributions of mass density of small areas (one sq. mm area) of a number of commercial paper samples, and reached the opposite conclusion, that the variances of such distributions were considerably greater than would be expected from random deposition. Norman and Wahren [48] used a less direct technique which involved the calibration of an optical, rather than radioactive, device, and obtained results supporting Wrist's conjecture.

Table 4.2 lists some of Corte's data; Fig. 4.12 illustrates the results of Norman and Wahren.

Cruder but very direct methods may be used to assess the uniformity over

TABLE 4.2

Variances of the mass density distribution of a range of papers (after Corte, Ref. 47)

| Type of paper | Basis weight $(g\ m^{-2})^2$ | Wetness (°SR) | Variance of MDD $(g\ m^{-2})^2$ | | Handsheets |
|---|---|---|---|---|---|
| | | | Machine-made papers | | measured |
| | | | Calculated [a] | Measured | |
| S [b] | 16.6 | 16 | 1.67 | 8.42 | 3.12 |
| S | 23.1 | 44 | 2.51 | 10.80 | 3.27 |
| S | 29.6 | 21 | 4.51 | 9.00 | 7.81 |
| K [c] | 40.8 | 85 | 6.32 | 8.40 | 7.61 |
| S | 41.9 | 30 | 5.33 | 16.82 | 6.74 |
| S | 42.1 | 28 | 10.70 | 30.25 | 12.61 |
| K | 42.4 | 85 | 6.32 | 10.80 | 7.92 |
| S | 50.0 | 15 | 4.92 | 15.22 | 5.58 |
| K | 50.3 | 65 | 5.56 | 12.25 | 6.42 |
| S | 55.3 | 38 | 5.69 | 24.05 | 6.17 |
| K | 58.1 | 48 | 6.67 | 20.25 | 7.71 |
| S | 75.7 | 53 | 5.72 | 26.01 | 6.02 |
| K | 76.1 | 68 | 2.87 | 9.00 | 3.11 |
| K | 83.5 | 90 | 2.87 | 15.21 | |
| K | 93.3 | 37 | 7.00 | 10.25 | 8.22 |
| S | 124.2 | 18 | 13.05 | 134.8 | 15.02 |
| S | 126.6 | 23 | 11.31 | 37.20 | 13.32 |
| S | 130.3 | 26 | 9.92 | 42.30 | 11.84 |
| S | 139.4 | 13 | 17.88 | 74.00 | 20.88 |
| K | 143.5 | 40 | 13.89 | 38.70 | 15.54 |
| K | 156.4 | 37 | 17.65 | 136.9 | 20.03 |
| K | 215.8 | 35 | 20.66 | 106.1 | 23.02 |
| K | 246.8 | 42 | 14.88 | 106.1 | 16.04 |
| K | 261.8 | 24 | 17.62 | 237.0 | 19.71 |

[a] For the case of a random distribution.
[b] Soft absorbent papers.
[c] Printing, copying, tracing grades.

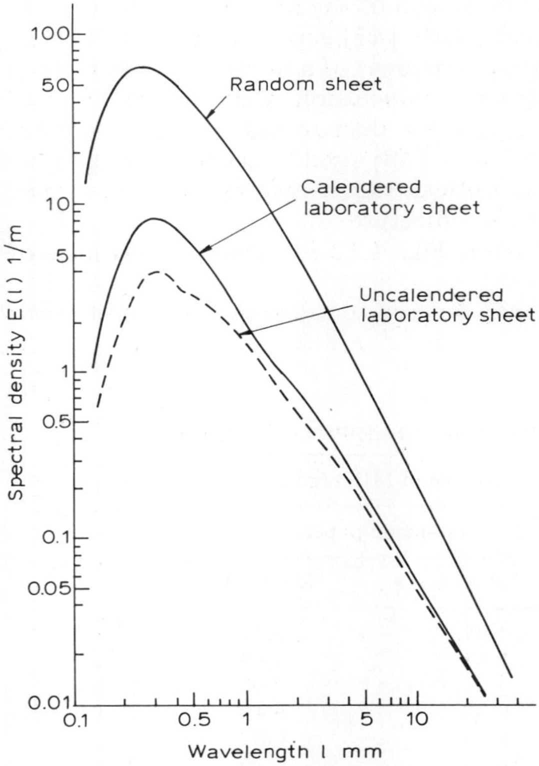


Fig. 4.12. A spectrum of the basis weight of paper, after Norman and Wahren [48].

areas rather larger than Corte's one square mm. A small proportion of the fibres are dyed a distinct shade and returned to the stock; their distribution in the finished sheet of paper may then be assessed by counting the number of such fibres which appear to cross a scanning line of a given length. Table 4.3 lists the results obtained on commercial samples of papers, with scanning lengths of 3 and 10 mm. The variance of these numbers is less than the mean value in nearly all cases; a random process of dispersion would have generated greater variability, the counts being distributed according to the Poisson formula, i.e. with variance equal to the mean. The length of the scanning line is similar to some of the "wavelengths" in Norman and Wahren's spectra, but the degree of improvement over the random case is much less, perhaps because a commercially made, flocculated paper, rather than a handsheet, was tested.

In the absence of direct observations of single fibres, or flocs being deflected towards light-weight areas during drainage, the controversy may be

TABLE 4.3

Variances of the number of tagged fibres observed crossing a scanning line (by permission of Wiggins Teape Research and Development Limited)

| Tabulating card grade | | Scanning length, 5 mm<br>Variance as percentage of the mean count | | |
|---|---|---|---|---|
| Reel 60 | MD | 91% | | |
| | CD | 86% | | |
| Reel 65 | MD | 79% | | |
| | CD | 105% | | |
| Lightweight coating stock | | Scanning length, 10 mm | | |
| | | Mean count | Variance | Variance as percentage of the mean |
| Reel A | MD | 0.567 | 0.446 | 79% |
| | CD | 0.686 | 0.654 | 95% |
| Reel B | MD | 0.510 | 0.520 | 102% |
| | CD | 0.724 | 0.611 | 84% |
| Reel C | MD | 0.438 | 0.418 | 96% |
| | CD | 0.495 | 0.526 | 106% |
| Reel D | MD | 0.452 | 0.400 | 89% |
| | CD | 0.562 | 0.484 | 86% |
| Reel E | MD | 0.486 | 0.433 | 89% |
| | CD | 0.619 | 0.521 | 84% |
| Reel F | MD | 0.745 | 0.553 | 74% |
| | CD | 1.158 | 0.804 | 73% |

Approximately one hundred fibres counted in each case.

considered unresolved. (It may be worth stating that it is of more than academic interest, since the result must affect the judgement of future designers of paper machinery.) Indirect evidence for the existence of preferential drainage comes from the observation that the common fault known as the "wiremark" must clearly originate from such an effect (the fibres depositing preferentially over distances equal to the wire mark, i.e. some fraction of a millimetre) rather than being caused by an "embossing" action of the wire upon an already-formed web [85]. Similarly, the best quality watermarks are made by preferential deposition of fibres over distances of several millimetres, on to an embossed wire cover of a cylinder mould machine — admittedly, at very low speeds.

The wire mark itself forms a characteristic feature of the structure of most papers, which originates in the forming process. It may be seen as an apparent impression of the upper surface of the fabric; Attwood and Parker [86] describe an elegant method of enhancing the visual impression of the mark by repeated photographic contact-printing of the sample, each print displaced by a mesh spacing. In effect, the method is a photographic construction of an autocorrelogram. The popular QNS/M optical formation meter produces the same information in a numerical form; the intensity of the wire mark is measured as the height of the peak on the graph, at a wavelength corresponding to the wire mesh. The mark may be more than just an optical effect, with the differences in basis weight between the light and heavy areas as much as 20—40%, of the average. Sectioning [87] and sheet splitting [88] also demonstrate that the mark penetrates the lower layers of the web, and may not be removed by subsequent pressing. Clearly, the mark originates in the process of forming the web in a way not yet understood, although some indications of the mechanism are now emerging from the results of work on the effect of the wire fabric on orientation (see p. 200). It may be worth noting that the wire mark is not always a fault — it may be deliberately induced in order to increase softness and absorbency [89].

Whatever the cause of the lack of uniformity on small scale — disruption of the web, flocculation, or wire mark — it will be seen in most cases that the light areas exhibit another well-known structural feature of paper; the pinholes. The name is self-explanatory: on examining a sheet of paper against the light, a number of small pin-points can usually be seen. It is not certain whether they form a distinct and separate population, or simply represent the extreme values of the distribution of pore sizes. Corte [90] reports that such distributions may be bi-modal in some cases, with a peak in the large-size range indicating the existence of a distinct population of pinhole-size pores (see Fig. 4.13).

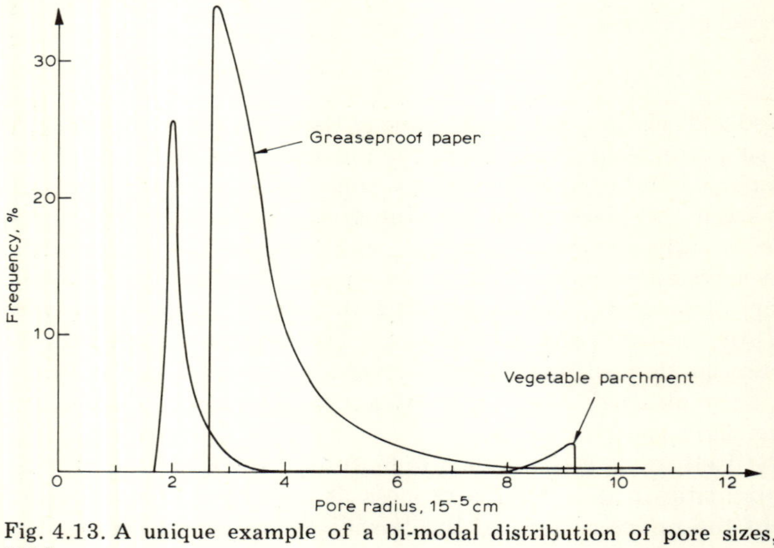


Fig. 4.13. A unique example of a bi-modal distribution of pore sizes, after H.K. Corte [90].

Pinholes present a serious problem in those areas of use of paper, where its barrier properties are essential, e.g. in packaging or in use as a coating substrate. Their incidence is measured by a variety of empirical methods, which rely on the penetration of liquids; or on measuring the minimum pressure of air which will force a bubble through a wetted sample; or merely on visual counting. Their causes are many, but in the absence of even an objective method of measurement, the knowledge of how they form belongs to the craft rather than science of papermaking — which, in this author's view, in no way detracts from the value of such knowledge.

### *4.3.3 Orientation of fibres in plane*

Apart from the lack of perfect uniformity in the distribution of mass over the plane of the paper, a sheet of paper will usually exhibit another typical structural feature: in general, more fibres will be found aligned preferentially in the "length" or machine direction, or more strictly a larger proportion of the total fibre length is so oriented. It is important to realise that this effect contributes to "grain" i.e. to the anisotropy of the mechanical properties of paper, but does not fully account for it: the effects of shrinkage in drying are probably at least as important as those of fibre orientation.

The exact measurement of the degree of fibre orientation is surprisingly difficult nor does it lend itself easily to automation. Strength anisotropy is not a good measure of fibre orientation, with the possible exception of the zero-span tensile test. It is claimed that when the jaws of the instrument are sufficiently close together, each fibre crossing the line of rupture is broken, so that the read-out is a measure of fibre orientation, but the technique is difficult and involves an uncertain extrapolation of the results from the nominal, to the effective zero span [91]. X-ray and polarised light techniques have also been used to measure the structural anisotropy of samples of paper, but their complexity has rendered them unsuitable as a means of tracing the causes of the orientation during the forming process.

The most obvious, and frequently-employed method, is to include in the furnish a small proportions of tagged fibres — for instance, through dyeing that proportion with a substantive dye, or with a fluorescent bleach — or else, to add a quantitiy of different, easily distinguishable fibres, for instance chopped synthetic staple. The orientation of these fibres in a particular direction may then be assessed in two ways:

Danielsen and Steenberg [92] described a device for viewing such fibres through a graduated protractor; each fibre is assigned to a particular direction of orientation to within ±4 degrees. The results are plotted in the form of a polar co-ordinates graph, some examples of which are shown in Fig. 4.14. The method is simple and suitable for rapid counting, but relies on an arbitrary assignment of a direction to fibres, any one of which may be flexed in several directions.

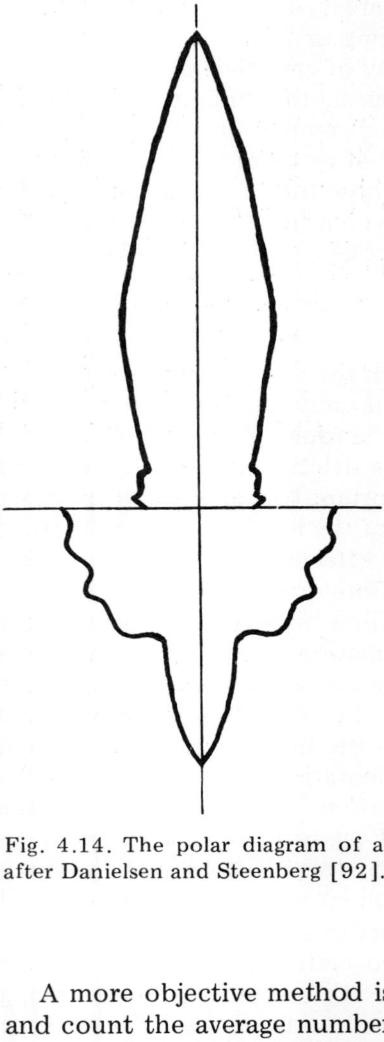

Fig. 4.14. The polar diagram of an observed distribution of fibre orientation in paper, after Danielsen and Steenberg [92].

A more objective method is to draw a scanning line in the given direction, and count the average number of fibres which appear to intersect it. Rotating the scanning line through 180° produces a polar diagram, but the mathematical interpretation of the results is far from simple. An approximate solution is given by Corte and Kallmes [93]. This assumes that the shape of the polar diagram is an ellipse which clearly is not the case, but it suffices to form an assessment of the eccentricity of the shape of the polar diagram. The complete solution is given in the appendix of ref. 93.

Both methods of course include the same assumption: (1) that the visible tagged fibres are a representative sample of all the tagged fibres and; (2) that the tagged fibres are a representative sample of the stock. The first may be assured by transparentising the web, but the second does not appear to have been considered; for instance, this author is not aware of a published study of whether long fibre fraction tend to be oriented more or less than the short ones.

At least three causes are known to result in fibres being deposited, and/or oriented preferentially in the machine direction. Firstly, the jet of stock flowing through the narrow slice channel will tend to orient the fibres in the direction of the flow. Bryant and Moss [94] obtained cine films of this phenomenon occurring at papermaking consistencies, but not through the mechanism observed by Mason [62] in the case of rotating single fibres; instead, whole flocs were seen to stretch and elongate. Figure 4.15 shows polar diagrams of the orientation of fibres in a transparent model slice, obtained by the author at speeds of some 150 m per min; observe the similarity of the diagrams to those recorded by Danielsen and Steenberg [92] and especially the development of a "horn" under conditions of high turbulence. Parker [95] observed a similar effect of a preferential alignment of some fibres across the direction of flow.

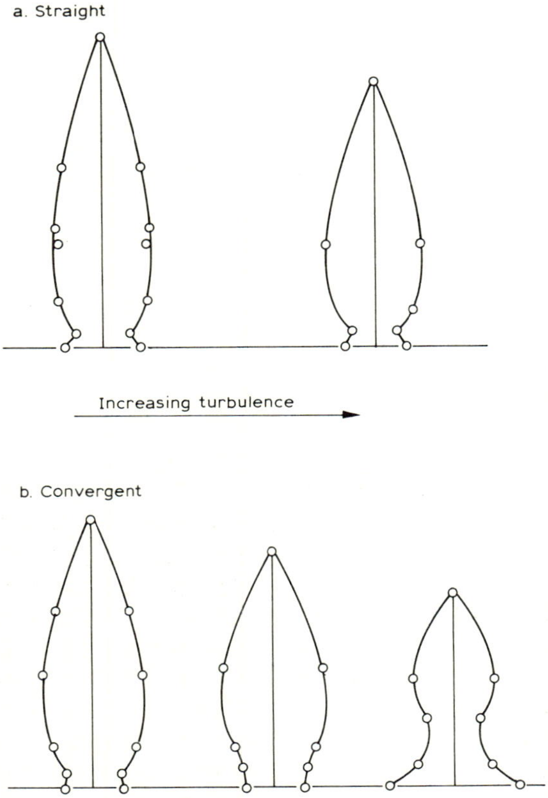


Fig. 4.15. The polar diagram of an observed distribution of fibre orientation in a slice.

It is not known to what extent the orientation of fibres in the slice jet survives the impact on to the wire. There a second effect takes place, which is probably of greater importance in controlling the orientation of fibres in the formed web: a "combing-out" action between the partly-formed web, and the slurry still flowing above it. Again, no rigorous description of this phenomenon is known, but it seems plausible, by analogy to say, the action of the textile carding machines. Where the effect is deliberately induced, extremely high degrees of orientation result, ratios of 10 : 1 and 15 : 1 having been repeatedly achieved, especially with high dilution formers, whether of the inclined wire, or of the rotary vacuum former types. On the Fourdrinier wire table, the effect is virtually impossible to avoid since (as explained earlier in the chapter) the speeds of the slice jet, and of the wire nearly always differ, however slightly. As the result, the wire side of a Fourdrinier paper is almost invariably oriented in the machine direction — which serves as a convenient means of distinguishing the two sides, but induces complex strains during drying, and hence a tendency to curl. Shaking the wire induces a degree of sideways orientation, probably by the same mechanism. Finger and Majewski [96] observed waves of diagonally oriented fibres resulting from combined effects of shake, and of stepwise drainage.

Thirdly, it has been frequently observed that a degree of fibre orientation may be induced by the topology of the upper surface of the wire fabric. If most of the depressions point in the direction of the flow, a greater degree of orientation will result, than if the top surface is ridged mainly in the cross-flow direction [28,29,97]. The extent of this effect occurring in practice is not known and is now being studied.

### *4.3.4 Orientation of fibres out-of-plane*

If the non-uniformity of paper is obvious to a visual inspection and its anisotropy in plane demonstrable by the most cursory examination, the third important feature of the structure of paper requires only the least magnification to become apparent: the constituent fibres are arranged each virtually parallel to the plane of the sheet, or only inclined to it by a small angle. This statement is often expressed by the phrase "layered" or "laminar structure" but it need not imply the existence of discrete layers — merely the observation that any given fibre may be seen not to penetrate any appreciable proportion of the thickness of the web. Some fibres lie disposed along the top surface of the web, some at an intermediate depth, and some near the wire side along the whole length of each fibre, and this in spite of the fact that each fibre length exceeds the thickness of the web by some two orders of magnitude. Figure 4.16 shows a typical cross-section of a sample of paper; contrast it with the visibly out-of-plane, three-dimensional arrangement of fibres in the subcutaneous layer of pig skin.

The layered concept is topological rather than geometrical, as best illustrated by Page [13]: an "interwoven" structure of three fibres is never seen, only the "layered" one (see Fig. 4.17). The difference between these two types of structure may be quantified by laborious microscopical observations of the numbers of apparent over and under crossings of tagged fibres in a transparentised sample: Radvan et al. report [13] a high degree of layering in all examined samples.

The causes of the layered structure are so fundamental to the papermaking process that they fall virtually beyond the control of the papermaker. Firstly, the dilution of the stock necessary to minimise flocculation is such that the depth of stock at the beginning of the formation is — unlike the thickness of the finished web — much greater than the length of individual woodpulp fibres — say 10—15 mm compared with 1—4 mm. With longer syn-

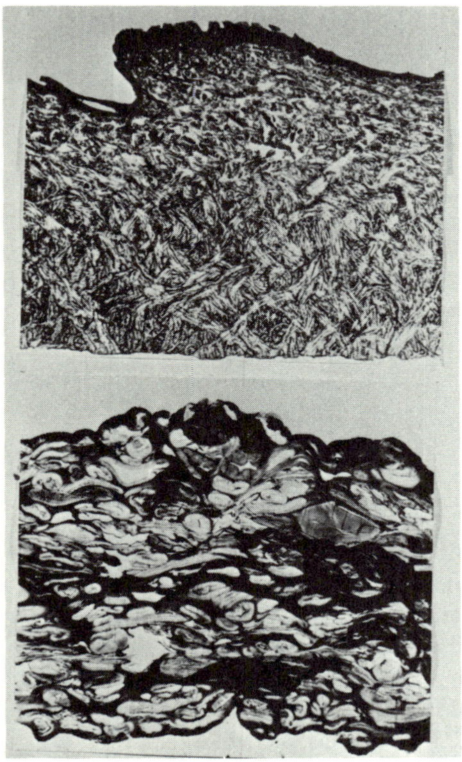

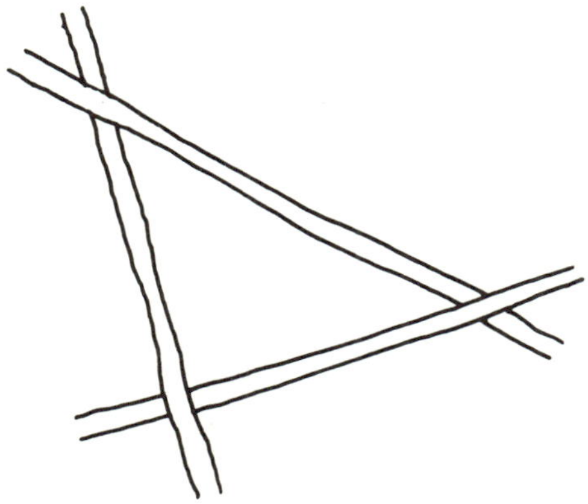

Fig. 4.16. Typical cross-section of paper (Courtesy of Wiggins Teape Research and Development Ltd.), and of pig skin, after J.W.S. Hearle [103].

Fig. 4.17. Three fibres in an "interwoven" arrangement, after D.H. Page [13].

thetic, or non-wood fibres, the dilutions have to be disproportionately greater (recall that the tendency to flocculate increases as the square of the length/diameter ratio). For instance to form a 100 g $m^{-2}$ web of say 6 mm, $1\frac{1}{2}$ denier viscose rayon fibres at an optimistic 0.5 g per litre would require a pool of stock with a depth of 200 mm — i.e. much greater again than the fibre length. The drainage now proceeds by filtration rather than by thickening, i.e. a discrete mat of fibres separates and grows by accretion, each newly added fibre being deposited on top of those already inside the mat. The layered structure is the necessary result and paradoxically, the longer the constituent fibres, the greater the degree of layering.

Long fibres will also tend to produce a more layered structure by a purely geometrical effect, and irrespective of the degree of dilution of the stock. Thus, consider two geometrically-similar arrangements of short and long fibres, all of the same diameter.

| | Shorth fibres | Long fibres |
|---|---|---|
| Fibre length | say, $L$ | $KL$ |
| Corresponding areas | say, $A$ | $K^2A$ |
| number of fibres in the corresponding areas | say, $N$ | $N$ |
| ∴ basis weight in the corresponding areas at similarity | $NL/A$ | $(NL/A) \div K$ |
| ∴ number of fibres in the corresponding areas at equal basis weight | $N$ | $N \times K$ |
| ∴ number of layers in the corresponding areas at equal basis weight | say, $M$ | $M \times K$ |

Secondly, the jet of stock issuing from the slice may be presumed to be already layered. No direct observations of this effect have been reported, but it may be inferred from the observed length-direction orientation by shear within the slice as discussed above. The same mechanism may be expected to orient both fibres and flocs, in-plane.

The two effects — dilution and orientation in the slice — are each sufficient to produce a layered structure of the formed web. Thus, a necessary condition of achieving a different 3-D arrangement is that the stock be handled at very high concentrations right up to the moment of formation and secondly, that it be maintained at a high intensity of small-scale turbulence. Just such a set of conditions are obtained within the slice of a High

Consistency Former [75] where a jet of high consistency stock (say 30 g per litre) is forced through a very shallow (say 3 mm deep) long channel. The small dimensions ensure small eddies, so that the network is broken into small flocs of 1—2 mm size; and the high consistency prevents larger flocs re-forming before the jet is delivered to and drained on, a Fourdrinier wire table. Cross-sections of the resulting web suggest a degree of out-of-plane orientation and the physical properties of the paper exhibit the expected increase in Z-direction strength, and in the ease of water removal in the presses.

No other example of a paper web has yet been known to exhibit a substantial degree of out-of-plane orientation, but scattered reports of slight departures from a layered arrangement of fibres are reviewed in ref. 98.

### *4.3.5 Distribution of loadings and fines in depth*

The operation of drainage of the stock, and formation of the web, has also a classifying effect, as the large fibres, and the fine fragments and mineral loadings are not distributed uniformly in depth. Majewski [97] described how various methods of formation result in different distributions: slow continuous drainage, as in a laboratory sheet machine, will tend to concentrate the fines in the lower layers, merely through the filtering action of the forming web. In contrast, the fast and turbulent discontinuous drainage on a Fourdrinier will wash out the fines from the wire side, leaving the top side relatively rich in fines content. Groen [99] measured a number of such distributions in greater detail, and observed several finer effects: for instance, the plot of filler content vs. the depth is not a straight line, but shows a plateau extending throughout much of the depth, with a sharp drop near the wire side. He attributed the latter to the washing action of the drainage elements, especially the table rolls. The presence of a dandy roll may cause a steep increase in concentration of the filler near the top surface (Fig. 4.18).

Both these investigations used the method of carefully splitting off successive layers of the web with an adhesive tape, a method both laborious, and causing a disruption of the structure of the web. More delicate methods have been developed since, but the results obtained largely confirm Groen's findings. Lehtinen [87] sectioned paper in very thin slanting (some $7\ g\ m^{-2}$) layers by means of a microtome. Parker [88] developed a method for uniformly freezing thin layers of paper before splitting them off. The latter method, in particular has been much used in illustrating the structural differences between webs formed on the Fourdrinier, and on the newer twin-wire machines; and also, the more uniform distribution of fines which may be achieved through the use of the gentler-acting drainage elements, such as hydrofoils.

Mays et al. [100] used the technique to investigate the effects of filler properties on their distribution within the web. Light-weight fillers of

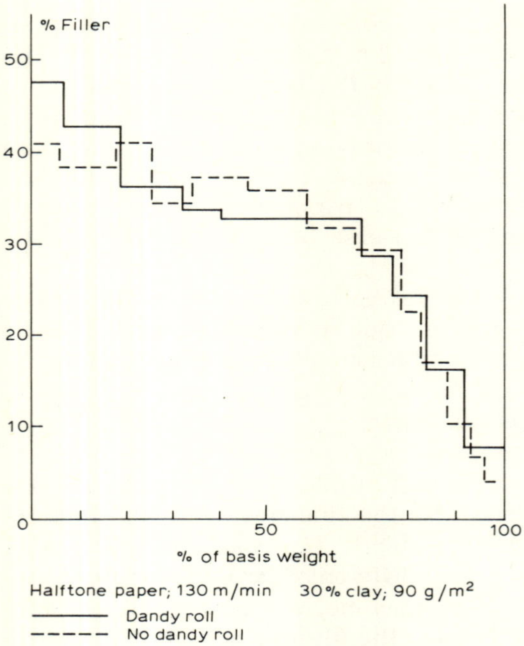


Fig. 4.18. A typical distribution of filler in the thickness of paper, after L.J. Groen [99].

extended shapes, and those which may be expected to form weak bonds with cellulose, such as sodium silicoaluminate will be more evenly distributed throughout the depth, than will heavy, inert powders such as filler clay, and especially barium sulphate.

Han and Chang [101] studied the distribution of radioactive-tagged fines within the paper web. Maximum concentration is usually observed near the top layer, but may be displaced downwards if the latter becomes "saturated". They concluded that structural theories are not suitable for describing the retention of fine particles.

## 4.4 REFERENCES

1 H.F. Rance, in F. Bolam (Ed.), Formation and Structure of Paper, Transactions of the Fundamental Research Symposium held at Oxford, 1961, Tech. Sect. Brit. Paper Board Makers Assoc., London, 1962, p. 1.
2 A. Lehtinen, N. Ryti, P. Mauranen and P. Aaltonen, Pap. Puu, 47 (1965) 673—675.
3 H. Meyer, Tappi, 54 (1971) 1426—1450.
4 S.T. Han, Tappi, 45 (1962) 292—295.
5 E. Cowan, Pulp Pap. Mag. Can., 38 (1937) 85—91.

6 W.E. Bennett, Proc. Tech. Sect. Brit Paper Board Makers Assoc., 36 (1955) 359—374.
7 P.E. Wrist, Pulp Pap. Mag. Can., 55 (1954) 115—119.
8 G.I. Taylor, Pulp Pap. Mag. Can., 57 (1956) 115—119.
9 I.F. Hendry, J. Mardon, H. Scruton and D.M. Wilkinson, Proc. Tech. Sect. Brit. Paper Board Makers Assoc., 33(3) (1952) 603—653.
10 G. Burkhard and P.E. Wrist., Pulp Pap. Mag. Can., 57 (1956) 100—118.
11 J.V.M. Calver, P.I.R.A. Bibliography No. 665.
12 N.J. Lindberg, Paper and Timber (Finland), 34 (6—8) (1952).
13 B. Radvan, C.T.J. Dodson and O.G. Skold, in F. Bolam (Ed.), Consolidation of the Paper Web, Transactions of the Fundamental Research Symposium held at Cambridge, 1965, Tech. Sect. Brit. Paper Board Makers Ass., London, 1966, p. 189.
14 J. Bergstrom, Norg. Skogind., 31(5) (1977) 133—138.
15 W.L. Ingmanson and R.P. Whitney, Tappi, 37(11) (1954) 523—534.
16 W.L. Ingmanson, Tappi, 42(6) (1959) 449—454.
17 W.L. Ingmanson and B.P. Andrews, Tappi, 46(3) (1963) 150—155.
18 C.K. Meadley, l. c. [13], p. 216.
19 S.T. Han, Pulp Pap. Mag. Can., 70(9) T134—46 (1969) T134—146.
20 W.L. Ingmanson, B. Andrews and R.C. Johnson, Tappi, 42 (10) (1959) 840—849.
21 H.G. Higgins and J. De Yong, in loc. cit. Ref. (13), p. 242.
22 R.L. Jones, Tappi, 46(1) (1963) 20—28.
23 T.A. Arnevik and T. Helle, in Fibre-Water Interactions in Paper Making, Transactions of the 6th Fundamental Research Symposium held at Oxford, 1977, Fund. Res. Comm. Tech. Div. Brit. Paper Board Ind. Fed., 1978, p. 315.
24 B. Wahlstrom and G. O'Blenes, Pulp Pap. Mag. Can., 63(8) (1962) T405.
25 I.T. Pye and E.H. Sinder, 62nd Annual Meeting, Tech. Sect. CPPA — Preprints Book B.
26 W. Brecht, H. Erfurt and W. Volk, Das Papier, 10 (1965) 395—402.
27 J.D. Boadway and T.J. Gray, Pulp Pap. Mag. Can., 64 (1963) T14—18.
28 T. Helle, Seminar Papers, Tech. Div. Brit. Paper Board Ind. Fed., (Dec. 1977).
29 W. Kufferath, Seminar Papers, Tech. Div. Brit. Paper Board Ind. Fed., (Dec. 1977).
30 R. Estridge, Ph.D Thesis, Inst. of Paper Chemistry, Appleton, WI, U.S.A., 1961.
31 R.W. Abrams, Ph.D. Thesis, Inst. of Paper Chemistry, Appleton, WI, U.S.A., 1964.
32 W.H. Cuffey, Tappi, 40(6) (1957) 190A—196A.
33 G. Burkhard and P.E. Wrist, Pulp Pap. Mag. Can., 55 (1954) 188—200.
34 H. Weiss and H. Dahl, Das Papier, 21 (1967) 766—768.
35 W. Muller-Rid and G. Pausch, Zellst. Pap., (10) (1957).
36 A.E. Beechers and R.A. Bareiss, Tappi, 53(5) (1970) 847—852.
37 K.A. Cutshall, Trans. Tech. Sect. CPPA, 1 (3) (1975) 77—81.
38 R.M. Tong, Ph.D. Thesis, Cambridge Univ., 1975.
39 J.R. Parker and J.B.A. Epton, Pulp Pap. Mag. Can., 78 (1977) 53/56.
40 S.G. Mason, A.A. Robertson, G.A. Allen and C.W. Walker, Pulp Pag. Mag. Can., 55 (1954) 97—108.
41 Van der Meer, Tappi, 37 (1954) 502—511.
42 L. Reiner and D. Wahren, Sven. Papperstidn., 74(9) (1971).
43 K. Wolf and D. Egelhog, Pap. Technol., 15(2) (1974) 77—82.
44 J.R. Parker and B.A. Epton, Tappi, 59 (1976) 114; Ibid., 60 (1977) 90—93.
45 C.J. Moen, Tappi, 60(10) (1977) 116—119.
46 K.A. Cutshall and J. Mardon, Appita, 28 (1975) 252.
47 H.K. Corte, Das Papier, 24 (1970) 261—274.
48 B. Norman and D. Wahren, in F. Bolam (Ed.), The Fundamental Properties of Paper Related to its Use, Transactions of the Fundamental Research Symposium held in Cambridge, 1973, Tech. Div. Brit. Paper Board Ind. Fed., publ. Benn, London, 1976, p. 7.

49 S. Schroder and O. Svenson, Sven. Papperstidn., 68(2) (1965) 25—33.
50 D. Attwood, Pulp Pap. Mag. Can., 68 (1967) T517—525.
51 M.S.Yalin, Proc. ASCE J. (Hydraulics Divn.), (5) (1964) 105—119.
52 R. Culver and J. Mardon, Appita, 13 (1959).
53 W.R. Debler, Tappi, 44 (1961) 589—591.
54 C.S. Yih and A.C. Spengos, Tappi, 42 (1959) 398—403.
55 D. Gustafson and J.D. Parker, Sven. Papperstidn., 67 (1964) 545—554.
56 C.S. Yih and S.J. Macnamara, Tappi, 46(3) (1963) 204—208.
57 M.P. Alston, I.E. Goodhew and J. Chapman, Pulp Pap. Mag. Can., 57(6) (1956) 125—128.
58 O. Andersen and L. Steen, (discussion) l. c. [1], p. 771.
59 A. Erspamer, Pap. Trade J., 110 (1940).
60 F.H. Gundersen, Nor. Skogind., 9 (1947).
61 F.D. Broadbent and H.A. Harrison, Proc. Tech. Sect. Brit. Paper Board Makers Assoc., 21 (1940).
62 S.G. Mason, Tappi, 37 (1954) 494—501.
63 O.L. Forgacs, A.A. Robertson and S.G. Mason, in F. Bolam (Ed.), Fundamental of Papermaking Fibres, Transactions of the Fundamental Research Symposium held at Cambridge, 1957, Tech. Sect. Brit. Paper Board Makers Ass., Kenley, 1958, p. 447.
64 D. Wahren and R. Meyer, Sven. Paperstidn., 67 (1964).
65 B. Steenberg, N. Thalen and D. Wahren, l.c. [13], p. 177.
66 H. Westra, Papierwereld, 17 (10) (1963).
67 M.Y. Chang and A.A. Robertson, Pulp Pap. Mag. Can., 68 (1967) T438—444.
68 J.D. Boadway, Pulp Paper Mag. Can., 57 (1956) 185—189.
69 G. Gavelin, Pulp Paper Mag. Can., 55 (1954) 191—200.
70 J.J. Bikerman, Foams, Springer Verlag, New York, 1973, p. 151.
71 J.R. Barnard et al., Can. J. Chem. Eng., 39(2) (1961) 94—97.
72 H. Seibert and K. Bahr, Das Papier, 24 (1970) 331—335.
73 O. Huber and J. Weigl, Das Papier, 24 (1970) 823—833.
74 B.W. Attwood, Pap. Technol., 7 (1966) 367—370.
75 K.J. Grundstrom et al., Tappi, 59 (1976) 58—62.
76 M. Judt, Das Papier, 13(3/4) (1959) 46—54; (19/20) (1959) 483—496.
77 D.W. Manson, Tappi, 49 (1966) 425—429.
78 Crown-Zellerbach Corpn. Br.P. 1 049 675, 1966; Compagnie Industrielle de Textiles Artificiels, etc., Fr. Pat. 1 327 082, 1963.
79 N.J. Parrat, Composites, 1 (1969) 25—27.
80 A.J. Hentschel, in O.A. Battista (Ed.), Synthetic Fibres in Papermaking, John Wiley and Sons, London, 1964.
81 T.W. Kidner, Pap. Technol., 15(6) (1974) 346; B. Radvan and A.P.J. Gatward, Tappi, 55 (1972) 748—751.
82 W. Muller-Rid and H. Nisser, Das Papier, 20 (1966) 791—798.
83 G. Duffy et al., l. c. [23], p. 195.
84 P.E. Wrist, l. c. [1], p. 839.
85 F. Luhde, Pulp Pap. Mag. Can., 67 (1966) T381—392.
86 D. Attwood and J.R. Parker, Pap. Technol., 3 (1962) 435—443.
87 A. Lehtinen, Pulp Pap. Mag. Can., 65 (1964) T61—64.
88 J.D. Parker and W.C. Mih, Tappi, 47 (1964) 254—265.
89 Kimberly Clark Corpn., Br. Pat., 994 661, 1961.
90 H.K. Corte, l. c. [63], p. 301.
91 W.F. Cowan and E.J.K. Cowdrey, Tappi, 57 (1974) 90—93.
92 R. Danielsen and B. Steenberg, Sven. Papperstidn., 50 (1947) 301—304.
93 H.K. Corte and O. Kallmes, l. c. [1], p. 13.
94 L.A. Moss and E.O Bryant, Pap. Trade J., 106 (1938) 46—57.

95 J.D. Parker, Private communication.
96 E.J. Finger and Z.J. Majewski, Tappi, 37 (1954) 216—224.
97 Z.J. Majewski, l. c. [1], p. 749.
98 B. Radvan, l. c. [48], p. 137.
99 L.J. Groen, l. c. [1], p. 697.
100 R.K. Mays et al. Tappi, 49 (1966) 99A—102A.
101 S.T. Han and N.L. Chang, Tappi, 52 (1969) 688—693.
102 J.D. Parker, The Sheet-Forming Process STAP No. 9 Tappi, 1972.
103 J.W.S. Hearle, CIBA Rev., 65 (1) (1965) 9.
104 S.T. Han, Pulp Pap. Mag. Can., 70 (1969) T134—T146.
105 A.A. Robertson and S.G. Mason, Pulp Pap. Mag. Can., 50 (1949) 103—110.

*Chapter 5*

# CONSOLIDATION OF THE PAPER WEB

H.F. Rance

## 5.1 INTRODUCTION: THE MEANING, NATURE AND PURPOSE OF CONSOLIDATION

### *5.1.1. Meaning of consolidation*

The word "consolidation" as such is seldom used in the general technology of papermaking, and apart from its appearance in the title and contents of the 1965 Fundamental Research Symposium on "Consolidation of the Paper Web" [1] it is seldom found in the relevant literature. Yet, it comprises logically the central processes of the papermaking sequence and it is the general concept underlying the practical operation and effect of the processes of drainage, pressing, drying and calendering which follow wet end deposition of the pulp stock in the form of a continuous web.

The relative absence of literature references to consolidation as such is however more than counterbalanced by the flood of references — especially over the last twenty years or so — to the essential cause and accompaniment of most of the consolidation of the paper web, viz. the removal on the paper machine of water from that web. There have been International Symposia on Water Removal at the Presses and Dryers; International Water Removal Symposia; and a host of allied papers in the leading technological journals of the papermaking industry.

Equally, there has been a flood of literature references to the process (calendering) which causes consolidation of the dry sheet of paper, rather than references to dry consolidation or compaction as such.

The reasons for this emphasis upon the processes which bring about consolidation, rather than consolidation itself, are not hard to seek. The paper industry is highly capital-intensive, and its main economic driving force is increase in process efficiency aimed towards increased output and lower cost. Although consolidation of the web is (after beating) the main determinant of paper quality, the main constraints upon achievement and variation of consolidation lie within the massive pieces of equipment that bring about the desired consolidation. As a consequence, the main technological focus of attention is the equipment itself, and the mode of operating that equipment for maximum efficiency and optimum effect.

This bias within the technology of the industry is good reason for making consolidation as such as far is possible a focus of attention in this chapter, thus seeking the underlying scientific factors which determine equipment design and operation, and the quality of paper made by that equipment.

The Oxford Dictionary records the following definition for "consolidation":

"Combination into a compact mass, single body or coherent whole". It also records the following definitions of the verb "to consolidate":

(1) "To make solid: to form into a compact mass."

(2) "To make firm and strong: to strengthen."

(3) "To combine compactly into one mass, body or connected whole."

(4) "To become solid or firm: to combine or unite solidly or compactly."

Here, explicitly or implicitly, are listed all the practical and theoretical processes involved in the consolidation of the paper web. There are the processes of *making* and *becoming solid* or *firm:* there is the process of *forming:* and there is the process of *combination* (implicitly of component elements) into a *connected* or *coherent* whole. These processes clearly encompass the several types of structural change that transform the weak, soft, pseudo plastic wet web, containing some 80 per cent of water, into the hard, solid dry sheet that we call paper, containing some 95 per cent of cellulose fibre (leaving aside the non-cellulosic constituents).

### *5.1.2 Nature of consolidation*

One can distinguish between different phases and modes of consolidation within this sequence. On the drainage wire, at the couch and at the presses there is *forming* and compaction of a pseudo-plastic mass. At the presses and on the driers, there is *compaction* of component elements, in one or more dimensional planes. On the driers and in the draws, there is *combination* together of component elements to form a connected whole. At the calenders, there is a further phase of compaction, by actual compression of the web which is already a coherent solid body. Throughout the sequence, as a consequence of these effects, there is the *making*, *becoming solid*, *becoming firm* of the material that eventually emerges as a dry, solid sheet of paper.

There are two main categories of modes of consolidation: those caused primarily by the removal of water, whether by gravity drainage, suction, or evaporation: and those caused primarily by the application of external forces. Examples of the two categories are frequently superposed one on the other, and are sometimes inevitably coexistent, but they are logically separable.

Water removal results in a succession of phases and modes of consolidation from the drainage wire at the wet end to the reel up at the dry end. The stock issuing from the slice of the paper machine generally contains about 100 parts of water to one part of fibre, although it already has at this stage a genuine network structure. When it leaves the drainage wire, it contains generally about 4 parts of water to one part of fibre, and has by this time been formed into a coherent body with a well-defined structure that clearly foreshadows the structure of the final sheet. This phase of consolidation has been brought about by compaction of the constituent elements through removal of free water, brought about partly by gravity drainage, partly by suction, partly (in some cases) by applied positive external force; aided throughout by surface tension.

The mode of consolidation through water removal induced by both pressure and suction continues in the wet presses, also aided by the compacting force of surface tension. Next, in the early stages of the drying section, surface tension compaction initiates consolidation during evaporation of the free water. Then, after the free water has been removed, the water in the cellulose fibres begins to evaporate, and a new mode of consolidation enters: a fibre-to-fibre combination through hydrogen bonding, which is the key to making *firm and strong the interconnected whole.* At the same time, a new mode of web contraction enters, based upon contraction of the component fibres.

Externally applied forces are secondarily involved from the start of the sequence of water removal. First, on the drainage wire, gravity and suction applied by table rolls, scraper foils and suction boxes compel the removal of free water: and on some modern machines involving two drainage wires (one each side of the wet web), a squeezing pressure is applied to expel the water. Later, at the wet presses, force applied by the press nip on the "sandwich" of the felt plus wet web squeezes water from both web and felt; suction applied through holes in the press rolls can assist removal of this water; the water remaining in the felt/web sandwich is then to a degree redistributed between felt and web by competing recovery forces; and after separation of the felt and the outgoing web, water is expelled from the felt by further nip pressure or suction before it returns to rejoin the incoming wet web. Still later, on the dryers, pressure of the dry felts assists heat transfer from the hot cylinders to the wet web, and also promotes transfer of residual free water from web to felts.

Over and above these secondary involvements of externally applied pressure in the water removal sequence, such applied pressure also assists consolidation directly. At the wet end, there is a degree of moulding and compacting of the wet web itself, by the use of dandy rolls on the drainage wire, within the wet presses, and by the use of so-called smoothing presses which mould the surface of the wet web. On M.G. drying cylinders, pressure rolls force the malleable wet web surface on to a smoothly polished surface, where the forces of adhesion to the surface, coupled with hydrogen bonding of the cellulose fibres, produce the typical glazed surface of M.G. paper. On Fourdrinier machines, smoothing rolls placed at successive intervals along the machine repeat the surface compaction. Finally, dry calendering closes the sequence by applying high nip pressure which compacts the dry web and to a degree "polishes" the surface.

The successive interactions of these several phases and modes of consolidation result in a complex total situation. In practice, there is much empiricism involved in the design and operation of the machine and its component parts, but both design and operation have benefited considerably from some scientific understanding of each factor in this complex sequence of events. Such scientific understanding can, in the first instance, be divorced from the design and operation of particular papermaking machines. One can consider, in isolation, each factor involved in the consolidation process: and having reached some understanding of all the factors, one can aim for at least a qualitative synthesis of them, which can be applied to papermaking *in toto*.

### *5.1.3 Purpose and desirability of consolidation*

There are two quite distinct purposes of consolidation, both important economically, but sometimes opposed to one another in terms of feasibility and desirability.

The first, central purpose of consolidation is the production of a paper sheet designed to fulfil the needs of its end use. In the case of such paper as toilet tissue, kitchen wipe or blotting paper, a low degree of consolidation is required, to ensure absorbency. In the case of a writing or printing paper, a fairly high degree of consolidation is required to ensure controlled acceptance of ink, with strength and durability in use. In the case of greaseproof or tracing paper, a still higher degree of consolidation is required to ensure oil resistance in the one case, transparency in the other. These are but selected examples from the wide range of degrees of consolidation which ensure that the character of the end product matches the needs of its use.

The second purpose of consolidation is central to the needs of the papermaking process, and can on occasion be opposed to the needs of the product. As such it is a purpose of expediency, rather than a purpose of ultimate intent. It relates to the need to be able to transfer the web, stage by stage from slice to couch, from couch to press, and press to driers, and from calender to reel up, and to be able to remove water at each appropriate stage so as to achieve the degree of consolidation required by the needs of end use.

Web transfer as such, especially in the early stages, cannot be carried out without a substantial degree of consolidation, even though that consolidation may to a degree sometimes conflict with the needs of product end use. In the case of paper products of a very soft, absorbent nature, this results in an ever present conflict between consolidation for the purpose of papermaking, and consolidation in relation to end use. A similar example can be found in the case of printing and writing papers at the dry end of the machine, where excessive calendering carried out as a process of local corrective consolidation may be required to ensure production of acceptable reels of paper, despite the fact that it may cause a degree of bulk consolidation undesirable for the needs of end use, by reducing the rigidity of the sheet. Sometimes different degrees of consolidation within the thickness of the sheet are desirable. For example, the ideal paper for certain printing applications or office copying machines may be one with a high degree of surface consolidation and a low degree of internal consolidation.

The need to remove water to achieve desired or necessitated consolidation itself imposes local variations and defects in consolidation which are undesirable for product end use, but often necessitated by production needs or deficiencies. Such local defects include wire mark from the wire drainage, shadow-marking and felt mark from the wet presses, calendered high spots and other end results of wet-end flocculation. These are all due to consolidation variations derived from the water removal processes.

It is clear from this consideration that consolidation as such is neither desirable nor undesirable. The *required degree* of consolidation is *desirable*, and any deviation either of overall degree or of local variation is undesirable. The reader must keep this well in mind when reading the rest of this chapter. Some cited authors, and the writer himself, may sometimes seem to imply that consolidation as such is desirable. This is not so. Consolidation is the *modus operandi* of all papermaking, but the objective must always be consolidation controlled to a pre-designed pattern and degree, without unwanted deviation or variation.

### *5.1.4 General comments on presentation of Chapter 5*

Inevitably, the contents of this chapter must to a degree overlap the contents of the preceding chapters in this volume, and the contents of those to appear to Volume 2. Also inevitably, understanding of the contents of this chapter depends to a considerable degree upon an understanding of the contents of the preceding chapters, which consider in detail the underlying determinants of factors and forces involved in the several phases and modes of consolidation. This applies especially to Chapter 3, and for this reason the next section of this chapter has been deliberately designed to bring out the general relationships between fibre treatment and consolidation, essentially encompassing the interactions between this chapter and Chapter 3.

For general reading, as a background to this chapter, references [1] to [7] are recommended. These will also later be the subject of more specific reference on specific elements of the subject of this chapter.

The amount of published literature relating to the successive phases and modes of consolidation of the paper web is far too great to permit comprehensive reference. Furthermore, much of it concerns empirical aspects of design and operation of machine components, coupled with factors of economic constraints. It is often difficult to identify considerations of fundamental significance in some of these publications, yet their contents in some cases provide an important source of information from which the underlying scientific principles of consolidation can be deduced.

In this situation, rather than overwhelm the reader with an indigestible mass of references, the writer has aimed to confine references (a) to review articles and conference proceedings which are themselves sources of more extensive lists of references than are practicable in this volume; (b) to outstanding contributions of major significance to the theme, including some of historic interest, which have laid the foundations of present-day understanding; and (c) to papers which stimulate new thought and ides and which may be expected to lead to further advances in understanding.

Over and above literature references, the writer aims to present the theme in discursive fashion, in an attempt to relate the several phases and modes of consolidation organically to the ultimate purpose of the papermaking sequence, viz. the production of an artifact designed for practical use in a variety of applications. It must never be forgotten that sheets of paper, in their wide variety of type, shape, geometry and physical structure, are deliberately designed and brought into being to perform a variety of practical functions. The science of papermaking, however fundamental it may be, ultimately serves to facilitate the control of machine design and operation, and hence to facilitate deliberate control of the many variables, so that the designed type of paper is produced to match the specified demands of usage. This consideration must be dominant, when dealing with consolidation of the paper web, since the modes and degree of consolidation determine more than anything else, the type and properties of the ultimate sheet.

## 5.2 GENERAL RELATIONS BETWEEN FIBRE TREATMENT AND CONSOLIDATION

Having stated that consolidation is the main determinant of the type and properties of the paper sheet, it must now be further stated that mechanical treatment of the fibre by so-called beating or refining is the main determinant of consolidation. Observably, it far outweighs paper-machine design and operation in its effect upon consolidation, and indeed has a considerable influence upon machine design and operation also, since these have to conform in large degree to the type of stock from which the paper is made. Well before the fibre dispersion hits the drainage wire, the degree of consolidation of the final product has already been largely pre-determined by beating treatment whose effects can only marginally be changed or reversed by anything that can be done on the paper machine.

Chapter 3 has already discussed at length in this volume the subject of fibre treatment by some form of beating or refining, which results in a greater or lesser degree of unravelling of the internal structure of the cellulose fibre, together with some fragmentation and fibre shortening. Fibre selection, dealt with in Chapter 2, does also have an influence upon web consolidation, and upon the consequent character of the paper sheet, both directly and through its interaction with the beating or refining process, but it is far outweighed in importance by mechanical treatment as such.

Both within and outside the paper industry, the word paper usually brings to mind a sheet made from moderately beaten stock, as encountered in printing and writing papers. Yet even in ordinary household and office use, a much wider variety of papers is commonly encountered, ranging from such extremes as toilet and handkerchief tissue, on the one hand, to dense, semi-transparent types such as greaseproof paper, glassine and tracing paper, on the other hand.

By volume of manufacture and usage, printings and writings, together with paper and board for packaging, certainly dominate the industry. Nevertheless the value, and the economic and social significance of the other types of paper are substantial. Even more important, in the context of this chapter, is the significance of the more specialised types of paper in broadening the spectrum of paper types in terms of both structure and physical properties; and it is only by considering the full spectrum of types, from toilet tissue to tracing paper, that the significance of consolidation and of the fibre treatment by which it is largely determined, can be fully understood.

Soft toilet tissue, like the once familiar grade of blotting paper, is made from fibres which have been subjected to a minimum of beating: it is visibly a loosely felted aggregate of almost intact cellulose fibres. It is highly porous and absorbent, with an internal air space substantially greater than the volume of the cellulose component. It is soft and weak, as a consequence of the looseness of the structure and the relatively low area of inter-fibre bonding. At the other extreme, tracing paper, greaseproof paper or glassine is a highly consolidated, dense aggregate of unravelled fibres, fibrils, and fibre fragements made from highly beaten stock: it is largely impermeable to air and to non-aqueous liquids, because of the very small air space, which is virtually nil in the case of highly calendered specimens (whose apparent specific gravity can approach that of cellulose itself (ca. 1.5)). These papers are also mechanically hard and relatively strong in terms of tenacity.

The substantial difference in character between a paper made from lightly beaten stock (such as soft toilet tissue, or blotting paper) and a paper made from highly beaten stock (such as natural tracing paper) is paralleled by a corresponding difference in the behaviour on the paper machine of the wet webs from which these papers are made.

The web from lightly beaten stock exhibits during drying a cross-direction linear contraction of only 1 to 2%. The web made from highly beaten stock can exhibit a cross-direction linear contraction exceeding 10%, most of which takes place in the dryer section of the machine. This phenomenon as well as the differences in character between the dried sheets, must be fully explained in any comprehensive understanding of consolidation.

The structural differences in the finished paper derived from differences in degree of beating of the fibre stock can be characterised in terms of various secondary single value measurements such as air permeability, or rate of oil penetration, or alternatively in terms of apparent specific gravity (pseudo density) of the sheet, derived from thickness and weight measurements. Measurement of the density of the sheet is rather imprecise, because of surface roughness and compressibility, and at high density levels it is relatively intensitive to structural change, as compared with measurement of such secondary properties as air permeability or rate of oil penetration. Nevertheless density, or apparent specific gravity, is a real structural property of the sheet as such, not a second order characteristic involving interaction with another material such as air or oil. It is the most direct measurement of consolidation as such.

If one compares toilet tissue with tracing paper, in terms of this measurement, the spectrum of variability of consolidation of the paper sheet is made evident. The one has a density less than 0.5: the other has a density approaching 1.5. The one is soft, absorbent and opaque: the other is hard, impermeable and transparent.

How does beating bring about this dramatic contrast in character and properties? It does so through a number of superimposed effects which influence in succession all of the main modes of consolidation down the paper machine. First, beating both shortens the fibres and renders them flexible and malleable by breaking internal bonds between their fibril components (so-called "bruising" or internal fibrillation of the fibres). As a consequence, they are more readily able to conform to forces (whether externally applied, or derived from surface tension within the web-system itself), which tend to pack the fibres more closely together. Unbeaten fibres are relatively long and rigid, and their strength provides recovery forces which tend to reverse any packing together that involves distortion of individual fibres. Beaten fibres are generally shorter and more malleable, and tend to stay in the positions imposed by applied forces.

Second, because of the malleability of the fibres both singly and *en masse*, the surface tension envelope which helps to compact the sheet during removal of free water can remain more or less intact, depending on the degree of beating and intrusion of air is inhibited or delayed.

Third, as beating continues to the point where the individual fibres are progressively unravelled into their fibrillous components, the structural restraint to bringing the components together is reduced, while the effect of surface tension acting on the individual component increases. The practical result of these effects is that webs of highly beaten stock shrink both in thickness and to a lesser degree transversely during removal of free water, well before fibre-to-fibre bonding enters the scene.

Fourth, by removal of inhibiting layers in the fibre structure, and by progressive unravelling and fibrillation of that structure, beating makes more and more surface area available for hydrogen bonding after the surfaces have been pushed and pulled close enough together by external forces and by surface tension. At this point, after free water has been removed, evaporation of bound water (dissolved or otherwise intimately associated within the cellulose fibres) causes the fibres to shrink, and that shrinkage is transferred to the web through progressive fibre-to-fibre bonding.

This general account of the causal relationship between beating and consolidation during water removal is today generally accepted, but the historic origins and development of the thesis tend to be forgotten. In fact, the significant role of surface tension was not explicitly formulated in the literature until the publication of Campbell's work [8] in 1933, which showed how progressive beating leads to improved formation and densification of the sheet. That pioneer work was largely obscured for many years by much confused and rambling speculation about so-called "hydration" of cellulose during beating, which failed to distinguish between the logically separate events of physical fibrillation, surface tension compaction, and bonding of cellulose surfaces through molecular bonds. The work and writings of A.J. Stamm exemplified by [9], clarified the distinctive role of bound water contained within water-swollen cellulose fibres, and facilitated the quantification of this phenomenon. The Fundamental Research Conference held in 1950 under the auspices of TAPPI and the Technical Section of CPPA, on the "Effects of Fiber Characteristics on Paper and Papermaking Properties" included relevant papers by Lewis [10], Gallay [11], Stamm [12] and Forman [13] which further developed the theme. A similar conference held in 1954 on "Fundamentals of the Paper Machine" included a paper by Rance [14] which developed a concept of wet web consolidation and shrinkage in relation to beating, which distinguished between inter-fibre web shrinkage produced by surface tension, and intra-fibre web shrinkage produced by fibre shrinkage communicated to the web by fibre-to-fibre bonding.

During the last twenty to twenty-five years, the general thesis relating web consolidation to beating has been further discussed, developed and refined by many workers, using the more refined techniques of examination and experimentation that have become available. These developments, and also some details of the above cited references are discussed in later sections of this chapter.

For some valuable general surveys of the subject in the present context, the reader is recommended to study the paper by Higgins and de Yong [15], presented at the 1961 Fundamental Research Symposium, and the more recent paper presented by Atack [16] at the 1977 Fundamental Research Symposium, in addition to the detailed consideration of the subject by Emerton in Chapter 3 of this work.

## 5.3. THE INFLUENCE OF WET-END FORMING ON CONSOLIDATION

Chapter 4 contributed by Radvan on Forming the Paper Web has inevitably already dealt to a considerable extent with the consolidating aspect of forming: the two matters are inseparably entwined.

Here attention will merely be drawn briefly to the key factors determining consolidation in the forming zone, and the main modes of consolidation brought about by those factors. The key factors are clearly the distribution of fibres within the stock approaching the wire, the flow of stock onto the wire, and the flows of water from the forming wet web due to gravity, to suction induced by table rolls or scrapers, and to pressure applied by converging drainage wires in the case of two-wire machines. The dominant mode of packing in the forming zone is determined by the fibrous character of the individual component of the web. It becomes very obvious, when one examines either a wet web or a dry sheet of paper, that the vast majority of the fibres lie essentially parallel to the plane of the sheet. This is an inevitable consequence of their shape, combined with the mode of deposition and drainage from a water suspension on the wire.

The degree to which these fibres (which may be some millimetres long) penetrate (albeit at a shallow angle) through the thickness of the sheet (which is only some hundredths of a millimetre), was investigated by Radvan et al. [17], as presented at the 1965 Fundamental Research Symposium on Consolidation of the Paper Web. All of the examples they examined showed a highly layered structure in the thickness of the sheet, which may be attributed to the mechanism of formation by filtration of successive layers on the Fourdrinier machine.

In addition to this dominant pattern by which the fibres lie almost entirely parallel to the plane of the sheet, and generally in superimposed layers within the plane of the sheet, there is the subordinate mode of packing caused by forces which produce a greater or lesser degree of fibre orientation, generally preferentially in the machine direction. A high degree of such orientation would clearly produce a dense packing, and would have considerable effect upon modes of consolidation at subsequent stages down the papermachine. Some papers, such as those intended for manufacture of paper twine or of tissue for cable manufacture take advantage of a high degree of fibre orientation to produce a very high machine direction strength. Generally, however, machine direction orientation of fibres in undesirable, expecially since it compounds with other subsequent sources of planar anisotropy to produce a degree of such anisotropy in the finished paper which in most cases is unwanted.

The other subordinate factors (i.e., subordinate to the planar disposition of the fibres within the sheet) which influence consolidation of the paper web during forming, are the degree of dispersion of fine material, and the degree of fibre dispersion (inversely, the degree of flocculation).

It is clear that a graded aggregate of coarse, fine and finer components can lead to a higher degree of packing during forming, which will also aid subsequent modes of consolidation. The use of mixed furnishes of long and short fibres to promote an evenly packed structure in papermaking is a practical example of this. Also beating, by providing a spectrum of degrees of fragmentation, can progressively promote closer packing which affects all modes of consolidation from the drainage wire onwards. In the extreme case of natural tracing paper, it contributes significantly to the ultimate achievement of a virtually solid sheet, with virtually no air space.

As for fibre dispersion, this is a *sine qua non* for a well-consolidated sheet as well as for a well-formed sheet. By definition, flocculation introduces deviations in packing which affect consolidation adversely at every stage of papermaking. It leads to overconsolidation at the flocs, and underconsolidation between the flocs, and must be regarded essentially as a structural fault which disturbs all the successive stages of papermaking, and leaves its compounded pattern of blemishes in the final sheet. Flocculation is yet another natural consequence of the shape of the fibre: its avoidance is one of the primary objectives for the forming of paper, and for both consequent and subsequent even consolidation of paper.

## 5.4. CONSOLIDATION AND WATER-REMOVAL BY WET PRESSING

### *5.4.1 Historic and general background*

The process of wet pressing dates from the early history of handmaking of paper. Open "sandwiches" of formed wet sheets and pieces of damp wet felt upon which these sheets had been couched from the drainage wire of the hand mould were assembled in a pile, and subjected to progressively increasing pressure normal to the plane of the sheets, within a screw press. During this operation, water poured from all four sides of the oblong mass, which was generally reduced towards one half of its original thickness by the time the pressing was complete. After removal of the compacted mass from the press, the individual sheets of paper had been compacted to a degree where they were strong enough to be peeled manually from the underlying felts, to be hung up for loft drying. Sometimes they were peeled and transferred to fresh damp felts for a second pressing *en masse* before loft drying.

In such batch wet pressing, the felt served (a) as a transfer medium from the drainage wire on which the sheet had been formed; (b) as a temporary reservoir for water forced vertically under pressure through its surface from the surface of the adjacent paper sheet(s); (c) as a permeable medium through which the water could be expressed laterally to flow away externally. Depending on the character of the wet paper sheet, some water was also expressed laterally through the sheet itself, but this was (generally) very small compared with the quantity expressed laterally through the (generally) thicker and (generally) more permeable felt. Indeed, any attempt to press wet paper sheets without felts results in "crushing" by disruption of the fibrous structure through excessive lateral flow of water through the wet web, quite apart from the almost insuperable difficulties of handling the wet web before pressing, and of separating adjacent webs after pressing.

The advent of the continuous papermaking process, pioneered by Robert, Fourdrinier and Donkin brought about inevitable changes in the wet pressing process. The process itself was still necessary to consolidate the wet web. In its earliest form in the continuous process, it simply involved passage of the wet web on the carrier wire on which it had been formed, through a pair of plain squeeze rollers. Later, the continuous wet web was pulled off the wire at the couch roll, to be transferred to a continuous band of felt (the so-called "wet felt", to distinguish from the "dry felt" of the dryer section). The open sandwich of wet paper web on the damp wet felt was then passed through a pair of plain squeeze rolls, the most elementary form of the continuous web press as a separate component.

This change from the screw press to the continuous roller press left unchanged the essential character of the pressing operation, although it introduced substantial changes in the temporal and dimensional parameters.

With the screw press, there had been a succession of turns of the screw, each involving a new surge of water transversely from wet web to felt, and then laterally through the felt to drain away down the outer faces of the compacting mass. During each turn of the screw, there was a dynamic situation, as in the continuous press, but the incremental rate of pressure increase was low, and consequently there was plenty of time for lateral transfer of water through the area of the felt (despite the large size of sheets made by professional papermakers), without building up excessive hydraulic pressure within the paper web which might have caused disruption by excessive lateral flow through that web.

In the continuous plain press, even at low speeds, there was now a much more rapid movement of water from wet web to felt and laterally throught the felt to external free flow. At the same time, the distance through which water had to flow laterally (whether through felt or to a much lesser degree through wet web), was substantially reduced. It was now only the few inches of the nip instead of the several feet of the sheet dimension. Thus a dimensional reduction in the distance of the flow compensated for an increase in the rate of the flow, allowing much higher rates of water removal without damage to the paper web through increased hydraulic pressure.

As the technology of papermaking developed, the plain press was improved by the use of rubber coverings on the press rolls, by varying the configuration of the felt run, by improving the quality of the felt in terms of permeability and evenness of surface, and by provision of nips to re-dry the felt on its return to rejoin the incoming wet paper web.

Apart from these improvements, the process remained largely unchanged until economic pressures brought about the dramatic acceleration in paper-machine speeds which occurred between the 1920's and the 1970's. In the 1920's, two hundred feet a minute was quite a good speed for papermaking. In the 1970's, two thousand feet per minute is fairly modest, and 5000 ft. per minute or more is a reasonable target for certain light weight grades. This substantial speed increase has brought major design challenges to every part of the paper machine, and not least to the wet press section. The main effect of speed increase on the wet press has been to increase the rate of water flow (and hence hydraulic pressure) without any decrease in the distance for lateral flow through the felt within the press nip. Indeed, the increase in paper-machine width which has generally accompanied increase in machine speed, has necessarily been accompanied by increase in diameter of the press rolls, which has increased the distance for lateral flow through the felt within the press nip before emerging into free flow, thus compounding the dilemma.

The only way out of this dilemma has been to provide for transverse removal of water from the felt, in place of lateral removal. The first major step for providing transverse removal was the introduction of the suction press. In this, the shell of the lower roll with which the felt is in contact is drilled with a pattern of holes into which can pass the water which has been forced into the felt by the nip pressure. There is still a transverse flow through the felt over the so-called "land" areas of the roll, but this dimension is relatively small, and there is an effective release of the hydraulic pressure built up by the combination of nip pressure and high speed.

Since the introduction of the suction press, many other refinements and improvements to the same end have followed, including the provision of plastic or metallic woven wires lying between the felt and its corresponding roll, whether plain or suction, and the provision of resilient layers which are porous, grooved, or embossed, covering the surface of this roll. Except in the manufacture of certain board grades, the roll directly in contact with the paper web itself is generally smooth and hard, to ensure that the top surface is moulded smoothly.

The purpose of all these expedients is the same: viz. to provide a release from excessive hydraulic pressure, a reservoir for expressed water, and a channel for external disposal of that water.

### *5.4.2 Early speculation and experiment*

Until the 1950's, development of wet pressing was dictated almost entirely by empirical considerations. In 1947, however, Campbell [18] devoted considerable attention to the removal of water in wet pressing, in a paper entitled "The Physics of Water Removal". This paper was largely a discussion of the theoretical considerations affecting water removal in papermaking, and contained detailed theoretical appendices on the permeability of pulp webs to water, the suction action of table rolls, the compressibility of pulp, and the removal of water by compression of mass. He referred to the division of mechanically applied pressure in wet pressing between hydraulic pressure and mechanical stressing of the wet web. He derived a formula giving the increase in density due to mechanical pressure as a function of the product of pressure and time of application and the product of the square of the basis weight, the square of the specific surface, and the viscosity of water (with which the increase in density varies inversely). He attributed "crushing" of the wet web to a separation of the fibres caused by an excessive rate of escape of water at the nip. He also pointed to decrease in water viscosity through temperature increase as the only way of increasing water removal in a given installation, and predicted a 2.5% increase in rate of water removal per degree Centigrade rise in temperature.

In 1954, Nissan [19] presented a paper to the TAPPI/CPPA Fundamental Research Conference on "Fundamentals of the Paper Machine", entitled "The functions of the Felt in Water Removal on the Papermaking Machine". This paper was partly concerned with wet pressing, partly with drying over hot cylinders, and contained much emphasis on felt quality as a determinant of drying efficiency. It contained only marginal reference to the consolidation aspect of wet pressing, and was largely speculative in content, rather than a presentation of experimental data: and indeed some of its findings have been subsequently contradicted (though not necessarily disproved) in detailed experimental studies on the mechanism of wet pressing. Nevertheless, it ranks with Campbell's paper as a pioneer contribution of major significance for the subject, in that it was one of the first attempts to bring scientific logic to bear upon an empirical process until then based entirely upon trial and error.

Nissan began by presenting photographic evidence that the wet web passing from a suction couch and approaching the first wet press contained a more or less regular pattern of wet spots in a less wet continuous network, the wet spots corresponding to the holes of the suction couch, where re-absorption of water from the wire to the paper had apparently taken place. He next observed that at the entry to the nip, the paper sheet and the felt were compressed and flooded with water. Nothing could be observed within the nip, beyond the evident compression of the felt. After the nip, the felt observably recovered its thickness. The paper web was photographed emerging (a) from the nip of a plain press (no suction), and (b) from the nip of a suction press.

In the case of the plain press, the surface of the emerging wet web was uniform in appearance, and no longer exhibited the pattern of apparently wetter areas corresponding to the holes of the suction couch: the spotted pattern had been completely obliterated. In the case of the suction press, the sheet once again revealed a pattern of dark spots on a light continuous network, apparently corresponding to the holes of the suction press roll.

Nissan argued that at the entry to the nip, both the felt and paper were compressed, with observable extrusion of water. He considered that most of this water had come from the felt (presumably having been brought back from the outgoing side of the press nip) and stated that little of the extruded water was likely to have come from the wet web itself, citing the observations of Sedoff et al. [20] concerning the retention of water by the wet web under very high pressures. In the nip itself, he stated that both felt and paper must be considered to be saturated with water, and that at this stage the total quantity of water in the system had been reduced by (a) water expressed backwards to free flow; and (b) in the case of suction press, water removed by aid of suction through the shell holes.

After the nip, both wet web and felt expand, and the total quantity of water was then considered to be partitioned between web and felt by (1) elastic forces in the felt sucking water from the wet web into the felt, and (2) surface tension forces displacing air from capillaries.

From these considerations, Nissan proceeded to deduce the desired characteristics of a wet felt liable to aid water removal; high compressibility, rapid elastic recovery, high capillary absorption, and high permeability to water. In calculations based on earlier work on liquid flow into porous materials, Nissan deduced that the total volume of water abstracted from the sheet by the felt at the outgoing side of the nip is proportional to the square root of the ratio of surface tension $y$ to the viscosity $\mu$ of the water. This ratio increases with temperature (as shown in Table 5.1, cited from Nissan [19]). It is indeed a known fact that higher wet web temperature at the presses does assist water removal. Thus Stamm [21] reported in 1943 that increasing the wet web temperature by 20° F increased the dryness after the press by 2—3%.

In 1959, Bergstrom [22] published the results of laboratory studies on removal of water from wet webs and wet felts pressed alone and in contact together, in different combinations, in a laboratory punch press to which loading could be rapidly applied and withdrawn. Bergstrom's results did not appear to accord with Nissan's theories. The final water distribution between

TABLE 5.1

Relation between temperature, viscosity and surface tension of water (from Nissan [19])

| $t^\circ$ | $(y/\mu)^{1/2}$ |
|---|---|
| 100 | 101 |
| 130 | 114 |
| 160 | 127 |
| 190 | 138 |
| 210 | 142 |

felt and paper was in Bergstrom's view most logically explained by the thesis that during compression (corresponding to a wet press nip) water was forced from the wet web into the felt, and from the felt into the supporting wire screen and perforated plate, which in Bergstrom's equipment simulated the corresponding elements in a suction press shell carrying a "ventilating" wire screen. In the recovery phase, corresponding to the post-nip phase of felt and wet web expansion in a wet press, some of the water in the felt appeared to be re-absorbed by the wet web, and some of the water in the wire screen to be re-absorbed by the felt. He suggested that this re-absorption is caused by capillary action, perhaps aided by suction induced by felt expansion.

Bergstrom's interpretation of his findings were thus diametrically opposed to Nissan's theories: both in respect of the removal of water at the ingoing side of the wet press nip, and in respect of the removal (or re-absorption) of water at the outgoing side of the nip.

White and Berdux [23] also reported in 1960 work carried out with punch press equipment similar to Bergstrom's. For a given press loading, they always found the web after pressing to be wetter when it was pressed in contact with a felt, than when it was pressed alone. When the web was held between two felts, it emerged wetter still. Insertion of impermeable sheets of cellophane between the web and the felts resulted in water contents for both felts and web the same as if they had been pressed separately. They concluded that there is a transfer of water from felt to paper during the expansion phase, when web and felt remain in contact.

In 1961 Sweet [24] presented evidence, based upon studies carried out some years earlier with an experimental press section, which supported the general findings of Bergstrom [22] and White and Berdux [23]. The experimental press section was run at 45 f.p.m., which slow speed (in Sweet's view) facilitated the study of the phenomena involved without changing their fundamental nature. The wet web used throughout was derived from re-wetted wet strength tissue apparently made from lightly beaten stock.

Sweet's experiments were deliberately designed to prove or disprove the opposing views already cited, regarding the removal of water from the wet web at the ingoing sides of the press nip.

To check whether water is removed from the web by the felt, at the outgoing side of the nip, Sweet passed through the nip (with the felt) a number of wet webs of increasing thickness made up from multiples of re-wetted sheets of free beaten tissue. He argued that, if the felt absorbs water from the wet web on the outgoing side, then the effect should be greatest with the lowest weight wet web. Conversely, if the wet web absorbs water from the felt, he argued that the reverse should be true. Figure 5.1 shows his prediction of the trends of results which he expected from the two opposing mechanisms.

The solid curve in Fig. 5.2 shows what he obtained in practice by pressing different thicknesses of wet webs carried on a wet felt. The shape of this

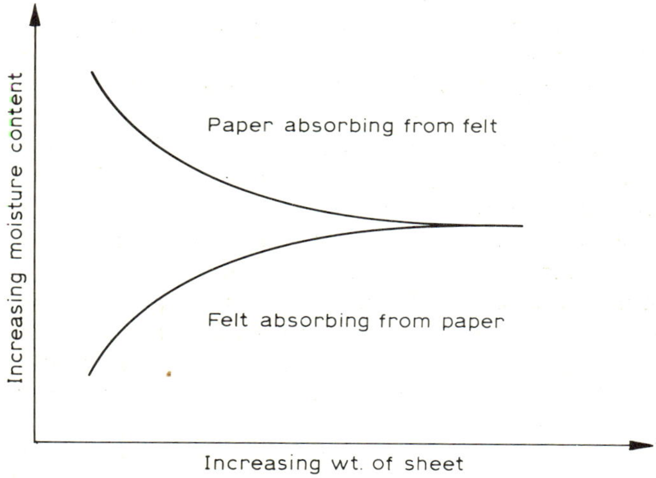


Fig. 5.1. Predicted patterns of flow of water from paper web to felt and from felt to paper web (from Sweet [24]).

curve, obtained from plotting a number of superposed sheets against wet web water content is similar to the shape of the solid curve in Fig. 5.1, which he had predicted as the consequence of the wet web absorbing water from the felt. The upper (discontinuous) curve in Fig. 5.2, depicts similar data obtained by passing the wet webs between two felts: Sweet argued that the

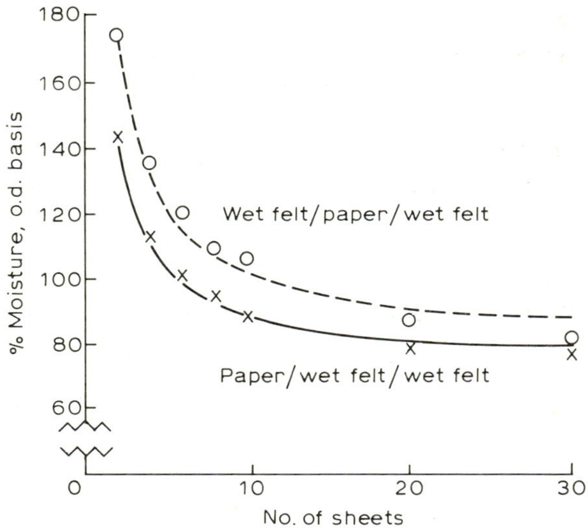


Fig. 5.2. Water contents after pressing of different thicknesses of wet webs (from Sweet [24]).

high levels of water content in this case are consistent with re-absorption of water from the two felt surfaces instead of one surface.

Having thus disposed to his satisfaction of the question of redistribution of water between felt and web on the *outgoing* side of the nip, Sweet turned his attention to the *ingoing* side of the nip.

To check whether or not water flows from the wet web to the felt on the ingoing side, Sweet ran his press with and without an impermeable sheet of plastic material (Saran) between the wet web and felt. The experiment was done with a range of thicknesses of wet web, made up from superposed layers of re-wetted tissue. Figure 5.3 shows the water content of the outgoing webs of varying thickness, with and without the intermediate layer of Saran. Clearly, the absence of the Saran sheet allowed substantial increase in water removal, and since he was already satisfied that this was not occurring at the outgoing side, Sweet logically deduced that it was occurring at the ingoing side.

Sweet's results have been subjected to critical comment in more recent times as will be discussed on p. 240, but in advance the writer must point out certain weaknesses in his work. First, wet webs made up artifically from superposed sheets of pre-dried, re-wetted paper, are far removed from genuine newly-formed wet webs. This is especially so when they are made from wet strengthened sheets, which by their nature cannot revert to the character of never dried unbonded sheets, since water-resistant bonds have already been formed. Second, soft absorbent tissue made from very free beaten stock (as implied by the origin of Sweet's paper — a tissue supplied

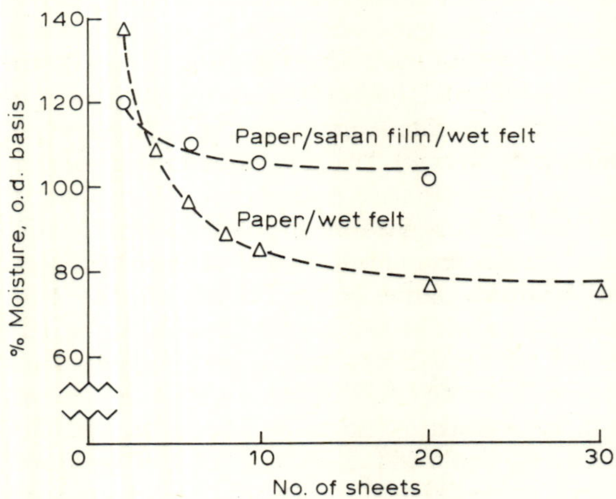


Fig. 5.3. Effect of interposed Saran film on water contents after pressing of different thicknesses of wet webs (from Sweet [24]).

by Scott) represents an extreme case of paper where structural resistance to water flow is very low, and where the recovery forces from compression of the fibrous network are very high (especially in the presence of wet strength bonds). Third, Sweet did not consider the possibility of any alternative explanation other than re-wetting of the wet web, for his observed inverse correlation between water content and increasing sheet weight. Fourth, he ascribed his deduced re-wetting to capillary flow, without any consideration of any alternative mechanisms such as suction induced by recovery from felt compression.

### *5.4.3 Formulation of broad concepts on the dynamics of wet pressing*

In 1960, Wahlstrom [25,26] published a massive record of results of observations derived from an extensive study of two suction presses on a medium speed newsprint machine. In reviewing his observations and postulates, and also those of later contributions to the subject, it is helpful now to define the successive physical phases which observably occur during the passage of a felt plus wet web sandwich through a press roll nip. Figure 5.4 taken from Wrist [27], depicts in conventionalised manner (not true to scale) the generalised situation. The following commentary on this figure is for convenience quoted from Wrist. "Phase 1 extends from the point at which one or more of the elements contact the surface of the rolls, i.e., at the line LL′ in the figure. It terminates at the line MM′ at which the zone of solid continuity between the two rolls begins. MM′ is usually considered to be the start of the nip proper. The way in which the different elements come together in the first phase can vary.

Phase 2 extends from the boundary of solid continuity MM′ to the point of closest approach in the nip NN′.

Phase 3 extends from the nip centerline NN′ to the end of the zone of continuous contact OO′ which is normally considered the end of the press nip.

Phase 4 extends from the line OO′ to the points PP′ at which the structure leave the roll surfaces.

Phase 5 extends for an indefinite period beyond PP′ and covers any interactions between the structures beyond the press or in the case of felts or wires, any conditioning which might occur before they return to the ingoing phase 1 for the next pressing cycle."

Wahlstrom's interpretation of his observations on the two newsprint suction presses referred to above, was as follows (using the numeration and lettering of Wrist's conventionalised diagram).

Wahlstrom assumed that at the start of Phase 2, neither felts nor paper web were saturated (certainly true for newsprint), although the paper had a much higher water content, and therefore needed less pressure to become saturated. During phase 2, as web and felt approach the nip centreline NN′,

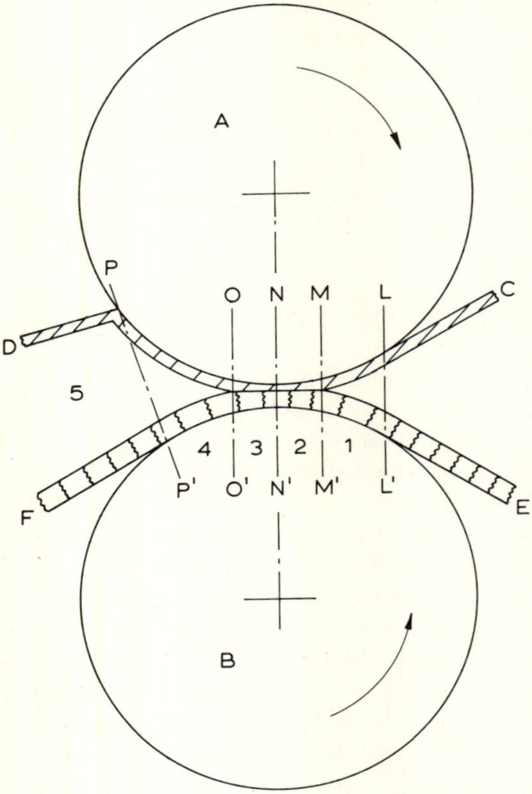


Fig. 5.4. Schematic drawing indicating successive phases in wet pressing (from Wrist [27]).

the converging rolls exert a compressive force on the web felt sandwich which is balanced by the sum of the mechanical stress in the solid network and hydraulic pressure in the liquid phase. Mechanical stresses precede hydraulic pressure, until the web becomes saturated. Thereafter, hydraulic pressure in the web causes water to move from web to felt. In turn, the felt becomes saturated, and a water flow within and from the felt commences. With a plain press, this flow must be backwards, against the forward movement of felt, and web. Consequently, water is discharged at point M, to run down the rotating lower roll, against the forward rotating movement of that roll.

Such backward flow can only take place under high pressure, with low felt resistance to lateral flow, and at relatively low speeds. If the speed and hydraulic pressure are too high, or the felt permeability too low, then excessive lateral flow will occur in the wet web before transfer to the felt, thus causing internal disruption of the wet web.

With a suction press, the drill holes can relieve the hydraulic pressure. The distance for the water to travel over the "land" areas is relatively small, so that excessive lateral pressure gradients do not build up so as to cause wet web disruption.

At NN′ the ratio of water to fibre was judged by Wahlstrom to be at a minimum for the sequence depicted in Fig. 5.4. Regarding the next phase (3), Wahlstrom followed Bergstrom [22] and Sweet [24] (whose then unpublished results were at that time already known to him) in assuming that a reverse flow from felt to wet web begins to occur: also that there is a reverse flow from each drill hole in the shell to the felt. Both wet web and felt were considered to expand allowing entry of air and formation of air-water interfaces, which leads to water movement within the capillaries of both felt and wet web. At the same time, there was said to be water movement from the coarse capillaries in the felt to the much finer capillaries in the paper wet web.

In 1964, at the first International Symposium on Water Removal at the Presses and Dryers, (held under the auspices of TAPPI and CPPA) Wrist [27] presented an excellent review entitled "The Present State of our Knowledge of the Fundamentals of Wet Pressing", from which Fig. 5.4 and its explanatory commentary on p. 227 of this section have been taken. Wrist's paper is essential reading for a grasp of the subject of this section, and for an understanding of the fundamental factors involved, especially since it contains useful reference to structural considerations within this phase of consolidation, as well as a comprehensive bibliography and commentary on papers dealing with all aspects of the possible mechanisms for water removal at the wet presses.

After reviewing published theories and experimental evidence on press operation which have already been discussed, Wrist turned attention to certain underlying factors which influence the effectiveness of that operation: notably the mechanics of compression and recovery of fibrous mats and felts, and the mechanical and hydraulic pressure gradients within compressed fibre mats.

Compression of wet fibre mats has been studied over many years by a large number of workers, including successive teams at the Institute of Paper Chemistry (Appleton, WI, U.S.A.). The subject is highly complex, as evidenced by a fairly recent paper by Han [28]. Wrist accepted the simplest among a variety of expressions for describing the relationship between applied pressure ($P$), and mat density or consistency ($C$).

$$C = MP^N$$

where $M$ and $N$ are compressibility coefficients ($N$ may be called compressibility power to distinguish it from $M$). This relationship was used by Campbell [18] and with various refinements has been the basic expression used by the workers at the Institute of Paper Chemistry, e.g. Ingmanson et al. [29], [30] and Jones [31].

It should be noted that $N$ is the percentage change in mat density corresponding a unit percent change of pressure: it is thus an index of compressibility. Data cited by Wrist show that the value of $N$ for a woollen felt is typically 0.2: for wet sheets made from wood pulp, it ranges from 0.3 to 0.45. $N$ seems to be unaffected by beating, an important consideration for consolidation of the web, although it has to be noted that all these data referred to static tests in which sufficient time was allowed for near equilibrium to be reached.

Wrist noted that compressibility under dynamic conditions is less well documented. Turner et al. [32] found that beating had no effect on $C$ at equilibrium, but considerably affected the rate at which equilibrium was reached. Mats made from unbeaten fibres reached equilibrium under pressure very rapidly: mats made from well-beaten fibres could take over 30 min to reach equilibrium.

Wilder [33], cited by Wrist, found that in repeated compression/recovery cycles, $M$ changed substantially towards a constant value after about six cycles. The changes in $N$, which also moved towards a constant value after about six cycles, were relatively small. Wrist points out the possible significance of this finding in practical wet pressing. Whereas the felt will become mechanically conditioned towards constant values of $M$ and $N$ after a short running in period, the fibrous wet web goes through the first cycle at the first press, its second at the second press and so on. This may have some bearing on the ability of a second press to remove water from the wet web, with a press loading similar to that used on the first press.

Significant in the context of the effect of beating on wet pressing are the results of an early paper (1939) by Seborg et al. [34]. These workers studied a still virtually unknown but very important aspect of wet pressing of fibrous webs, viz. the recovery phase after compression. Their findings confirmed the above cited reports that while beating did not affect the equilibrium compressibility of the fibrous mat, it did substantially affect the rate of reaching equilibrium. In addition, they found that beating considerably reduced the degree of recovery of sheet thickness after compression.

Wrist [27] commented on the often neglected complication of hydraulic pressure gradients in a fibrous mat under compression. The compressive load on the mat is balanced by the mechanical resistance of the network to compression, and by hydraulic pressure built up in the liquid between the fibres. Both factors change through the thickness of the mat as compression takes place. Wrist cited the theoretical case of a fully saturated fibrous network contained between an impermeable piston and a highly permeable screen. When pressure is applied by the piston, it creates initially a hydraulic pressure which causes water to flow from the screen, at the same time compacting the fibres adjacent to the screen. Some of the pressure initially carried by the liquid is transferred to the compacting fibrous network. The flow of water relative to this network is nil at the piston, and increases towards the wire mesh. Consequently, the proportion of the load carried mechanically by the network increases towards the screen, and the proportion carried by the hydraulic pressure decreases towards the screen. At the same time, the degree of compaction of the network increases progressively from piston to screen, just as in the case of fibrous sheet formation under gravity drainage.

The problem is still more complicated in an actual press nip, where there is inevitably a lateral component of water flow. This is especially so in the case of a plain press where the water cannot escape transversely, but it also applies to the land areas of a suction press. Most of the lateral flow is through the felt, but the dynamics of the press nip dictate that there must always be a lateral component, however small, through the paper as well.

The so-called "wedge effect" in a wet press nip also requires consideration. When a wet felt enters the nip, it encounters a pressure gradient which tends to force water backwards against the movement of the felt, this flow being resisted by viscous drag of the felt fibres. With increase in speed, the hydraulic pressure increases. As a consequence, the load borne by the solid phase of the felt structure decreases, so that the felt expands and causes the nip rolls to separate. Thus the water in the felt has the dynamic effect of wedging the two rolls apart. One consequence is that the water content in the felt leaving the press increases as speed increases.

This phenomenon has been studied by Yih and McNamara [35], Asklof et al. [36] and Means [37]. All these workers confirmed the fact and significance of the wedge effect whereby speed increase causes more water to be carried through the nip, but they differed in their interpretation of the data. Asklof et al. reached the somewhat surprising conclusion that at nip centre the water is passing through at a higher speed than the felt and the roll surface. They further concluded that this speed difference, which can be as high as 20%, increases with both speed and press loading.

While the "wedge effect" is clearly only another manifestation of the overall pattern of mechanical and hydraulic forces discussed by Wahlstrom and others, it is so obtrusive in its effect upon press operation and efficiency as to deserve more profound study. Among other things, it could have a significant bearing upon the vexed quesion of redistribution of water on the outgoing side of the nip. If water is in effect extruded forward by this wedge effect, between phase 2 and phase 3 (Fig. 5.4), then we need to understand what happens to it in phase 3, where it must slow back to the felt speed. It is reasonable to expect that it should influence any pattern of transverse flow between felt and web.

In his review paper [27], Wrist considered at length the question of water re-distribution in phase 3, referring among others to the work of Nissan [19],

Bergstrom [22]. White and Berdux [43] and Sweet [24] which has already been considered above. In addition, he cited experimental and machine observations by Welch [38] and Woodside [39] which support (though sometimes on fairly flimsy evidence) the concept of rewetting of the outgoing wet web in phase 3 by transfer of water from the wet felt. He also added some new experimental evidence derived from high speed pilot plant studies, on several aspects of wet pressing, including the re-wetting concept. Among other findings, he reported that increase in basis weight in nearly every case resulted in wetter sheets from the press. Since this is contrary to Sweet's findings at low speeds, he concluded that at his higher speeds, resistance to water flow during compression rather than the equilibrium compression of the sheets made from the least beaten stock used in his experiments was the dominant factor controlling water removal, and that it overshadowed any subsequent re-wetting effects. He found that apart from press loading, the main determinant of final water content was the nature of the fibrous furnish. A neutral sulphite semi-chemical (N.S.S.C.) web was very resistant to water removal, a relatively free kraft fibre web being least resistant. Removal of the fines fraction from the N.S.S.C. increased its wet pressability considerably, and reduced the tendency to "crush".

Wrist judged that the re-absorption of water by the wet web from the felt in the expansion phase was well established, but he seriously questioned whether Wahlstrom's proposed transfer mechanism, viz. capillary flow under surface tension can apply to phase 3, although it may well be correct for any absorption occurring beyond the nip when felt and web remain in contact. In his opinion capillary transfer is too slow to take effect in phase 3. He suggested instead the possibility of a film split phenomenon between two rough surfaces with different scales of roughness as an explanation for phase 3 re-absorption.

Further reference must be made at this point to work in the 1960's on the subjects of "shadow marking" and "crushing", phenomena of importance to the consolidation aspect of wet pressing.

"Shadow marking" has already been referred to when discussing Nissan's paper [19]. It is the term used to describe the optical pattern in the paper web, observable both during drying and in some cases in the dried paper, derived from the pattern of holes in the drilled shell of the suction press. This optical pattern can be particularly troublesome with certain speciality papers such as natural tracing paper, where it constitutes a major defect. Visual evidence suggests that the "hole areas" of the wet web emerging from the press have a higher water content than the "land areas". Redfern and Gavelin [40] tested the respective "hole" and "land" areas of observably "shadow marked" sheets of dry paper, seeking physical differences that might be correlated with the visual pattern. They found several clearly measurable differences between the two areas. First, fibres were more aligned in the machine direction over the "land" areas than over the "hole" areas.

Second, the surface of the paper over the "hole" areas was raised above that over the "land" areas. Third, the areas over the holes had higher air permeability than those over the holes. Fourth, the "land" areas had 10% higher tensile strength than the "hole" areas. Fifth, the intensity of "shadow marking" increased with press loading. Redfern and Gavelin concluded that "shadow marking" was caused by a lesser degree of web consolidation over the "hole" areas than over the "land" areas, resulting in a lower degree of bonding, with concomitant effects on tensile strength and opacity.

This conclusion has a significant and important concomitant, for it demonstrates the true function of pressing as a major determinant of intermediate web compaction, in the absence of which the later stages of compaction, produced by surface tension forces and hydrogen bonding cannot be fully effective. The "land" area represents the ideally press-compacted sheet, from the viewpoint of consolidation. The "hole" area represents the imperfection in that compaction introduced by the necessity to provide transverse relief from hydraulic pressure and means for transverse disposal of expressed water, at high speeds of papermaking. It is clear that the so-called "fabric press", which provides an intermediate porous layer of woven mesh between the felt and the press roll (whether drilled as a suction roll or not) offers a means of evening out this sharp contrast between the perfectly and imperfectly compacted areas.

The observations of Redfern and Gavelin also suggest that in the "land" areas there is a certain degree of disruption of the original disposition of the fibrous network, in that the component fibres appear to have become more aligned in the machine direction by lateral flow of water. This may well be regarded an incipient stage of the phenomenon which exhibits itself in grosser form as "crushing" at the presses, especially on plain presses.

From the foregoing discussion of the disposition of hydraulic forces in the wet web, and the pattern of flow of water within the web and felt, it is clear that excessive speed and/or nip pressure, coupled with inadequate means of relieving the pressure and providing outlets for the water, can cause the wet web to stretch on the ingoing side of the nip and within the nip. This causes web disruption, which can build up to the point of fragmentation, if the forces increase excessively. This is the phenomenon called "crushing": presumably the term has arisen because nip pressure is one of its main determinants. Lyall [41] investigated "crushing" in the case of thick boards, where it results in shear failure at one or more planes of the sheet. He observed that when "crushing" occurs, the centre of the sheet passes throught the nip at a speed higher than that of the roll surface: clearly another manifestation of the extrusion of water noted in connection with the "wedge effect" (p. 231).

Felt permeability is of prime importance in determining the conditions under which "crushing" will not occur. For this reason, felts intended for first presses generally have open structures. By the same token, the provision of woven fabric layers between felt and roll also provides relief from hydraulic pressure, and hence from the danger of "crushing".

*5.4.4 Growing doubts regarding the dynamics of wet pressing*

In 1968, several papers of significance for the subject were presented at the 2nd International Symposium on Water Removal at the Presses and Dryers held under the auspices of TAPPI and CPPA. In addition to the already mentioned paper by Han [28] on Compressibility and Permeability of Fibre Mats, they included an updated version of "Our Present Understandings of the Fundamentals of Pressing", this time presented by Wahlstrom [42], a contribution by Nilsson and Larsson [43] which in addition to its experimental content further refined Wahlstrom's analyses of events in wet pressing, a paper by Morrison [44] entitled Felt Capillary Structure and Water Removal, a paper by Schiel [45] on Nip Geometry of Transveral Flow Presses, and several papers relating to specific press types and designs, with emphasis on the use of woven fabrics as "ventilating" media between felt and roll.

Turning first to the paper by Nilsson and Larsson [43], the aim of the work presented was to develop a mathematical model of the flow conditions in a pressed web, based largely upon measurement of compressibility of pulp mats, and measurement of flow resistance in pulp mats. These measurements were carried out quite simply by using apparatus already described by Asklof et al. [36].

Their work had the praiseworthy objective of concentrating upon events within the paper web itself. They pointed out that much interest had been concentrated in investigation of movement of rubber in the nip and on flow conditions in the felt, but the paper web itself, which they rightly declared to be "the real subject of pressing", had been largely neglected. They felt this was especially important in the case of so-called transversal flow presses (those where the use of fabric reservoirs etc. permit transversal relief for the expressed water). The writer considers that the same considerations apply to all presses, since flow from paper to felt is always largely transversal: the difference between plain presses and transversal presses lies primarily in the direction of movement of the expressed water within the felt, not within the paper.

Nilsson and Larsson's attempted mathematical model unfortunately failed to produce simulations matching practice, so they concluded that their basic assumptions regarding the source of dynamic pressure in the system were wrong: and that there must be source(s) of dynamic pressure other than simple hydraulic pressure, perhaps arising from inertia, the structure of individual fibres, and/or viscoelastic behaviour of the fibres.

The general concept of events in the press, used in developing their model, was essentially based on Wahlstrom's account of these events [25,26], but contained some refinements, including the introduction of new ideas about events post mid-nip. Their total concept of events is depicted in Fig. 5.5 which indicated projected curves for pressure distribution and paper web thickness at different phases in the nip. Three separate pressure distribution curves are shown, one for the total nip pressure, one for the fibre structure pressure, and one for hydraulic pressure. They postulated that the fibre-structure pressure curve reaches its maximum and the paper web thickness its minimum, at a point after the mid nip point. Indeed, their phase 3 (not to be confused with Wrist's phase 3 in Fig. 5.4) was defined as the phase between mid nip (more accurately the maximum point of the nip pressure

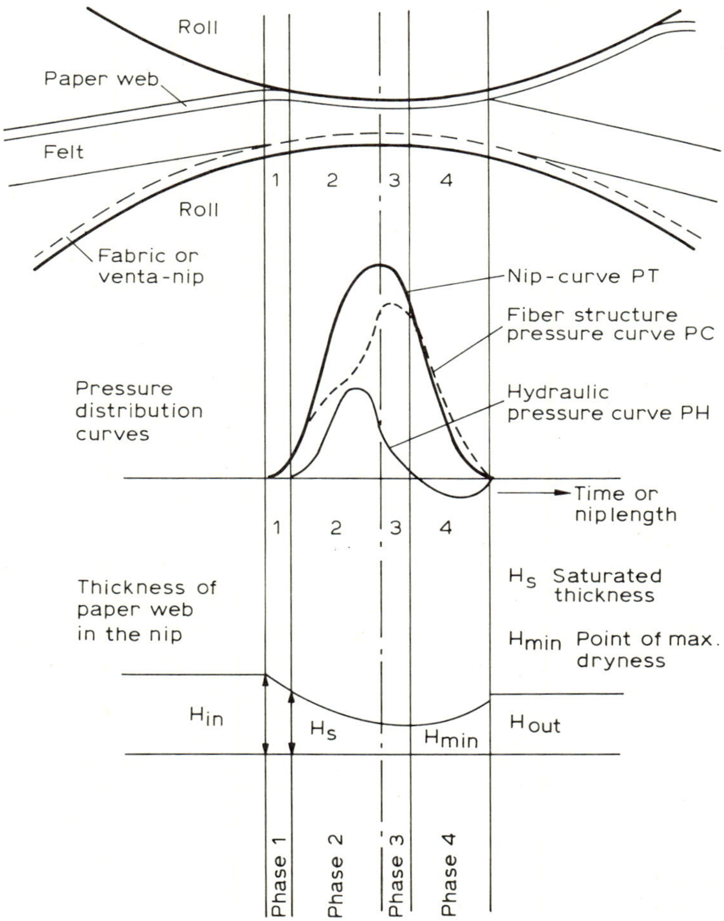


Fig. 5.5. Schematic drawing indicating successive phases in wet pressing (from Nilsson and Larsson [43]).

curve) and the maximum web dryness point. This added element to the account of events within the nip appears to be essentially speculative, although not necessarily for that reason untenable. In any case, it was accepted enthusiastically by Wahlstrom, who in his review paper [42] presented at the same symposium, accepted Nilsson and Larsson's depiction of events, and incorporated Fig. 5.5 in his own review as the basis for re-statement of his own concepts.

Like Wrist's 1964 paper [27], Wahlstrom's 1968 review paper [42] constitutes a valuable source of background information and bibliographical references. It set out to cover only new knowledge since 1964 (including that included in other papers presented at the symposium) but in fact developed into a far ranging review which took in much pre-1964 material, and which was largely dominated, perhaps understandably, by refinements, extensions and re-exposition of Wahlstrom's own concepts as already presented in 1960 [20,26]. In addition to the refinement added by Nilsson and Larsson [43], Wahlstrom discoursed at length on these concepts and their consequences for press and felt design and operation. It is interesting to note that phase 3 (introduced by Larsson and Nilsson), was stated by Wahlstrom to be "an expanding portion of the nip" in which the felt takes up all the expansion, since the paper gets still further compressed: there is thus "a vacuum in the felt". This, said Wahlstrom, forces air and water to enter the felt from underneath through the fabric or grooves, and there is also "a fluid flow of water still active as a transfer mechanism from paper to felt . . . . Capillary forces help transfer water from felt to paper".

Despite the lack here of a clear statement as the proximate cause of this transfer of water web to felt, the writer finds it difficult to distinguish between this concept during Nilsson and Larsson's phase 3 and Nissan's 1954 concept [19] whereby on the outgoing side of the nip, water was considered to be sucked out of the paper web by compression recovery forces of the felt. This is not to deny that there may be a subsequent reversal of flow, changing to a flow from paper to felt, as visualised by Wahlstrom in the later rewetting of phase 4, but his new account of events in phase 3 constituted a rather unexpected admission that there is in fact some kind of "competition" between felt and paper for water transfer in successive phases, during their respective recoveries from compression.

It is impossible to summarise or adequately review Wahlstrom's 25,000 word review, but some key statements need recording. "The optimum felt should give a perfectly uniform pressure, lowest possible flow resistance in the fluid flow region, and a smallest possible re-wetting in the outgoing part of the nip" (his p. T351).

"In essence, pressing then involves compression of the wet fibre structure (mat), fluid flow through paper and felt, fluid flow in a two-phase system due to capillary forces and gross pressure differences, film splitting and uniformity of pressure application (his p. T351).

"The ingoing part of the nip is normally by far the most important determining dryness of the outgoing paper. However, in the outgoing nip, rewetting occurs which can be of sizeable proportions. The amount of water transferred by rewetting can be in the order of up to 35 g m$^{-2}$ (no source cited for this figure — HFR), which, especially for light weight sheets, is extremely important . . . . It is an interfacial phenomenon with the transfer only dependent on participating structures and their states of compression and wetness; but not on factors like basis weight, speed, etc." (his p. T352).

"Permeability of felts is at least a hundred times larger than that of paper. With transversal flow presses, which have reduced flow distance radically, flow resistance through felt is normally much less than flow resistance in paper" (his p. T355).

In pages T362—364 of his review, Wahlstrom reverted to detailed argument about rewetting and the possible mechanisms causing it. He considered at length the three possible mechanisms: (i) pressure differential between felt and paper; (ii) film splitting, as proposed by Wrist [27]; and (iii) capillary transfer. He eventually concluded, as in earlier publications, that capillary transfer is the most likely cause. Nevertheless, in the course of his considerations on this matter, he referred to reported major variations in the dimensional expansion of the wet web after mid nip. Whereas he (Wahlstrom) had reported 50—70% expansion, Wilder [46] reported only 10—20%. He also referred to indications that re-wetting occurs only at the first part of the expanding phase, and is limited by the availability of water in the interface.

These observations, coupled with Wahlstrom's earlier cited admission that post nip transfer of water from paper to felt takes place during felt expansion in Nilsson and Larsson's phase 3, suggests that despite his rather authoritative analysis of the subject, Wahlstrom was far from confident about the precise pattern of water transfer on the outgoing side of the nip. Indeed, he concluded:

"Our knowledge of factors affecting rewetting is still limited both experimentally and with respect to theoretical prediction. Work must be done to refine theoretical models and to collect data on capillary forces in paper and felts as a function of structure, saturation and compression".

In a more general conclusion, he stated, with justice,

"The importance of a better understanding of pressing is amply demonstrated by large improvements realised in reduced moisture content after the presses, achieved by applying modern press concepts. The key has been to reduce flow resistance in felts by using transversal flow presses. Fabric presses, sleeved presses and grooved presses have all produced large reductions compared to plain and suction presses used before. Dryness improvements of 2 to 10% have been reported". (His p. 368.)

In a paper presented at the same symposium, Morrison [44] concentrated attention on the influence of felt structure upon water removal at the wet press. He started by recapitulating the consequences of Wahlstrom's concepts of wet pressing upon felt properties, stating that "The Wahlstrom Theory postulates that:

(A) The between-yarn voids within a felt should be relatively large to allow transfer of water through the felt at the ingoing nip of a press.

(B) The within-yarn capillaries (particularly at the surface of the felt) should be as small as possible. In this way, the capillary force of the felt would more nearly equal that of the capillary force of the paper web; thus miminising the transfer of water from the felt to paper web at the outgoing nip.

(C) A compressible but resilient felt is essential to provide a narrow nip width (high specific nip pressure), and yet allow rapid recovery to facilitate redistribution of water from the between-yarn voids to the within-yarn capillaries.

(D) The lengthwise yarns should be large and as far apart as possible to minimise the resistance to water flow in the machine direction. This aspect of the theory is somewhat dependent on the physical nature of the yarn if we consider the difference between hard or soft twist (large and small capillaries)."

Briefly, Morrison's work with a variety of felts corroborated requirements A, B, C and D as stated to be postulated by Wahlstrom. The observed relationship between felt structure (capillary size) and water removal established early in life continued through felt life. It was also found that by separating sheet from felt immediately, instead of allowing the sheet to stay with the felt approximately six feet, a 1—2% increase in sheet dryness was realised. This is a rare recorded example of proven rewetting in practical operation, as opposed to laboratory work, or theory: but it explicitly refers to an unduly prolonged contact between felt and wet web located well after Wrist's phase 3 or Nilsson and Larsson's phase 4. There is no reasonable doubt that this work reinforces Wahlstrom's conclusions as to desirable felt characteristics: it is less clear as to how much support it gives to the detailed underlying concepts from which those conclusions were drawn.

In another contribution at the same symposium, Schiel [45] described experiments to determine and define pressure distribution in the press nip and the sharpness of the nip under static conditions. He also carried out dynamic press investigations in machine runs. His work showed that a very distinct optimum pressing condition exists for every type of ingoing sheet. Even slight decreases in pressing time below the optimum level made very sharp increases in nip load necessary in order to regain equal dryness. On the other hand, if pressing time was too long, the influence of apparent rewetting also had to be compensated for by nip load increases.

*5.4.5 Critical comments and a return to basic considerations of the dynamics of wet pressing*

In 1975, an International Water Removal Symposium was held under the auspices of the Technical Division of the British Paper and Board Industry Federation. There were indeed many relevant papers published between this and the preceding 1968 symposium in Canada (some fifty or so could be listed), but as on previous occasions the symposium itself provided both a summary of the continuing work, and a forum for presentation of stimulating new work.

Westra [47], under the title "A new Constribution to Press Nip Analysis", discussed the work of Nilsson and Larsson [43] referred to above and suggested that the failure of their model to match practice was due to a mathematical mistake. Westra presented a new version of the model, which he claimed could predict quantitatively the result of pressing up to the point where rewetting starts. For the rewetting phase, only semi-quantitative results were obtained. The rather inconclusive summary to this paper pointed to the practical difficulties in the way of obtaining measurements of moisture ratio or hydraulic pressure in the nip. In this sense, the paper made a valuable negative contribution in indicating how far speculation in this subject has been allowed to develop into detailed mathematical theorising without any firm factual basis, or any hope of practical verification at the present time.

The British Symposium also contained some useful papers biased towards the engineering and felt design aspects of press operation [48,49], but special comment is appropriate for a most significant, well-prepared, and refreshingly sceptical contribution by Heller et al., entitled "Back to Basics in Wet Pressing" [50]. Their work was based on a technique of pressing handsheets on small felt swatches in a slow speed (3—60 cm $sec^{-1}$) press nip. It was found that in a plain press the felt always comes out of the nip drier than the paper. It was suggested that this may reflect the fact that major resistance to water removal arises from the sheet rather than the felt. Also the paper was wetter leaving the presses as the speed of pressing increases, and as the freeness of the stock was reduced, (i.e. as beating was increased). Coarse, rough felts were found to produce wetter paper than fine-textured smooth felts. This was considered due to more uniform transmission of the press load to the sheet with the smoother felt. Experiments with the multilayered sheets showed direct evidence of a substantial dryness gradient through the thickness of the pressed sheet. Studies with hydrophilic felt fibres and a study where a membrane filter was inserted between the wet sheet and the felt suggested that capillary competition was not a major factor in rewetting for the conditions of this study.

In discussion of their results, the authors considered their significance in relation to currently held theories of wet pressing, and started by surveying the past literature.

They stated that, in their view, little had been unequivocally established regarding the basic mechanisms which are operative in this vital part of the paper machine, and emphasised that with the many different mechanisms of mass and momentum transfer involved, the predominant mechanism(s) will depend upon the particular conditions of the observations. Their review of the literature had revealed a number of papers where the reliability of the observations was either questionable or not well documented, and where the authors of the papers tended to compensate by loose interpretation and unsupported speculation. They judged that the relevant literature in the field of wet pressing presents a confusing picture.

The authors continued by presenting an excellent summary of Wahlstrom's original concept of the successive phases between entry and exit of web and felt. They added that much of the emphasis in the literature for 15 years past had focussed on the rewetting aspects of Wahlstrom's theory, and surprisingly little had focussed on the entry side of the nip. Regarding rewetting, they emphasised the point which seems to be almost universally ignored (although it is sometimes implied in passing, as in Westra's paper), *that the evidence for rewetting is only indirect, and must remain conjecture* until some means is devised to measure the actual moisture content of a paper web in the mid-nip region.

They went on to review the three mechanisms that have historically been adduced to explain rewetting, for they did not allow their comments on the conjectural nature of the phenomenon to lead them to deny the possibility of its existence. They subjected the published work of Sweet [24], already discussed at some length by the writer, to critical comment. In particular, they commented that Sweet's reasoning in interpreting his results was only valid if the resistance to water flow in the paper web was zero or very small in relation to other factors. They pointed out that this might be so for the predried, wet strength tissue used by Sweet, but was unlikely to be true for any normal papermaking furnish. Busker [51] and Nilsson and Larsson [43] were cited as maintaining that the major controlling variable in wet pressing is the resistance to water flow out of the paper itself. (The writer must add that it is the dominating variable above all else, with highly beaten furnishes, which are so seldom referred to in the literature.) In further analysis of Sweet's results, they commented on the fact that two felt/paper interfaces had not, as should be expected, doubled the apparent degree of rewetting. They maintained that none of Sweet's data unequivocally proved either the fact of rewetting or the validity of the capillary competition as a re-wetting mechanism. They added that the question of the time available for rewetting had already led to serious criticism of Wahlstrom's theories of capillary induced rewetting. Wrist [27] and Busker [51] had referred to the slow action of capillary force, in relation to current machine speeds. Wilder [52] had contended that there was ample time for capillary transfer, but the author's examination of Wilder's data showed certain mathematical errors which considerably affected the magnitude of the viscous forces. They could not accept the validity of Wilder's conclusions, and concluded their review of literature references to rewetting by stating that there is inadequate evidence for a definitive answer, and that if there is rewetting then all three mechanisms — capillary transfer, film splitting and differential pressure on expansion, may occur simultaneously.

The authors went on to deplore the virtual absence of literature reference to the important aspect of recovery of both felts and wet webs from compressive deformation, a lack earlier emphasised in this Section. After then considering with favourable comment the positive contributions of Nissan [19], Wahlstrom [42], Wrist [27], Schiel [45], Busker [51] and others, the authors concluded their rather devastating literature survey with the statement — somewhat generous in the circumstances: "As these, and other studies show, headway is being made in developing a basic understanding of wet pressing fundamentals. However, much more needs to be done in a careful well-documented manner before a complete understanding can be developed".

The body of their paper described the meticulous care taken in their own well-documented experimentation. By choice of wet webs ranging from free beaten tissue, through moderately beaten kraft, to wet beaten glassine furnish, they demonstrated attention to their own warning of the need to cover a wide variation in major variables. Their results were presented with-

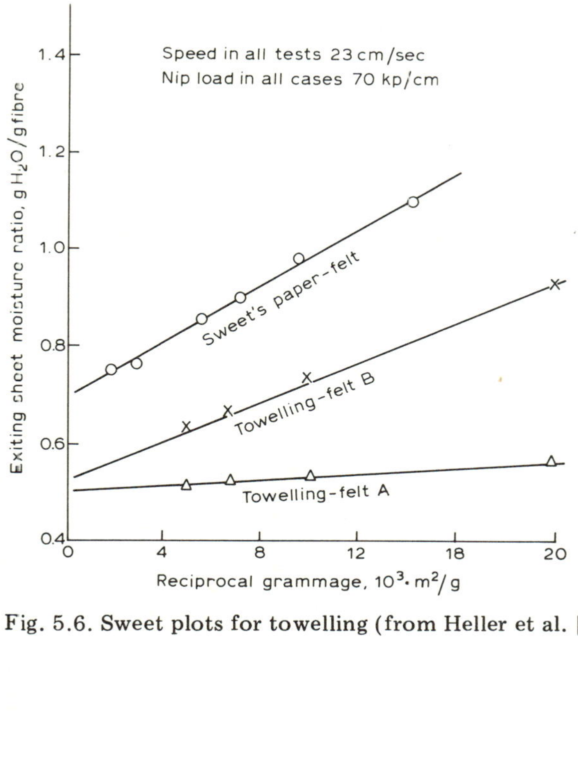


Fig. 5.6. Sweet plots for towelling (from Heller et al. [50]).

out pre-conceived intent of interpretation. They followed the mode of presentation already indicated by Sweet [24], relating existing sheet moisture ratio to basis weight and reciprocal basis weight, but declined to be led into unilateral interpretations.

The main body of their results was concentrated into two figures, which contain so much material of significance to the subject that they are reproduced here in Figs. 5.6 and 5.7. Their most striking feature is the dramatic effect of change of web furnish upon all data and derived curves. Their "towelling" was a pre-dried wet strengthened sheet introduced in a successful attempt to stimulate Sweet's work. The data and curves for kraft and glassine extended the scope of the work through what might be called "normal" furnish, to a near extreme of wet beating. For a fuller consideration and discussion of these results, the reader is recommended to study the original text. Comment here will be confined to a few key questions.

What is the significance of the slope and axis intercept of the plots relating

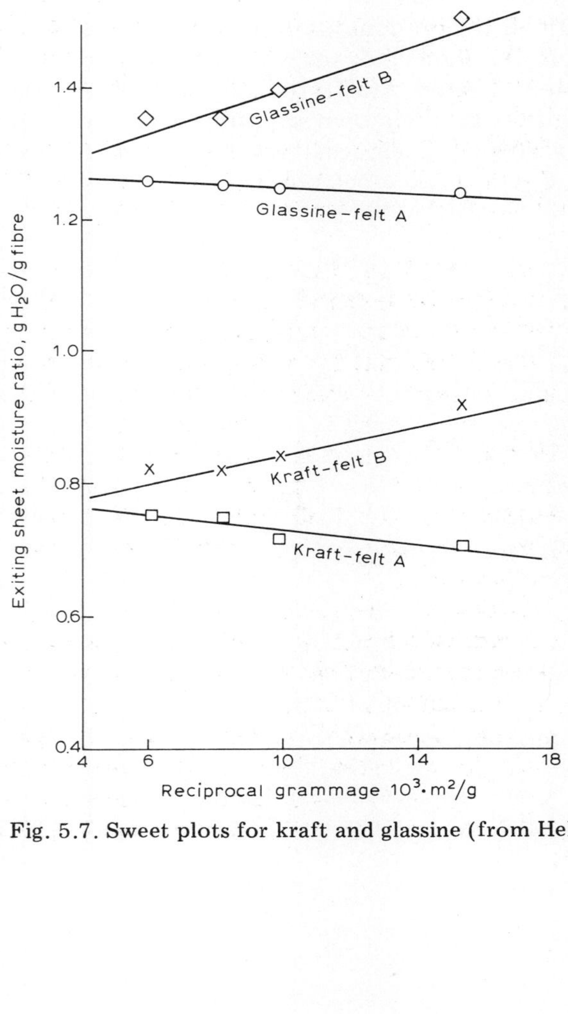


Fig. 5.7. Sweet plots for kraft and glassine (from Heller et al. [50]).

exit sheet moisture ratio to reciprocal basis weight? Sweet's interpretation, and that of most who have cited his work is that the slope of these plots is the amount of rewetting and that the axis intercept is the mid-nip dryness. This, the authors state with some restraint, seems to be an over-simplification. The writer would go further and say there is no evidence whatever for such an interpretation. It is only a speculation, which could be matched by alternative speculations. Cautiously, the authors suggest that the slope of such plots is related to the felt used, and that the axis intercept appears to be primarily a function of the type of wet web, since it is the same for both felts used. That is all that can be unequivocally stated. In any case, the slopes of the plots is sometimes positive, sometimes negative, depending on the felt used. The authors rightly emphasise that felt structure and surface smoothness, as well as sheet furnish, is a primary determinant of the relationship.

As far as the main body of data is concerned, the authors preferred to remain non-committal as to their implication regarding rewetting. However, the investigation into the dryness gradient through the thickness of the sheet, which showed a large and progressive change from top to bottom, showed no discontinuities at the felt/paper interface, such as one should expect if rewetting were occurring. Nor did the study of hydrophilic/hydrophobic balance in the felt produce any evidence which would support a "capillary competition" mechanism of rewetting.

The authors were content to leave most of their questions unanswered, and many possible interpretations undefined. This is an admirable characteristic of an altogether excellent investigation, soberly and sensibly carried out and reported.

Between the 1975 International Symposium and the time of writing, there have inevitably been further relevant publications, at the usual rate of six or seven a year. None has been of major significance, but several are worthy of study. Wiseman [53] published in 1976 an excellent "popular" account of wet pressing, which avoided too much back reference or mathematical interpretation and presented in a simple manner some of the key relationships in practical operation of presses, including some intriguing references to temperature effects. While following the general Wahlstrom interpretation of events in the press, he avoided polemics or any alternative speculation as to the underlying mechanisms of press operation, which would indeed have been out of place in his simplified exposition.

More recently, Wiseman [54] published a paper on the effect of press loading on water removal in low speed web pressing with a fabric press, in which he presented the results in accord with the conventional rewetting concepts, using the Sweet method of presentation. The results could however well have been interpreted otherwise, in the light of the paper of Heller et al [50]. As they stand, they are additional useful data to fill in the gaps in a very complex matrix, rather than affording evidence for one or other interpretation of pressing mechanisms.

So must rest a somewhat inconclusive account of the dynamics of wet pressing. It would be wrong to suggest that the work of the last quarter century has provided a reliable working model to enable us to predict wet press performance in terms of measureable variables. It would be equally wrong to contend that no progress has been made to this end. The words of Heller et al., cited on p. 241, offer a fair statement of the level of progress of the science or the art of the subject.

### *5.4.6 Consolidating effects of wet pressing*

The preceding sub-sections on wet pressing have been largely though not exclusively concerned with the dynamics of water removal rather than consolidation as such. The writer has felt it necessary to deal at length with the extensive body of reported work on the mechanism of water removal at the presses, since such removal is an essential requirement for this vital stage of consolidation, even though consolidation as such is often ignored, by those working in this field.

Wiseman [53] has gone so far as to say "Wet pressing must therefore be regarded primarily as a water removal process" in a paper which nevertheless carried an excellent photograph of a perfect example of localised web consolidation caused by pressure of the felt. Water removal can only be seen as the primary function of pressing, if it is the primary function of the whole papermaking process. In fact, the primary function of the papermaking process is to make paper from a suspension of fibres, and it does so through a succession of steps of consolidation from slice to reel up. It is also important to recognise that each step in the consolidation process puts its mark on the drying web, and that each step is the stepping-off point for the next step in consolidation. Until drainage compaction and wet press consolidation have taken place, the consolidation effects of drying and calendering cannot take place. In the fullest sense of the words consolidation is the primary objective of wet pressing, as of all other component processes of the paper machine. One hopes that more direct attention will in future be turned to this primary objective.

In simple fact, wet pressing increases the solid content of the wet web generally from about 20% towards about 45%. This increase in solid content is generally accompanied by a substantial compaction and collapse of the fibrous components of the web, and a corresponding substantial decrease in thickness of the wet web.

The compaction of the wet web by static pressing has been studied to a fair degree by a number of workers in addition to those already cited on p. 229.

Doughty [55—57] carried out in the earlier 1930's an extensive pioneer study of the relationships between wet pressing of handsheets and the properties of the dried sheets, using unbeaten spruce sulphite pulp. His results showed that sheet density, and tensile, folding and bursting strengths increase with increased force of pressing. Sheets made from shorter fibres were denser and stronger than sheets made from longer fibres. The amount of lateral shrinkage during drying increased rapidly to a maximum under low pressures, and then decreased at higher pressures. Doughty made the broad generalisation that wet pressing had the same kind of effect on dry sheet properties (essentially density and strength) as was produced by beating.

Brown [58] reported in 1932 results obtained from static pressing of hand sheets whose thickness dimensions were measured immediately after compression, and then at a certain time after removal of the compressive force. This seems to be the earliest record of attention being paid to thickness recovery of the wet web after compression, a matter of considerable importance both for the dynamics of water removal by wet pressing, and for the consolidating effect of wet pressing. Brown derived a numerical value which he termed the "spring back" of the pulp used in making the handsheet. He related this "spring back" to various properties of the subsequently dried paper sheet.

Seborg et al. [34] whose work has already been briefly mentioned in connection with the dynamics of water removal during wet pressing, reported in 1939 work based essentially on Brown's approach, but considerably more detailed and comprehensive. Their work stands out both as a rare example of and as a major contribution to our knowledge of this vitally important property of dimensional recovery of the wet web after compression.

They showed that as beating increased, equilibrium wet sheet density under pressure was approached more slowly; the sheet was more difficult to compress. On the other hand, the equilibrium sheet density after relaxation from pressure was approached slightly more quickly. The equilibrium sheet density under pressure was largely independent of the degree of beating, but the degree of recovery leading to reduction in density after removal of pressure decreased substantially with beating: recovery was both slower and much less in degree. Sheets made from very highly beaten stock showed virtually no recovery, whereas sheets made from unbeaten stock showed recovery generally equivalent to 35—40% of the compressed sheet thickness. Sheets made from unbeaten Southern pine kraft and Loblolly pine kraft showed recoveries even as high as 85—100% of compressed sheet thickness: clear evidence of the important role played by the stiffness of the fibres in causing recovery.

Breitveit [59] presented some intriguing findings in 1949, coupled with some speculation, regarding the effect of wet pressing upon the anisotropy of the paper web. He suggested that backward flow of water through the paper web in a plain press could orient fibrils in the machine direction to a degree which could significantly increase anisotropy of the ultimate strength properties of the sheet. Although this unusual suggestion is to large degree irrelevant to most papermaking today, since plain presses are now seldom used, it is interesting to note its conformity with the findings of Redfern and Gavelin [40] cited on p. 232 regarding the preferential machine direction alignment of fibres over the land areas as opposed to the hold areas, when using a suction press.

In 1954, McLaurin and Whalen [60] published a paper on the effects of wet pressing on sheet properties. They emphasised that wet pressing not only removes water from the wet web, it also plays an important role in determining the properties of the finished sheet. They stated that a typical Fourdrinier sheet enters the wet presses at about 16—20% solids and leaves at about 30—35% solids, (figures appropriate to paper machines in 1954). They carried out work relating pressure, number of pressings, pulp freeness and dry sheet properties for unbleached Douglas-fir kraft handsheets made from stock beaten to different degrees. They concluded that for the purpose of increasing density and strength through wet pressing, two or possibly three pressings, to a water content of about 50% gave optimum results.

Pye et al. [61] reported in 1966 the findings of a study on structural changes in paper on pressing and drying. Their work was primarily concerned with direct observations of changes in the structure of paper, using the scanning electron microscope. In addition to laboratory-made handsheets, samples of wet webs taken from various positions along the paper machine were observed, after having first been frozen and later dried under vacuum. Their impressive photo-micrographs showed how, under pressure, fibres are deformed plastically, especially at crossing points and asperities. The machine made samples included bag-kraft, bond paper, and newsprint.

The photographs of the bag-kraft web showed that the first press caused fibrils and fibril laminae to aggregate or collapse on to the surfaces of the larger fibres, which themselves also showed some degree of collapse, the whole structure becoming more compact. There were regions of pressure-induced distortions of fibres at crossings although, only after the second press, was a large number of fibres brought into really intimate contact.

In the case of the bond paper, the first press was seen to have considerably deformed the fibres, many having wholly or partially collapsed: the second press produced further deformation.

In the case of the newsprint, although the structure of the web was seen to have compacted significantly under wet pressing, with some bending of fibres to conform with each other, the stiff groundwood fibres (supported structurally by lignin) did not collapse and deform to the same degree as the more flexible chemical pulp fibres in the bag-kraft and bond papers.

No attempt can usefully be made here to reproduce examples of the photographs presented by Pye et al., since study of the full range presented in their paper [61] is required for one to see clearly the effects discussed by them.

Back and Norberg [62] in 1966 reported work on the effect of temperature on the compressibility of wet pulp pads (mechanical Asplund pulp) which included data for dimensional springback after deloading. The springback was highly dependent upon temperature. At a pad moisture ratio of 0.50, it decreased from 44% at about 10°C to 27% at about 90°C. At a pad moisture content of 0.80, it decreased from 18% at about 10°C to virtually nil at 90°C. The density and tensile strengths of the products from drying of the wet pads were correspondingly increased by increase in pressing temperature.

The conformability of the wet web surface during wet pressing is strikingly illustrated by electron photo-micrographs shown by Wiseman in a previously cited paper [53], and reproduced here in Figs. 5.8 and 5.9. These show the structural impression of a batt-on-mesh felt produced on a hot-pressed refiner groundwood sheet, under low and high magnification, respectively. They demonstrate not only the conformability of the web surface, but also the unevenness of the distribution of press loading, dependent on the character of the felt.

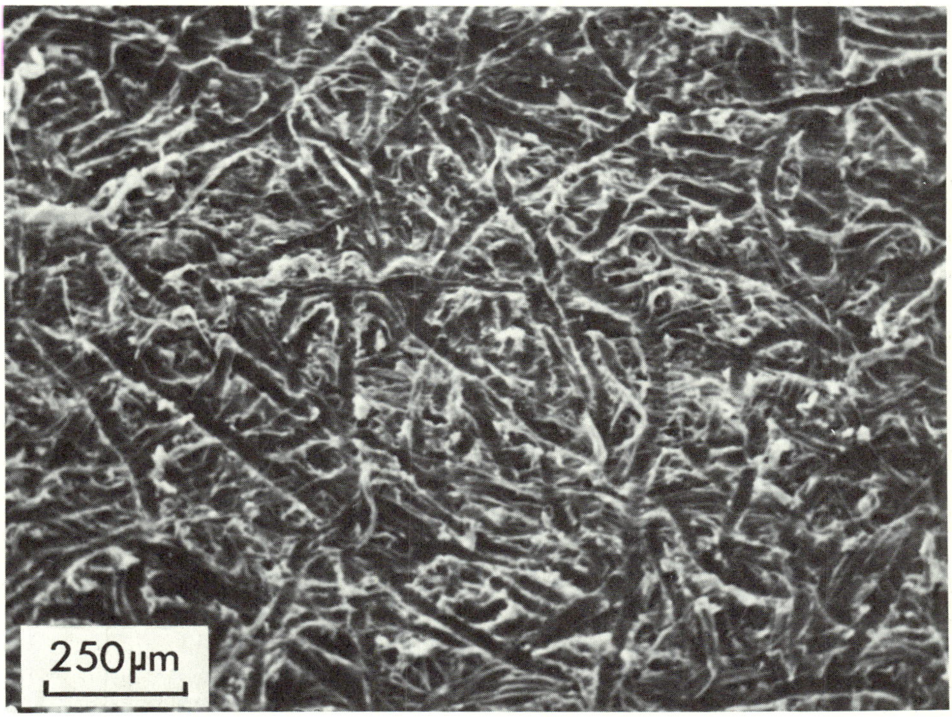


Fig. 5.8. Structural impression of batt-on-mesh felt produced on a hot pressed refiner groundwood handsheet: low magnification: showing the proportion of contact area (from Wiseman [53]).

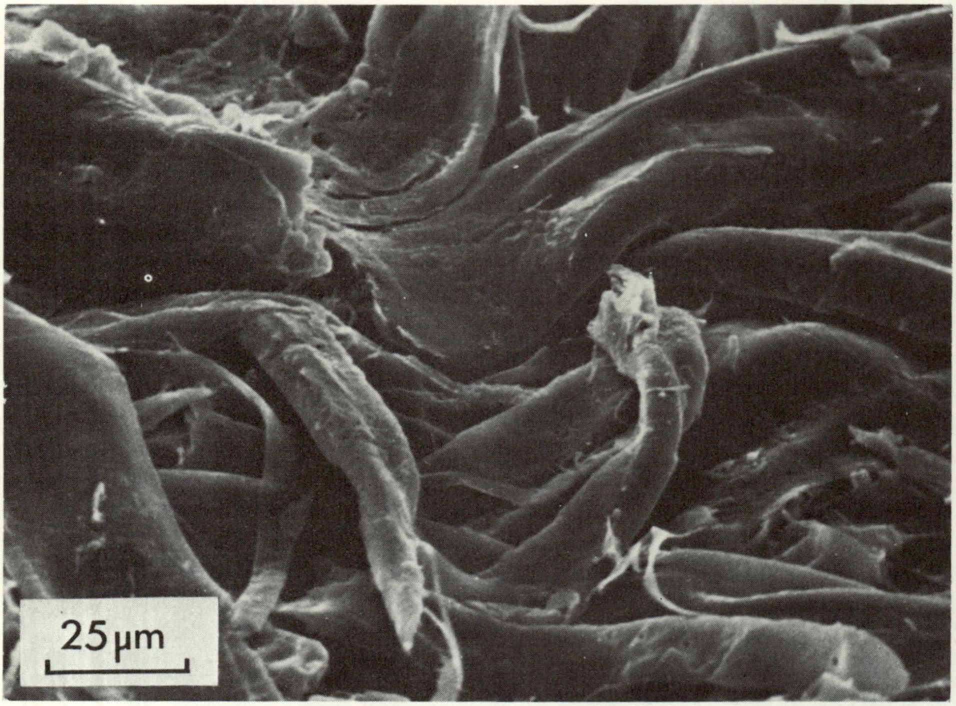


Fig. 5.9. High magnification of an area to right of centre of Fig. 5.8, showing how fibres have undergone plastic deformation as a result of intense local pressure (from Wiseman [53]).

Finally, a very recent paper on wet pressing by Oliver and Wiseman [63] was presented at the 6th Fundamental Research Symposium in the U.K. This incorporated some of the material from Wiseman [53], and enlarged on the significance of felt roughness in compaction of the wet web in wet pressing. Impressions of actual contacts between felt and the paper fibres in handsheets during wet pressing were produced by heating the press to 125° C. The felt impressions were examined under a scanning electron microscope, and values of $F$, the percentage of felt/paper contacts were estimated from low magnification micrographs. $F$ was found to be very low, varying from 25 to 33% for a variety of commercial felt structures.

The studies demonstrated that irregularities in the felt surface can deeply impress and permanently deform fibres in the sheet. It became evident that at the scale of papermaking fibres, wet press felts are very rough, and that the pressure exerted on the paper is far from uniform. The authors considered that substantially improved water removal might be achieved if a felt could be produced which would distribute the load more evenly over the surface of the paper, quite apart from the question of uneven surface compaction and distortion.

Wiseman entered the caveat that application of heat during pressing would inevitably exaggerate fibre and wet web deformability due to felt roughness. The results of Back and Norbert [62] already mentioned, on the reduction of springback caused by press heating are relevant in this context. At the same time Wiseman pointed to the above discussed work of Pye et al. [61] as evidence that such deformation does occur, albeit at a lesser degree, at ordinary pressing temperatures.

## 5.5 THE ROLE OF SURFACE TENSION IN CONSOLIDATION

Surface tension, coupled with the viscosity and strength properties of water, is an ever present determinant of wet web consolidation from the earliest stage of drainage on the machine wire, right up to the removal of the last fraction of water in the drying section. It is a primary factor in the suction exerted by tube rolls and scrapers on a Fourdrinier drainage wire; it helps to hold down the compacted formed web after drainage on the wire; it enters into practically every aspect of consolidation and water removal at the presses; it fixes to a large degree the compaction produced by pressing, and it influences the process of consolidation of the web in the early stages of the dryer section.

Over and above these effects, it cannot be too strongly emphasised that surface tension is a major factor contributing to the integrity of the wet web at all stages while free water remains, since it provides the main force which holds together the fibrous components which would otherwise be very loosely felted together as in a dry laid web.

Without surface tension, the fibres would not be drawn down together and held in a compacting mass on the drainage wire. Without surface tension, it would not be possible to transfer the wet web still containing about four parts of water to one part of fibre, from the wire to the press section. Without surface tension, it would be difficult to transfer the more compacted but still very wet web between the presses and from the last press to the dryers. Furthermore, it is largely the force of surface tension which prevents to a major degree elastic recovery of the wet pressed web after emergence from each press. Mechanical pressure brings about wet press consolidation, but surface tension, coupled with viscosity and strength of water maintains that consolidation until it can be fixed and further enhanced in the drying section.

At all stages, the force of surface tension is combating the force of elastic recovery of the wet web. In the case of webs made from free beaten stock, elastic recovery force at an early stage exceeds the surface tension force holding together the web as a whole, so that air can enter the web even while it is on the drainage wire: although even in this case, surface tension forces *between* fibrous components of the web still maintain a compacting influence. In the case of webs made from highly beaten stock, as in the case of natural tracing paper, elastic recovery may never exceed surface tension forces, which consequently maintain an "envelope" ensuring the integrity of the whole consolidating structure without any entry of air.

These factors were discussed by Campbell [8] in 1933 in a section of his classic paper on "The Cellulose-Water Relationship in Paper Making", entitled "A Physical Theory of the Beating Process". His insight into the sequence of physical events during papermaking was quite remarkable, and is still in general accord with the more refined and detailed understanding developed in more recent years. He visualised clearly the stage of compaction of the whole web induced by surface tension before the resistance of the fibrous networks to deformation breaks open the surface tension envelope to allow entry of air. He then visualised the next stage wherein individual fibres in the web are still drawn together by the surface tension of water remaining between them: and the final stage when the fibres have been drawn into contact, and are forcibly maintained in position by residual water menisci, so that molecular attraction forces can come into play at the cellulose-cellulose contact areas.

Campbell saw clearly that the magnitude of the surface tension force is independent of the dimensions of the fibres acted upon, but at the same time pointed out the dramatic dependence of the deflection that could be produced by this force, upon the dimension of the fibre or fibril. Thus a fibril with a diameter of one tenth of that of a wood fibre would be deflected 10,000 times as much, which is one reason why fibrillation of the fibre component structure substantially encourages compaction. He also logically deduced that the substantial increase in compaction caused by relatively small degree of beating, which can leave the fibres superficially intact, must be attributed to these fibres having been rendered more flexible, by what he called "internal fibrillation".

It was many years before Campbell's clearly expressed thesis became generally accepted, and expanded into a broad understanding which could incontravertibly be illustrated by both quantitative and qualitative observations. Stamm [12] discussed succinctly the significance of surface tension in papermaking, in the light of Campbell's thesis, emphasising the balance between surface tension and structural restraint, and quantifying the attractive force due to surface tension between fibres and fibrils of diminishing dimension. He stated that the attractive force between fibres $2 \times 10^{-3}$ cm in diameter is 87 p.s.i.; that between fibrils $2 \times 10^{-4}$ cm in diameter as 540 p.s.i.; that between fibrils $2 \times 10^{-5}$ cm in diameter is 2480 p.s.i. These high internal tensions between fibrils far exceed the normally applied external compression force on the paper machine, which is why beating can over-ride external factors as a determinant of consolidation once the fibrillous components have been brought close enough together. He also pointed out that the wet web shrinks appreciably before the fibre saturation point is reached because of a drawing together of the fibres as a result of surface tension forces. Below saturation or shortly thereafter, the fibres themselves begin to shrink.

In this same paper, Stamm dismissed the notion of hydration as a concept which confused the distinction between the gross physical effects produced by fibrillation and surface tension, and the subsequent physical effects of molecular bonding.

Lyne and Gallay presented a series of studies between 1949 and 1954 [64—66] dealing with the development of the strength and other properties of the wet web during drying. Both air drying and drying by pressing between blotters were employed to obtain a range of water-to-solids ratios. They deduced that surface tension forces were predominant in web compaction and strength development in the early stages of drying. Some of their data indicated a fairly critical demarcation at about 50% web solids content: for example the web breaking load and development of shrinkage tensions shown in Fig. 5.10, from [64]. Other of their data indicated an apparently critical water content at a web solids content of about 25%, at which some kind of fibre-to-fibre adhesion appeared to enter [66]. The authors were nevertheless careful to emphasise in this latter case that there was no sharp transition, and that surface tension forces continued to contribute to the developing properties of the web after their maximum value had been reached. They were also careful to distinguish between their concept of inter-fibre adhesion and the concept of inter-fibre bonding "as applied to the last portion of the drying web . . . these two applications of surface energy on strength development in the fibrous web are quite distinct in nature" [66].

In 1954, Rance [14] presented a paper on "The Effect of Water Removal on Sheet Properties: The Water Evaporation Phase". This paper was based upon both laboratory and paper-machine observations, coupled with data derived from previously published literature. On this basis, the author discussed a general thesis regarding web compaction during drying, involving two distinct and successive phases: first, compaction induced by surface tension: second, compaction induced by shrinkage of the fibrous components of the web during evaporation of bound water. The two phases were designated "Inter-fibre web contraction" and "Intra-fibre web contraction".

These two successive phases of web contraction must not be identified or confused with the successive phases of development of wet web strength reported by Lyne and Gallay. The relationship and overlap between these two related but differing sequences of events will be discussed in the section which deals with transition between surface tension consolidation and consolidation due to molecular bonding.

The central importance of Campbell's concepts for the surface tension phase in inter-fibre web contraction was acknowledged by Rance [14] whose

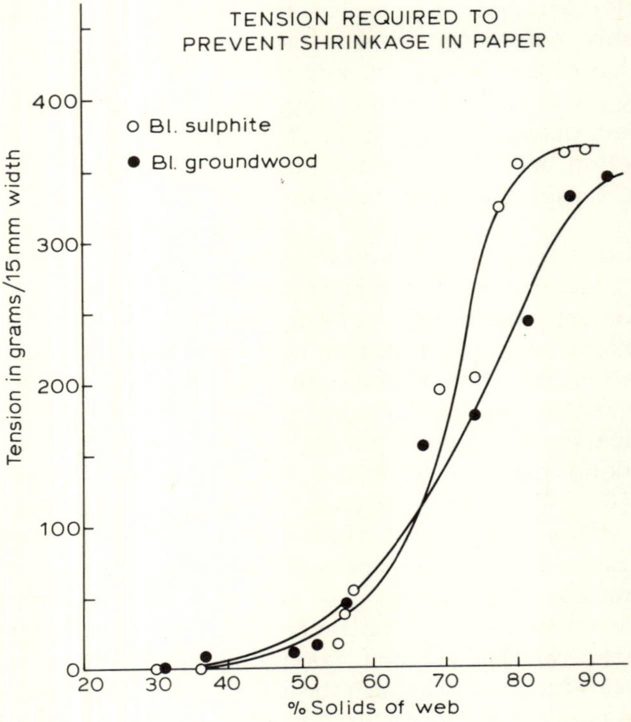


Fig. 5.10. Plots of tension required to prevent shrinkage, and of wet web tensile strength, against solids of drying wet web (from Lyne and Gallay [64]).

interpretation of this phase followed Campbell's views as expressed in [8]. It was visualised that at the beginning of evaporation, with a web made from very wet beaten stock, the only air/water interface is at the surface boundaries of the web, and in effect this interface acts as a contracting envelope, dragging the aggregate into a more compacted form. The fibres lying in the surface layer are pulled inwards to effect a compaction within the thickness of the web: at the same time, each of these surface fibres is subjected to a sideways pull towards its neighbours, so that there is from the outlet a slight lateral contractive force. Resistance to thickness contraction sets up a hydrostatic negative pressure within the water medium thus distributing the contractive forces evenly throughout the mass and reinforcing the lateral contractive force. Then the web contracts as a whole as though it were a composite body. With a web made from free beaten stock, air has entered the web at an early stage, well before the water evaporation phase, even before wet pressing. As indicated by Campbell the net contractive force acting on the web can in this case be considered as the aggregate of the contractive forces of the diminishing bridges between individual fibres. Surface tension was seen in both cases as a major determinant.

The permanence of the consequent contraction was seen on the one hand as dependent upon the ability of the fibres to conform by relative movement, which could give denser and irreversible packing without deformation of individual fibres. On the other hand, any contraction involving distortion of individual fibres was seen to require "fixing" by subsequent adhesive forces between fibres: outherwise it would be reversible after disappearance of the surface tension force.

From these considerations, it was deduced that irreversible surface tension contraction was least with long, strong, unbeaten fibres: and greatest with heavily beaten, fibrillated and fragmented fibres. This was in accordance with the observation made in 1950 by Glover and Pritchard, according to unpublished data cited by Rance [67] to the effect that a wet web made from wet beaten stock can contract laterally to a considerable degree, long before the water content is reduced to the point where the individual fibres begin to shrink. The same work showed that the water content, at which lateral contraction or lateral shrinkage tension begins, rose progressively as the degree of beating rises. To cite actual figures, the water content of the web had to fall to a value below 1 : 1 water: solids ratio before lateral shrinkage tension appeared in a web of unbeaten sulphite pulp, whereas it appeared at a ratio approaching 2 : 1 in webs of the same pulp beaten to 90° Schopper Reigler. Note that these water: solids ratios at which lateral shrinkage began were at levels which normally precede entry to the drying section of the paper machine. They confirmed the observable fact that lateral web shrinkage, apparently caused by surface tension (and probably reinforced by machine tensions on the paper machine itself) starts within the draws of the press section, well before the start of drying by evaporation. This is especially observable on a paper machine making paper from highly beaten stock.

Conversely, the start of lateral web contaction can be considerably retarded by physical hindrance, as shown in work published in 1954 by Pritchard [68] on the relation between lateral shrinkage of pressed sheets left to dry freely after pressing, and the hygroexpansivity of the dried sheets. The findings in that work relevant to the present subject matter are well summarised in Fig. 5.11, taken from Pritchard's paper. In this figure, the total linear web contractions during free drying after pressing at different press loads were plotted against final sheet density after drying, for each level of applied pressure (broken lines), and isometric lines were drawn each representing a different degree of beating. With free beaten stock, where web contraction enters at a fairly late stage, the load applied during pressing had little effect on lateral contraction, although it had a considerable compacting effects on the thickness of the sheet and hence on sheet density (note the horizontal isometric lines). With wet beaten stock, the load applied during pressing had little effect on thickness or final sheet density, presumably because the

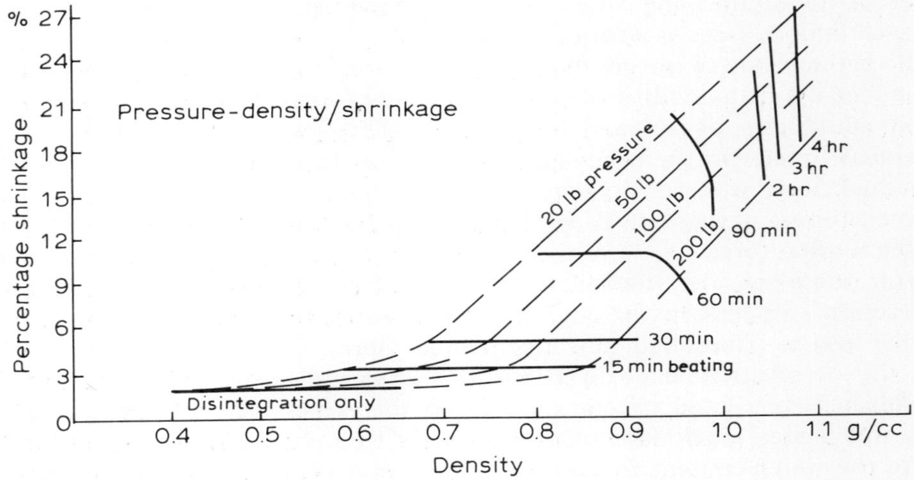


Fig. 5.11. Effect of Pressing on wet web density and wet web planar (linear) shrinkage at different degrees of beating (from Pritchard [68]).

internal contractive forces were great enough to over-ride other factors: but the pressure did effect considerable hindrance to lateral contraction of the wet beaten web, presumably by preventing the web contraction which otherwise would have occurred at the relatively high water contents during pressing.

These shrinkages recorded by Pritchard covered the whole period of water evaporation including that taking place after removal of free water: but the events depicted included also the whole of the phase of lateral contraction taking place in the presence of substantial amounts of free water, and therefore properly attributable to surface tension forces.

At the 1965 Fundamental Research Symposium on Consolidation, Robertson [69] presented data on the development of various properties in drying wet webs, derived from work akin to that of Lyne and Gallay [64—66], but accompanied also by a parallel study of water retention isotherms (involving the use of pressure to force water through pulp pads). The results of the wet web drying experiments were presented in the form of curves relating wet web strength, shrinkage etc., to the solids content of the web. He noted (see Fig. 5.13, p. 260), an inflexion point D corresponding to about 45% solids (55% web/water content) for never dried pulps and to about 60% solids (40% web/water content), for once dried pulps. He also plotted the water tension in his isotherm experiments against solids content, and obtained curves of similar shape, showing apparent inflexion at 40—50% solids. He concluded that the observed changes in the wet web properties (strength, shrinkage etc.) were initially related to the increasing tension. However, the testing of wet webs immersed in dioxane showed a development of strength that he attributed to some form of interfibre cohesion or friction, which begins to develop at about his noted inflexion point D. Also the effect of reducing surface tension at various stages of wet web drying (by use of surfactants) gave evidence of a new type of cohesion superimposed upon the surface tension effect after D.

Robertson was very cautious in his interpretations. He clearly visualised surface tension as the main cause of development of strength and shrinkage up to a web water content of the order of 50% (1 : 1 water: solids), but was non-committal as to the nature of the greater rate of development of these properties thereafter.

It should be noted that Robertson's inflexion point D, seen by him as a possible transition point from surface tension consolidation to some other mode of consolidation, was at about the same wet web water content as indicated by some work of Lyne and Gallay [64], but at a much lower water content than the inflexion point reported by Lyne and Gallay in 1954 [66], which corresponded to a ratio of 3 : 1 water to solids.

Further comments are made on Robertson's findings on p. 260, in the next section dealing with transition between modes of consolidation.

That surface tension forces initiate web shrinkage, at least with webs made from beaten pulps, is now generally accepted, and has not been challenged by any serious weight of evidence during the last twenty years. For completeness, however, there must be recorded a dissenting view. Tarnawski [70] published work in 1972 which led him to conclude that sheet shrinkage arises only from fibre shrinkage transmitted through the network by fibre bonds. He found the web shrinkage was independent of drying temperature, and concluded that shrinkage was not a function of surface tension forces as had been reported earlier. Byrd [71], accepting Tarnawski's conclusion, pursued the concept of web shrinkage energy as an index of network fibre bonding, but presented no new data supporting Tarnawski's conclusions.

It is possible that Tarnawski was working with sheets made from free beaten stock, where only a very small part of the wet web planar shrinkage occurs under the influence of surface tension. Even with wet beaten stock, a substantial part of the planar web shrinkage occurs after the removal of all free water. There is no reason to doubt the existence of an empirical correlation between web shrinkage energy and degree of fibre bonding, as proposed by Byrd: such a correlation must exist whether or not a part of the web shrinkage is attributable to surface tension, since the latter effect would affect only the pattern and precision of correlation without negating the fact of an empirical correlation.

Other workers, concentrating attention upon the intra-fibre component of planar shrinkage, have sometimes suggested that this is the only possible mode of planar shrinkage, and that planar shrinkage attributable to surface tension compaction does not occur. Since however, these workers have not adduced significant evidence to support this negative claim, the writer considers that citation of their comments would be purely polemical, and unjustified in this treatise which is essentially intended to review *positive* observations and relevant speculation thereon.

## 5.6 CONSOLIDATION IN THE DRYER SECTION OF THE PAPER MACHINE

The mechanisms involved in water removal at the wet presses have been discussed in considerable detail in this chapter, for although in a sense water removal at the presses is subordinate to consolidation at the presses, it is a process unique to papermaking. An understanding of its mechanisms cannot be found outside the literature of papermaking, and it has therefore properly been dealt with here in detail.

The case is different with the thermal drying of paper. The mechanisms of the drying processes used in papermaking are to a large extent common to the general body of engineering knowledge on drying, heat transfer, thermal efficiency and ventilation. Certainly, as Corte has made clear in Chapter 1, the peculiarities of the cellulose-water relationship introduce complexities beyond those encountered in the drying of many other materials, but he has to a large degree dealt with those complexities in that Chapter. It is therefore proposed to present here only a brief statement regarding the drying processes as such, and as regards this part of the paper machine to concentrate attention almost exclusively upon the state of papermaking science in relation to the consolidation of the paper web during drying.

In brief, the removal of water in the dryer section occurs through the following mechanisms. Considering a conventional sequence of felted drying cylinders, at the beginning of the train there first may be some removal of free water by capillary transfer from the surface of wet web to the dry felt. There is no precise information on this mode of transfer, and in any case its existence and extent have little significance for the main theme of this chapter. Second, there soon enters the main mechanism of transfer of moisture from wet web to felt, involving condensation within the felt of water vapour evaporated from one side of the wet web. Third, water is evaporated from the thus wetted felt into air in the pockets between the cylinders. Fourth, water is evaporated from one side of the wet web into air, during the passage of the web through the pockets between the cylinders, and from both sides during passage through any pocket between cylinder sections.

The efficiency of the drying sequence is dependent upon such factors as the applied pressure which forces the wet web into contact with the hot surface of the cylinder, the efficiency with which water condensed within the cylinder is physically removed, the nature and conditions of the carrier felt, and the ventilation of the pockets between dryers. These are all conventional problems of heat transfer efficiency and ventilation, and merit no further comment here, although the technical literature of the industry abounds with articles, both theoretical and practical, on these matters.

For the purpose of this chapter, the significant facts are that during this drying sequence first the residual free water, second the bound water from the cellulose fibres, is removed, mainly by evaporation, to reduce the water content from the entry level of generally about 60% (1.5 : 1 water to solids), down to the 5—6% of the emerging dry paper.

Similar considerations apply with less conventional methods of drying, including M.G. drying on a Yankee cylinder with or without pre-dryers or after dryers, so called through-drying, and drying by application of radiant heat or some more exotic form of heat application. These alternative methods of drying do affect the pattern of consolidation insofar as they involve more or less physical constraint upon the sheet itself, and different patterns of successive steps of constraint; but they do not change the fact that the last 55% or so of water to be removed from the wet web, comprising the residue of free water and the major part of sorbed water, is removed from the wet web largely by evaporation.

At some stage during this water removal, there is a transition from evaporation of residual free water, to evaporation of sorbed water. The nature of this transition, and the consequences of this transition, are of great importance for the consolidation process. The amount of volume compaction during drying is clearly very much less that that which has taken place earlier on the machine. At entry to the drainage wire, the wet web contains about 100 parts of water to one part of fibre: at entry to the wet presses, it contains about 4 parts of water to one part of fibre: at the entry to the dryers, it contains only about $1\frac{1}{2}$ parts of water to 1 part of fibre. Yet the removal of this last fraction of water has more profound effects upon the consolidation of the sheet (here to be distinguished from compaction as such) than all of the preceding stages. It brings about the final "fixing" of all the consolidation that has gone before. It brings the entry of a new mode of consolidation of over-riding importance, which by firmly joining together the component elements of the web structure, transforms it from a weak, easily disrupted assemblage of fibres, into the strong, usable, durable artifact, the paper sheet. Moreover, it brings about a new mode of compaction which, though small in extent compared with all the compaction that has gone before, has profound effects upon the properties of the final product.

The key to these drastic effects of the removal of the last parts of water is the transition from consolidation induced and aided by surface tension, to consolidation caused by cellulose-cellulose molecular bonding. This transition will be considered in the next section.

## 5.7 TRANSITION FROM CONSOLIDATION INDUCED BY SURFACE TENSION TO CONSOLIDATION INDUCED BY BONDING BETWEEN CELLULOSE SURFACES

The thesis advanced by Campbell [8] and Stamm [9] and developed by Rance [14] at the Fundamental Research Conference in 1954 involved a clear logical distinction between wet web contraction caused by surface tension, and that due to the shrinkage of the fibres themselves. Rance [14] rejected the view advanced by some writers that there is no logical distinction between the two phases, and that they are manifestations of the same phenomenon at different ranges of dimension (a view re-advanced in more recent times by Wrist [72]. He considered that this view ignored the essential difference in nature between the fibrous and molecular domains, and the gross difference in order of dimension between these domains, even after fragmentation and fibrillation due to beating. He recorded data relating lateral web shrinkage to moisture ratio for wet webs during drying on a paper machine of a web made from wet beaten stock. These showed (in round figures) that reduction of the water : solids ratio from 4 to $1\frac{1}{2}$ was accompanied by only $\frac{1}{2}$% lateral shrinkage of the wet web: reduction from $1\frac{1}{2}$ to $\frac{1}{2}$ was accompanied by 3% lateral shrinkage: removal of the last $\frac{1}{2}$ part of water was accompanied by nearly 4% more lateral shrinkage.

These data were presented in the form of a graph showing points of inflection at moisture ratios of $1\frac{1}{2}$ : 1 and about 0.35 : 1, (Fig. 5.12). It was sug-

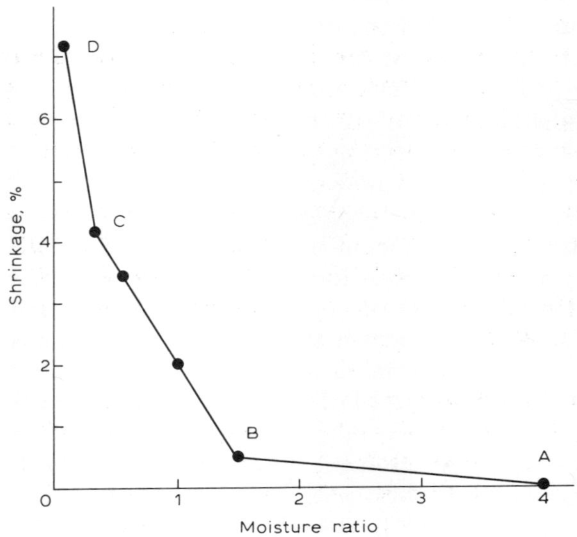


Fig. 5.12. Relation between water content and cross-direction shrinkage on paper machine of wet web made from highly beaten stock (from Rance [14]).

gested that A to B represented an initial stage of surface tension contraction largely within the thickness of the web, terminating when resistance to further thickness contraction sets up forces of planar contraction strong enough to overcome planar restraint. B-C was considered to correspond to a second stage of surface tension contraction, with the web contracting in both thickness and in area. C-D was considered to represent contraction communicated to the web by contraction of component fibres during removal of bound water, and consequent upon fibre-to-fibre molecular bonding (so called intra-fibre contraction). Communication of fibre shrinkage to the web through fibre-to-fibre bonds was considered to introduce local structural distortions through axial deformation in the crossing fibres. Such deformation was considered to involve compression of the crossing fibres in a sheet made from wet beaten stock, and a micro cockle in the case of a sheet made from free beaten stock. With a highly compacted sheet, "diametric shrinkage of the fibre could be accommodated only through axial compression of the fibres in the transverse lamina".

Figure 12 was criticised by Tydeman et al. [73] on the grounds that a smooth curve could fit the data presented just as well as a succession on straight lines indicating sharp points of inflexion. This is perfectly true, and indeed it is doubtful whether any modal transition would show up as a sharp transition in practice. There must inevitably be an overlap of modes, especially in the drying of a heteregeneous material such as paper: this comment applies equally to most of the reported data on this subject indicating critical transition points.

Nevertheless, it remains undeniable that Rance's data showed that the shrinkage per unit of water removed rises very sharply in the final stage of water removal, and that about one half of the total linear shrinkage occurred during removal of the last one half part of water (i.e. at web water contents below about 33%). The textual explanation given for this succession of events in 1954 by Rance as outlined above is still considered by the writer to be valid.

Despite their proper criticism of the detail of presentation of Rance's data, Tydeman et al. [73] confirmed his general findings, presenting data based on handsheet studies: these confirmed the essential features of the shape of the curve, with very low shrinkage rate at the wet end of the range, increasing and becoming very high during removal of the last 20—30% of the water.

Stamm's already cited estimates of up to 30% water on solids for fibre saturation [9], based on water vapour sorption and other studies, and supported by similar estimates from Heath and Johnston [74] and from Seborg [75] were accepted by Rance as the water content critical for demarcation between phenomena related primarily to the presence of molecularly bound water (notably web shrinkage promoted by intra-fibre shrinkage), and phenomena related primarily to the presence of free water (notably inter-fibre web shrinkage caused by surface tension). He emphasised that inter-fibre bonding marked the end of so-called inter-fibre web contraction, and the start of so-called intra-fibre web contraction, these two phases of web contraction being "sympathetically" linked through fibre-to-fibre bonding.

On the other hand, Lyne and Gallay's already cited work published at the same conference [66] on the development of wet web strength indicated a transitional water content of the order of 75% (3 : 1 water to solids). Some of their earlier work, already cited [64] on the development of both strength and shrinkage tension of the wet web had indicated a transitional water content of the order of 50% (1 : 1 water to solids).

Robertson, in his 1966 paper already cited in the last section, [69] reported evidence from studies on wet web strength development. They will now be discussed in more detail. Figure 5.13 is a schematic curve from Robertson, showing the development of strength as the wet web was progressively dried.

In Robertson's view, this figure should be interpreted as follows:

| | |
|---|---|
| AB | Web consolidation without air intrusion |
| B | Air intrusion followed by |
| BC | Removal of inter-fibre capillary water |
| C | First strength inflexion when removal of free inter-fibre capillary water is complete. |

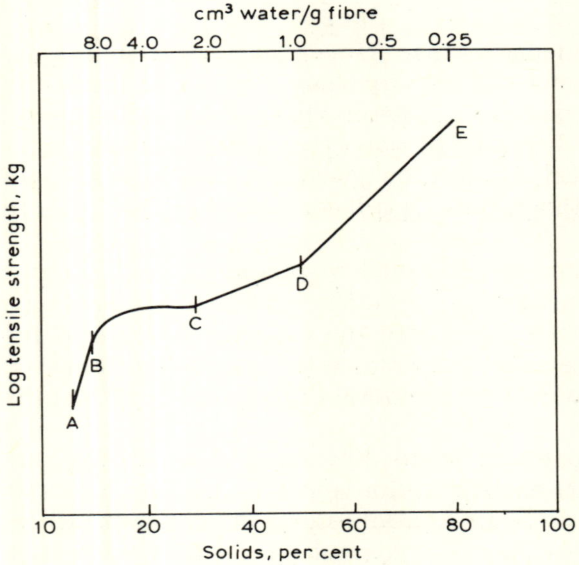


Fig. 5.13. Schematic curve showing the development of sheet strength as the wet web is progressively dried. A, at formation; B, at air intrustion; C, first strength inflexion; D, second strength inflexion (from Robertson [69]).

CD Removal of water associated with fibrillation and lumen.
D Second strength inflexion
DE Onset and development of interfibre bonding.

C was regarded not only as the end of removal of the last free inter-fibre capillary water, but also as the start of removal of the first of "associated" water. D was regarded as the start of inter-fibre bonding (though later in the text Robertson had cautious reservations as to the meaning of this phrase), at which also planar shrinkage in webs of unbeaten fibres became evident, and surface tension effects appeared to diminish. For never-dried chemical pulp, D occurred at a moisture content of about 55% (on total web weight). For once dried fibres, it was said to occur at a moisture content of about 40% (on total web weight).

Robertson's inflexion point C might appear to correspond to Lyne and Gallay's reported transition point [66] at a water content of 75%, though his interpretation was different from theirs. Similarly his inflexion point D (0.7—1.2 parts water to 1 part solids), seen by him as marking the possible transition from surface tension effects to some kind of inter-fibre adhesion or bonding, might appear to correspond to Lyne and Gallay's earlier reported transition point [64] at a water content of around 50% (1 : 1 water to solids).

All of these suggested transition points were considerably higher than the 25—30% water content (0.25—0.3 parts water to 1 part solids) accepted by Rance from publications of Stamm [9] and others on water vapour sorption, heat of wetting, etc., as a measure of the fibre saturation, and as marking the transition between web contraction induced by surface tension, and web contraction communicated to the web by fibre shrinkage.

Christensen and Giertz [76] in a review paper on cellulose-water relationships presented at the 1965 Fundamental Research Symposium on Consolidation discussed at some length the recorded literature on this subject, and made the significant comment that "the transition between the first and second phase as outlined here, is not sharp when drying paper, board and pulp sheets. In technical drying, the surface of the sheet will always dry faster than the centre and the surface thus will change to the second phase while the centre is still in the first, resulting in irregular drying. Certain papers show "skin" effects and the irreversible hornification of pulps shows up even at an average dry solids content of 60—65%, whereas the fibre saturation point is about 75% dry solids."

They went on the record results from some drying experiments where the drying was performed slowly by bringing all fibre to the same dry solids content, step by step. The wet webs were dried in equilibrium in atmospheres of different humidities, and they claimed that this technique enables the two postulated phases of drying to be separated. The opacity of the sheet at the successive stages of drying was used as the indicator of the effects of drying of a strong sulphite pulp. Figures 5.14 (from their paper) shows their results.

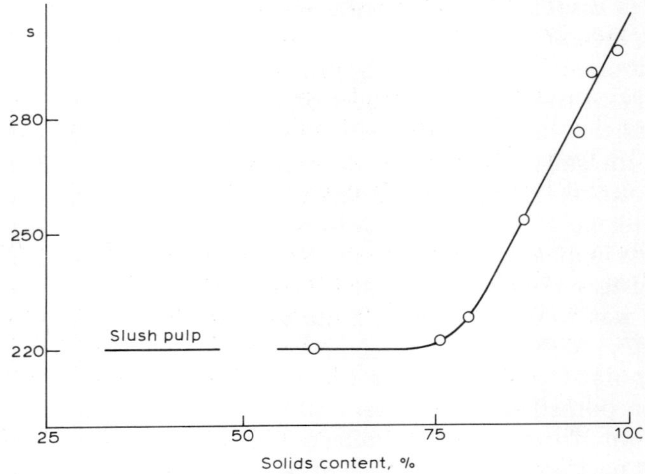


Fig. 5.14. Effect of drying on opacity of a wet web made from never-dried strong sulphite pulp, dried to equilibrium in atmospheres of different humidities (from Christensen and Giertz [76]).

The data show a distinct transition point at about 75% solids content (0.33 parts water to 1 part solids), a figure closely corresponding with the fibre saturation figure that had earlier been found for this type of pulp in sorption studies, heat of wetting, etc.

Nissan [6] in his recent "Lectures on Fiber Science in Paper" has also reviewed this subject, in a much wider context. He recorded that the fibre saturation value has been variously reported to be between 0.25 and 0.33 g water/g cellulose, although he commented as Stamm had earlier commented on the difficulty of obtaining this parameter accurately through sorption studies since capillary condensation follows adsorption without a break. He cited a saturation value of 0.25 g water/g cellulose reported in 1959 by Morrison and Dziecuich [77].

Stone, Scallan and coworkers carried out a wide range of studies between 1965 and 1972, aimed at determination of the water content of cellulose fibres at saturation point [78—84].

Their earlier work [78—80] used the technique of nitrogen adsorption at —195°C on pulps dried by solvent exchange, applying the Kelvin equation isotherm in the region of high partial pressures, where liquid nitrogen condensation within the pores is presumed to occur. This work was discussed in detail at the 1965 Fundamental Research Symposium [80].

Their nitrogen adsorption data suggested that in the fully swollen wall there are up to several hundred lamellae, each of the order of 100 Å thick, with a median separation of about 35 Å. During the drying, the lamellae appeared to draw together into thicker and thicker aggregates, decreasing the total pore volume, but leaving approximately the same median separation in the remaining spaces.

On drying from very high moisture content, there was little change in the pore volume (nearly 0.2 $cm^3$ $g^{-1}$) until a value of 50% (1 : 1 water to solids) was reached. Thereafter, the pore volume decreased progressively to a negligible level at dryness. Stone and Scallan therefore deduced that this 50% was fibre-wall saturation level.

Closing up of the inter-lamellar spaces during cycling was said to take place from the outer surface of the fibre, passing in towards the lumen. Re-wetting was said to take place through the lumen, with progressive separation of the lamellae outwards from the lumen. It was suggested that the pores which closed irreversibly — an event in the early stages of drying — were located towards the outside of the cell wall.

At water contents at and below 28%, they found that about one fifth of the water content was held within the inter-lamellar spaces, the remaining four fifths being apparently molecularly bound or dissolved within the lamellae themselves. From this, they deduced that the removal of water through drying appeared to reduce the inter-lamellar water and the intra-lamellar water in equal proportions, resulting in postulated shrinkage of each of the lamellae at the same rate as that of the closing up each of the apparent inter-lamellar spaces.

The authors found that drying of the swollen fibres caused about one half of the pores to close up irreversibly so that on re-wetting, the fibres took up substantially less water, as little as 30% (0.4 parts water to 1 part solids). Beating caused these closed pores to re-open to a degree dependent on the nature of the drying: it was only completely effective when the conditions of drying were mild.

Stone and Scallan used this same nitrogen adsorption technique to determine the pore volume of wood cooked to different yields by the kraft process [78] and found that the detectable pore volume (expressed in terms of g water in the pores to g solids) increased progressively from 0.01 g/g in uncooked wood to around 0.6 g/g (38% on total weight) at yields of around 50—60%.

In a later paper [81], Stone and Scallan expressed suspicions of the inadequacy of solvent-exchange drying for preserving exactly the swollen cell wall structure, and hence upon the reliability of the pore volumes deduced from nitrogen adsorption date. They considered that some of the pores in the saturated fibre wall were closed during solvent exchange, so that the measured pore volumes were too low.

This paper started with a general consideration of the types of water that can be held in a cellulose pulp pad, and referred to the various methods for determining fibre saturation point (F.S.P.) that had earlier been reviewed by Stamm; by plotting properties such as mechanical strength, electrical con-

TABLE 5.2

Fiber saturation point of kraft cooked spruce (from Stone and Scallan [81])

| Sample | Yield (%) | Lignin (%) | Fiber saturation point ($g\ g^{-1}$) | |
|---|---|---|---|---|
| | | | Pressure plate [a] | Non-solvent water |
| KO | 100 | 27.0 | 0.40 | 0.42 |
| K20 | 92.4 | 27.3 | 0.67 | 0.70 |
| K35 | 89.0 | 27.6 | 0.74 | (0.66) |
| K50 | 80.0 | 28.5 | 0.86 | 0.92 |
| K65 | 77.8 | 28.2 | 0.94 | 0.94 |
| K80 | 70.4 | 25.2 | 1.06 | 1.08 |
| K95 | 61.6 | 19.3 | 1.16 | 1.22 |
| K110 | 53.4 | 12.3 | 1.21 | 1.28 |
| K125 | 48.7 | 6.5 | 1.14 | 1.21 |

[a] Water retention at $p/p_0 = 0.9975$ (50 psi).

ductivity, or heat of wetting against moisture content and observing points of inflexion; or by measuring water sorbed by the material at various relative water vapour pressures. They pointed out that F.S.P. values of about 0.3 $g\ g^{-1}$ had been found by all these methods for a wide range of cellulosic materials. They went on to say that although the value of 0.3 $g\ g^{-1}$ has been widely quoted for many years, it has become increasingly apparent that it could not be correct. The only example cited in support on this statement was however their own deduced F.S.P. value of 1.0 $g\ g^{-1}$ derived from the above discussed work using nitrogen adsorption.

In attempts to clarify the situation further they now used two other methods. First, they carried out desorption isotherms of wood and kraft-cooked microtome cross-sections (specimens similar to those they examined in [78], using essentially the same technique as reported by Stamm but with considerable refinement to achieve accuracy at the high levels of partial pressure which have always been recognised as a source of imprecision (see p. 262). In addition to experimental refinements, they plotted their data with the function $(1 - p/p_0)$ presented on a logarithm scale, thus stretching out the scale in the sensitive high partial pressure region.

Their results are depicted in Table 5.2 and Fig. 5.15 * which shows a succession of inflexions from water content of 0.3 $g\ g^{-1}$ upwards. It can be noted that all the specimens showed a common inflexion point at about 0.3 $g\ g^{-1}$ water content on solids, but at other higher inflexion points the curves for

* This diagram has already appeared as Fig. 1.19 in Corte's chapter 1, but it is judged desirable to reproduce it again here, in juxtaposition with the textual commentary on it in this chapter.

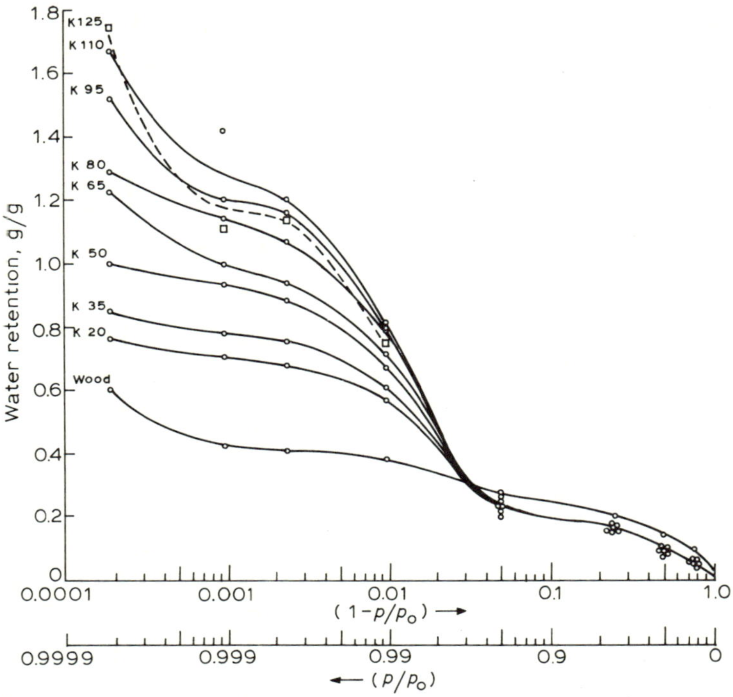


Fig. 5.15. Water desorption isotherms of wood and kraft cooked microtome cross sections (from Stone and Scallan [81]).

different specimens progressively diverged. Although this 0.3 g $g^{-1}$ inflexion point would appear to imply some fixed pattern of the behaviour at this point for the cellulose/water relationship, independent of external treatment, it was passed over by the authors who proceeded to calculate deduced values for the F.S.P. of the materials at the inflexion point around a value for $p/p_0$ of 0.9975, which would correspond to an effective capillary radius of around one micron, according to calculations based on the Kelvin equation. It is not clear as to why this inflexion point, and this level of partial pressure should have been selected and identified as indicating reliably the F.S.P. The values thus deduced are shown in column 4 of Table 5.2, from which it will be seen that a fully cooked wood pulp had an F.S.P. value about 1.2 g $g^{-1}$ (55% on total weight); uncooked wood had an F.S.P. of about 0.4 g $g^{-1}$ (29% on total weight).

The second method used by Stone and Scallan in this work was the method reported by Morehead [85] using dilute solutions of large water-soluble molecules of dextran to leach water out of saturated fibre pads. The consequent dilution of the dextran solution was taken as an indicator of the water taken out of the pads, thus facilitating calculation of the inaccessible water remaining in the pads, this latter being taken as a measure of the F.S.P.

The values thus obtained are recorded in column 5 of Table 5.2, and they show close correspondence with the F.S.P. values derived from Stone and Scallan's water vapour sorption technique and interpretation.

The authors proceeded to compare the values for F.S.P. by these methods with the values cited above from [78] for specimens from the same cooks, using the nitrogen adsorption method (Table 5.3).

In this Table they showed in column 2, the average F.S.P. values obtained by the sorption and solvent exclusion methods, in column 3 the pore volume obtained by the nitrogen adsorption method, in column 4 the difference between columns 3 and 4. They called the *total* pore volume from nitrogen adsorption the *macro-pore* volume and the difference the *micro-pore* volume.

Stone and Scallan concluded this paper by expressing doubt as to the usefulness and significance of classifying pores into macro-reticular and micro-reticular, in view of the arbitrary nature of solvent exchange, and referred to studies soon to be published which would show that "there is a continuous distribution of pore sizes in water swollen cellulose, ranging from virtually zero up to several hundred ångstroms".

In their 1968 paper on the same subject [82], Stone and Scallan pursued application of the method using accessibility of the cell wall to water-soluble macro-molecules. Here they developed further the concept that "the water must be distributed within the wall in spaces between the solid elements of the wall, and, for the sake of definition, whether these spaces occupy only

TABLE 5.3

Pore volumes for wood and kraft cooked pulps (from Stone and Scallan [81])

| Yield (%) | Total pore volume [a] FSP in (ml $g^{-1}$) | Macropore volume [b] (ml $g^{-1}$) | Micropore volume [c] (ml $g^{-1}$) |
|---|---|---|---|
| 100 | 0.41 | 0.01 | 0.40 |
| 92.4 | 0.68 | 0.04 | 0.64 |
| 89.0 | 0.70 | 0.08 | 0.62 |
| 80.0 | 0.89 | 0.27 | 0.62 |
| 77.8 | 0.94 | 0.33 | 0.61 |
| 70.4 | 1.07 | 0.54 | 0.53 |
| 61.6 | 1.19 | 0.63 | 0.56 |
| 53.4 | 1.24 | 0.55 | 0.69 |
| 48.7 | 1.17 | 0.57 | 0.60 |

[a] Average of pressure plate and nonsolvent water data Table 5.2.

[b] As measured by nitrogen adsorption (2) with graphical interpolation to permit comparison at the same yield.

[c] The difference between columns 2 and 3.

one water molecule, or a good many, they will be termed pores". They used solutions of polymers of a wide range of molecular weights, from glucose and other sugars, with M.W. of 180 and upwards, to the dextrans ranging from M.W. of 1400 to $24 \times 10^6$. (Molecular diameters thus ranging overall from 8 to 1600 Å.)

The results and contents of this paper are of considerable interest in the context of wood and wood pulp treatment involving a variety of solutes (pulping, bleaching, acetylation, etc.) but add little to the conclusions of Stone and Scallan [81] regarding the F.S.P. or regarding the transition point of hydrogen bonding which, it must be recalled, is the primary focus of attention in this survey of relevant work.

The paper by Stone et al. in 1968 [83] reported further work on water accessibility measurement by use of solute molecules ranging in weight from 180 to $24 \times 10^6$. In this paper, they moved significantly from the earlier implied concept that there is no critical demarcation within the range of pore sizes from the size of one water molecule up to several hundred Å. In this new work, they found that water of the order at 0.4 g/g on fibre weight was in all cases contained within pores smaller than 25 Å; that this amount of water was the same for kraft and sulphite pulps; and that it was completely unaffected by beating, drying or re-slushing. External influences could only change the volume of water contained in pores of 25 to 560 Å dimension.

In discussion on the results, this finding of a critical transition at 0.4 mil $g^{-1}$ (ca. 29% water on total) was considered in the light of findings from Stone and Scallan's earlier nitrogen adsorption studies to the effect that water in pores smaller than about 25 Å cannot be removal by solvents to leave spaces and surfaces detectable by nitrogen adsorption. The authors went on to say on their p. 692 "The reason for this is probably due to sorption forces extending out from cellulose surfaces and affecting all water molecules in the first few layers, and, since a 25 Å space is spanned by just three layers of water molecules on each of the opposing surfaces, it seems likely that all the water within the micro-reticular pores will behave differently from free water. That the affinity of cellulose for water may, indeed, extend over a volume of water such as that found in pores smaller than about 25 Å is shown by the differential heat of wetting and change in entropy which do not drop to zero until the water content of cellulose exceeds about 30%. This, according to the present data, is the water contained in pores smaller than about 22 Å".

The paper published in 1972 by Scallan and Carles [84] gave further data on F.S.P. determined by the solute exclusion method and showed that they correlated well with water retention values determined by centrifuging, up to a level of around 1.8 g $g^{-1}$. Thereafter the solute exclusion method gave substantially *higher* values than the centrifuge method, which seems to suggest that the former method must involve free water in fairly large cavities. Nothing new emerged regarding the significance of their F.S.P. values in relation to modal transition to fibre-to-fibre bonding during drying.

Page and Tydeman, at the 1962 Fundamental Research Symposium on Formation and Structure of Paper, presented "A New Theory of the Shrinkage, Structure and Properties of Paper" [86]. This was based upon microscopic observations of fibres under compression, and during shrinkage of sheets.

First, the lengths of dyed fibres incorporated in handsheets were measured before and after sheet shrinkage, the sheets being free to shrink in the length direction of the measured dryed fibres. It was found that even in the case of shrinkage as high as 12%, the longitudinal shortening of fibres was the same as the shrinkage of the sheet. There had been no relative movement of the

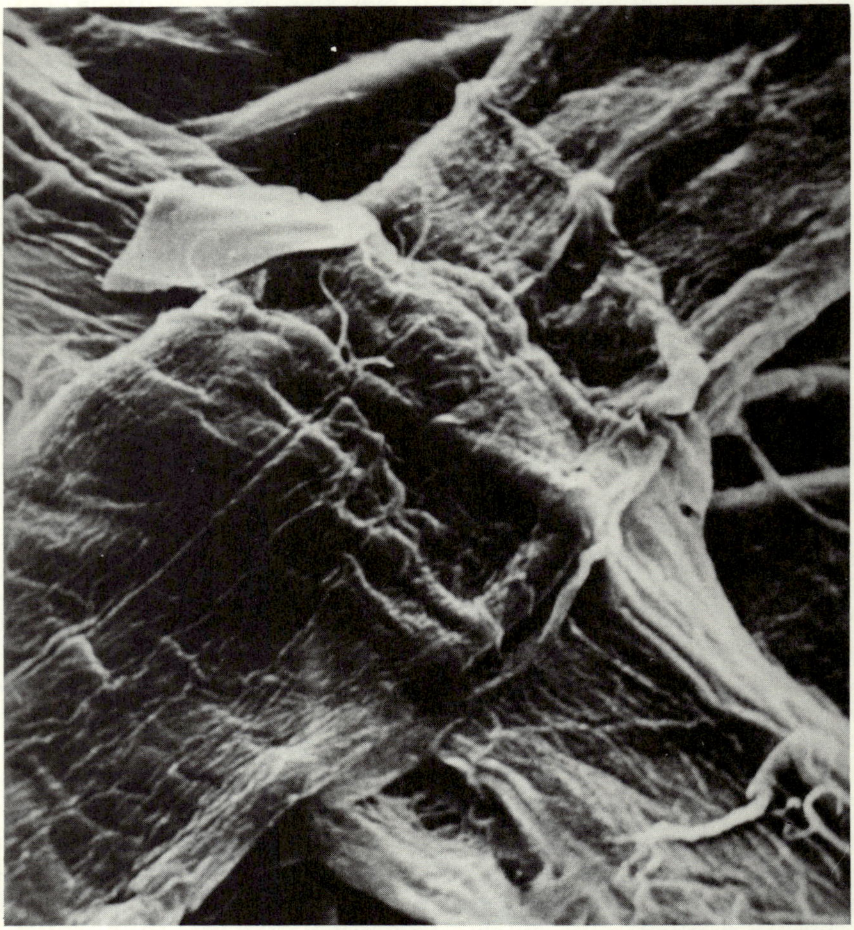

Fig. 5.16. Scanning electron micrograph of fibre-to-fibre crossing showing the compression folds produced during drying (× 2000) (from Page [87]).

fibres at the points of crossing. Thus it seemed that the dyed fibres had been forced to shorten longitudinally to an extent far greater than their longitudinal shrinkage, which is very small, said to be of the order of 1—2%. The transverse shrinkage of the fibres was reported to be in the region of 30%.

The authors proceeded to note that the greatest degrees of shrinkage are found in webs made from highly beaten stock (e.g. greaseproof paper), where there is no unbonded length of fibre to allow microcreping between bonds. They deduced that in such a case fibre-to-fibre bonding must be sufficiently strong before complete transverse shrinkage of the fibres for this transverse shrinkage to cause enforced longitudinal shrinkage of a crossing fibre.

To check this, they placed wet fibres on to a strip of wet regenerated cellulose which was then allowed to shrink in one direction only. It was found that independent of the direction of orientation of the fibre with respect to the cellulose film, it remained bonded to it along its length, and took up the shrinkage of the sheet. Along the fibres that were dried with their axes in the direction of large shrinkage of the cellulose film, wrinkles were observed, evidently due to local microcompressions.

The next step was to examine micrographs of the surfaces of freely-dried hand-sheets and of machine-made paper produced from highly-beaten stock, to seek such microcompressions at the fibre crossings. Many such microcompressions were indeed found, always in fibres oriented in the direction of sheet shrinkage, as exemplified in Fig. 5.16, taken from a later summary article by Page [87] on the same subject (see also Fig. 1.21 of Chapter 1, p. 49).

The authors proceeded further to maintain that since no relative movement of fibres was observed at fibre-crossing points, then in addition to the observed microcompressions at the crossings themselves, there must also be shortening of any unbonded segments, which occur in free beaten sheets. This enforced shortening could manifest itself either as kinks or as microcompressions. It was thought that transverse shrinkage of fibrils also was likely to be imposed on crossing fibres to which they bond. They concluded that the shrinkage of paper, presumably in the post surface tension phase (although this was not stated) is controlled by:

(1) The intrinsic potential shrinkage of the fibres.
(2) The resistance of the fibres to axial compression during shrinkage.
(3) The strength and extent of fibre-to-fibre bonding during shrinkage.
(4) Shrinkage forces associated with fibrillation.

In all essential respects, Page and Tydeman's findings provided unequivocal confirmation of the general concepts of intra-fibre shrinkage put forward by Rance [14], as is summarised on p. 251ff. Their elegant photo-micrographs put to shame Rance's crude diagramatic attempts to depict the postulated incidence of shrinkage in bonded networks, but they were nevertheless in complete accord with his textual concepts. By their nature, however they could not significantly touch on quantitative parameters governing onset of web-shrinkage caused by intra-fibre shrinkage, and communicated to the web through fibre-to-fibre bonds.

Page and Tydeman [88] presented a further paper on this same subject at the 1965 Fundamental Symposium on "Consolidation of the Paper Web". Many more micrographs were presented and analysed, in substantiation of the findings of the earlier paper, and much additional evidence was presented in support of the essential concept in the theory, viz. the concept of "adhesion before shrinkage" of the constituent fibres.

In particular, the micrographs gave elegant visual evidence of another necessary consequence of the shrinkage of fibres bonded at their crossing points: viz. the "necking" of these fibres in unbonded segments adjacent to the bonded area at the crossing. Rance [14] had earlier pointed out that, when bonded fibres try to shrink at crossings, the force of contraction of the one across the width of the fibre must be balanced by the axial resistance to contraction of the other, so that if (as might be expected) the two forces are roughly equal, a balance will be struck. If this is so, then despite axial micro-compression, the cross-direction shrinkage of each fibre will not be as great at the bonded area as at any adjacent unbonded area where there is no restraint to shrinkage. Page and Tydeman's observations showed very clear examples of the consequent necking phenomenon in unbonded segments.

The reader is recommended to study the original publication with its impressive array of photomicrographs, which cannot effectively be described or reproduced here. The net result of this further work was to reinforce strongly the evidence presented earlier, and especially to emphasise the significance of adhesion before shrinkage.

In this paper, Page and Tydeman also observed and photographed directly the course of events on the surface of a drying pad of paper formed on a porous plate by applying suction to a fibre suspension of 0.5% consistency. The water content of the pad was determined for different pressures at intervals in calibration trials. The micrographs presented in the paper illustrated two drying sequences, one for unbeaten pulp, one for beaten pulp. The sequence covered events from the first appearance of fibres projecting above the falling water level, down to the air-dry state. They gave visual indications of the start of bond formation involving progressive elimination of light-scattering faces, at the *surfaces* of pads of unbeaten pulp containing about 1 : 1 water to solids (45—50% solids). With this unbeaten pulp, from observations of recession of the water menisci, it was concluded that the surface layers of the pad began to lose continuous water at a pad solids content of 10—15%. With pads of moderately beaten pulp, the surface layer just receded below the highest fibres at 12% pad solids, the water appearing to be retained by menisci just below the uppermost layer of fibres. At 23% pad solids, there was still some suggestion of water menisci below the top layer: interesting evidence of the progressive influence of surface tension as beating time increases. Unfortunately, the work on beating was not completed at the time of publication.

Gallay [89] in a contribution to discussion when Page and Tydeman presented their 1965 [88] paper, reverted to consideration of the material presented by Lyne and Gallay [66] in 1954, as already discussed on pp. 251 and 260. He considered that the evidence now adduced by Page and Tydeman regarding fibre-to-fibre adhesion before shrinkage accorded with the evidence of the critical inflexion point in the development of wet web strength, reported by Lyne and Gallay to be at a wet web content of about 25% solids. He considered that at this stage the cellulose surfaces could well be "tacky" enough to promote adhesion, which he explicity distinguished from fibre-to-fibre bonding as such, and he went on to speculate that such an effect might well be manifest at even earlier stages in the web consolidation process. He also referred to the work of Robertson [69] already cited on p. 254 and 260 as evidence that some kind of adhesion or bonding begins at 50% solids content in the drying wet web made from never dried pulp, but he did not attempt to explain the substantial gap between Robertson's 50% (1 : 1 water to solids ratio) and Lyne and Gallay's 75% (3 : 1 water to solids ratio).

Scallan [90], in a paper presented to the 1977 Fundamental Research Symposium, reviewed the subject of accomodation of water within pulp fibres, with especial emphasis upon application of the solute exclusion method of investigation. To a degree, this involved some modification of his earlier published interpretations regarding fibre structure and porosity, but it added no new information regarding the transition point under consideration here.

Caulfield [91], in a paper given at the same Symposium, presented data relating sonic tensile modulus of wetted paper to the water content of the paper (N.B. not wet webs, but already dried paper): a plot of the logarithm of water content against sonic tensile modulus showed a straight line relationship up to a water content of about 76% of the wet paper (water-to-solids ratio about 3 : 1) (see Fig. 1.6). He interpreted these findings in terms of hydrogen bonding extending several layers from the cellulose surface, and progressively weakening up to the 76% water content. He did not present these wet strength data in terms of simple plots of strength against water content, which in fact would show the paper strength to have been reduced to one tenth of its dry strength at a water content of about 28%, and which could be held to indicate a transition at about that water content rather than around 76%.

The writer feels it desirable to make some independent comments on the evidence here accumulated, in an attempt to reconcile apparent discrepancies regarding the key question of the water content of the drying fibrous material at which effective hydrogen bonding between cellulose surfaces is initiated.

The first and most important thing to state is that determination of events at or closely approaching molecular level must involve considerations of parameters at or approaching molecular level, not considerations of parameters relevant either to the macro level, or the averages of the macro and molecular levels. Therefore, if one is concerned with the onset of hydrogen bonding between contiguous cellulose fibre surfaces, one should concentrate attention upon the state of the cellulose fibre surfaces and of the cellulose itself after macro changes at the surfaces and within the fibres are approaching or have given way to changes at the molecular level. This essential condition for determination of events at the molecular level has been obscured in much otherwise excellent work by mass observation of changes spanning both molecular and macro levels, which has prevented concentration at or approaching the molecular level.

Second, the subject has undoubtedly been complicated by some confusion of terminology, and perhaps also by some mis-interpretation of results. Nevertheless in the writer's view, there is nothing essentially incompatible between the different findings that have been recorded and discussed above.

Regarding terminology, a wide variety of adjectives has been used to describe the water involved in the total sequence of events. The term free water has been used with various meanings: sometimes to describe simply the water which runs away easily before the presses: sometimes to include all water which is not held by capillary condensation and molecular forces. The term bound water is sometimes used to cover only "molecularly-bound" water, including "surface-bonded" water, although some including Stamm, prefer to include some capillary condensation water as a component of bound water.

To a lesser degree, the words "bond" and "bonding" have been used with various meanings. Both Robertson and Gallay have used the term inter-fibre bond or bonding to indicate an adhesive force apparently distinct from true molecular bonding. The writer will restrict use of the phrase fibre-to-fibre bonding to hydrogen bonding at the molecular level.

Turning now to the question of experimental observation on this matter, we can distinguish between three types of approach.

First, there is the direct visual observation of the optical effects of bonding, as in the work of Page and Tydemann [88] and Christensen and Giertz [76]: in each case what was observed was the progressive reduction of light-scattering surfaces, indicating the emergence of bonded areas, not of course of individual bonds.

Second, there is the approach involving the measurement of the non-optical properties of an assemblage of fibres observed visually or through properties readily converted to visual observation, such as measurements of heat of wetting, electrical conductivity, shrinkage tensions, strengths of wet web, or vapour sorption (whether water vapour or nitrogen).

Third, there is the approach involving measurement of parameters only indirectly related to fibre or fibre-assemblage properties, as in the solvent exclusion technique, or the centrifuging technique.

It is clear that the first approach is the most immediately concerned with the central question under discussion, viz. the transition from consolidation induced by surface-tension forces to consolidation induced by molecular-bonding forces.

The second approach is certainly related to this central question, but indirectly so, since it generally depends upon interpretation of selected points of inflexion in plotted data implying changes in modes of behaviour within the fibre assemblage.

The third approach is indirect to a second degree, since it involves the interaction of the fibre assemblage with the parameters of an applied system of materials or forces.

The first of these approaches is the most precise at the immediate point observation, although it can be imprecise in relation to other parameters beyond the point of observation. Paradoxically, the third of these approaches is the most precise in a mean quantitative sense, but is the furthest removed from the local events under consideration. Whichever approach is used, the precision of its relevance to actual events has to be considered, in addition to precision at the immediate level of observation.

Measurements of such properties as heat of wetting, and electrical conductivity are probably among the least ambiguous, and most precise within the field under consideration. They have indicated that the critical transition point under consideration is around 0.3 g water per g dry cellulose.

Measurement of sorption isotherms are precise up to a partial vapour pressure of 0.98 after which they enter a region (well recognised by Stamm, Nissan and others) where definition as to what constitutes capillary condensation combats precision. Stamm and others up to about 1960 judged that sensible extrapolation of their curves above 0.98 partial vapour pressure gave a critical moisture value of about 0.3 g per g solids. Stone and Scallan have queried this and with their more refined techniques of both measurement and presentation have indicated points and zones of inflexion upwards almost indefinitely towards water contents of extremely high order. By selection of inflexion at partial vapour pressure of 0.9975, they determined critical moisture contents for fibre saturation of the order of $>1$ g water : 1 g solids, but one must state that far from resolving the ambiguities of such water vapour sorption studies, their work seems virtually to have negated the possibility of terminal precision in such studies.

Measurements of macro physical properties of drying wet webs can be potentially very misleading as indicators of the water contents appropriate to modal changes. This is especially so when the properties being measured are highly sensitive to moisture content. The main reason is the difficulty of avoiding moisture gradients through the wet web. As far as the writer is aware, Christensen and Giertz [76] are alone in having explicitly recognised this source of imprecision and in having deliberately avoided it. The writer is subject to correction on this point, but much other recorded work on the physical properties of drying wet webs seems to have assumed that the average moisture content of the drying webs is uniform, without special precautions having been taken to ensure uniformity. Without such precautions, the surfaces of such drying webs are bound to be considerably drier than is indicated by the mean web moisture content, thus giving the impression of the onset of hydrogen bonding at much higher moisture contents than is the case. The considerable increase in web strength that is produced by a relatively small amount of hydrogen bonding can considerably magnify this source of error.

This potential source of error can be even more evident in the case of visual observations of the surface of wet pads being dried by the deliberate withdrawal of water from the opposite surface, as in the work of Page and Tydeman [88]. It is clearly evident that in their work the upper layers of the fibrous pad were much drier than the average level of dryness of the pad as a whole. The paper stated that at the beginning of the sequence fibres were protruding above the surface of free liquid. This was an extreme case of differential dryness which must have persisted through the whole process of water removal: any fibres which were initially seen above the receding liquid in the early stages must have become much drier than the average pad dryness by the time they were photographed. The evidence of bond formation in their photographs was incontravertible, and most impressive: but there could have been no precise measurement of the moisture content at the actual physical location of the observed bonded area: one can only say that it occurred at a moisture content less than that of the drying pad as a whole.

Summing up so far, one can say that measurement of heat of wetting, and electrical conductivity have suggested water contents of the order of 0.3 g per g of cellulose as the transition point under consideration: that the evidence of water vapour sorption isotherms has been inconclusive; that the work of Christensen and Giertz on the opacity of drying wet webs has suggested that 0.3 g per g water-cellulose is the transition point; and that the work of Lyne and Gallay, Robertson and Page and Tydeman has indicated transition at a water content below about 1 g water per g cellulose, with the possiblity that it could be substantially below this figure.

There remains the work of Stone and Scallan et al. on nitrogen adsorption [78—80], solute exclusion [81—83], and centrifugal expression of water [84]. At first sight this important series of studies might appear to have rendered almost impossible a reconciliation of conflicting data and views, but a closer study suggests that their work contains within it both the keys to understanding the apparent divergencies and also a maturing of interpretation which itself has reconciled a number of associated and overlapping concepts.

Stone and Scallan's paper [80] presented at the 1965 Fundamental Research Symposium at first reading suggests a transition point for fibre saturation of the order of 100% water content on dry weight of fibre. It must be emphasised at once that the source of error attributed above to visual observations of the surface of pads, or of the macro-properties of webs, cannot apply to this work. Thick pads of cellulose fibres were used, and even if some premature drying had taken place at the pad surface, it would have had no significant quantitative effect upon their figures. Regarding the primary objectives of this work, the writer considers it to be well designed, well documented, and imaginatively presented. Regarding its peripheral deductions concerning fibre saturation levels and transition moisture contents for molecular bonding, however, it contains some internal inconsistencies, and the authors seem to have overlooked (at least at the time of presentation) one observation of central significance.

Because the total pore volume on the first drying starts to drop at about 50% web water content (100% on solids), the authors assumed that 50% was fibre saturation level. Having made this assumption, they stated:

> "While the sample is above cell wall saturation (about 50%), drying results in a loss of free water only. As drying proceeds, below the saturation value, the ratio of the water lost from the micro- and macro-reticular pores remains constant".

Analysis of the data in Table 1 of Stone and Scallan [20] shows that the statements in these two sentences are inconsistent. In fact, the ratio does not reach constancy of about 20% (the figure which the authors have accepted throughout their text), until the water content of the web has fallen to about 25%; at 50% web water content the ratio is still only about 15%. If the statement in the second of the above two quoted sentences is correct, then the fibre saturation value is 25%, not 50%, and the statement in the first quoted sentence is incorrect. Or *vice versa*.

Because of the near constancy of the 20% ratio at the lower water contents, Stone and Scallan built into their whole concept of fibre shrinkage and swelling the idea that the volumes of the swollen lamellae and of all the macro-reticular pores shrink and expand in constant proportion as water is removed from them. The writer finds this difficult to reconcile with the fact clearly demonstrated by Stone and Scallan, that a substantial proportion, about one half of the total of these inter-lamellar pores, entirely fail to open up on rewetting of the fibrous mat. This seems to contradict the notion of an inevitable "sympathetic" interaction between the substance of the lamella, and the volume of the macro-reticular pore, whereby water must flow from the one to the other, as the one or the other decreases or increases.

Without recourse to any particular structural models, is it not more consistent with the data simply to suggest that about one half of the pores detected by the nitrogen during drying are so intimately and sympathetically associated with the cellulose structure itself as to be in effect a part of that structure, and that their content of water should be regarded as an inseparable part of the total water of cellulose saturation? Conversely, it would seem logical to suggest that the water in the other half of the pores is not a part of the total water of saturation, and is in effect more akin to free water, since it appears to be influenced more by the forces exerted by the last fraction of free water than by the forces of the water molecularly bound in the cellulose itself. These pores do not appear readily to re-open, nor do they appear to share the property of the other pores in — as it were — "breathing sympathetically" with the molecularly swollen cellulose. On this view, one would deduce that the total water of saturation is about 25%, not about 50%, which would remove the inconsistency of the statements quoted above from Stone and Scallan: and such a figure would be also consistent both with those earlier derived from heat of wetting and electrical conductivity, and also with that derived by Christensen and Giertz [76] from carefully controlled web-drying experiments.

The writer is aware that some other information in Stone and Scallan [80], notably the fact that there appears to be no dimensional distinction between the reversible and irreversible pores, argues against the idea of considering the water in the irreversible pores as a component of free water. Nevertheless, this work clearly indicates: (a) that there is a transition in the behaviour of the material in the region of 25% total water content of the pads; and (b) that there are two quite distinct species of pores, detectable by the nitrogen adsorption technique, one of which is not intimately associated with the cellulosic material.

Turning now to the work of Stone and Scallan et al. using water vapour sorption, solute exclusion and centrifugal methods, it seems that their concept of fibre saturation points as such (F.S.P.) at first moved further away from the concept of any demarcation point between molecularly bound water and free water. Their measurements of F.S.P. also became more empirical in mode of measurement, culminating in the centrifuge method of water extraction: and their conceptual interpretations became at one stage more diffuse, almost to the point of appearing to deny any modal distinctions between types of water involved in later stages of the drying process, indeed between "pores" of dimensions between 4 Å and several hundred Å.

However, there emerged in one of their papers [83], an unequivocal indication of a very distinct modal transition, apparently first identified in terms of a pore size (25 Å) totally inaccessible to solute molecules. It appears also that this caused reconsideration of the nitrogen adsorption data, with consequent emphasis on the fact that pores less than 25 Å dimension are also inaccessible for nitrogen adsorption. This demarcation point was found to be the same for *all* specimens, and was unchanged by external influences such as pulping and beating. It corresponded to a water content of approx. 0.4 g per g cellulose: and coming full circle, the authors likened it to the water content of 0.3 g per g cellulose obtained from the heat of wetting experiments, which would be equivalent to about 22 Å pore dimension. They considered the water in pores below about this dimension to be controlled by sorption forces extending out from cellulose surfaces, affecting all water molecules in the first few layers, so that it would "behave differently from free water".

The authors thus clearly recognised that there is a modal distinction which others have recognised as a transition from molecularly bound water to free water, at around 0.3 or 0.4 g per g cellulose, even though they continued to use the phrase "fibre saturation point" (F.S.P.) for a different condition involving substantial extra quantities of water in a free state.

Clearly F.S.P. as measured by these authors is a concept of considerable value in a defined technological context. There is perhaps a case for distinguishing between this F.S.P. which varies from specimen to specimen according to pre-treatment, and a "cellulose saturation point" which appears to be a basic, unchanging property of the cellulose as such.

This transition point which the writer will term C.S.P., acknowledged by Stone et al. [83] as indicating a demarcation between molecular-bound water and free water, appears to correspond with the modal transition point accepted by Campbell, Stamm and others over many years, as indicating demarcation between consolidation of the drying paper web, induced by surface tension, and consolidation induced by hydrogen bonding, when cellulose surfaces have been brought close enough together.

There seems to be no major weight of evidence against the view of those authors, indeed there seems to be substantial new evidence to support their view that the onset of hydrogen bonding during the drying of paper occurs locally at a water content of around 30% expressed in terms of dry weight. The word "locally" must be emphasised because on a paper machine, with repeated high temperature contact at alternate surfaces, there must be local hydrogen bonding while average water contents of the web are still well in excess of this 30%. Indeed there is probably a repeated sequence of bond formation, breaking and re-formation, of progressively diminishing amplitude as water is driven alternately from one surface to the other as the web progresses towards its finally dried state. The reality of events on the machine is far removed from the ideality of a steady progressive and homogeneous drying sequence, as Christensen and Giertz [76] have so rightly indicated.

## 5.8 HYDROGEN BONDING: THE FINAL ACT OF WET CONSOLIDATION

The ultimate determinant of consolidation of the wet web is the formation of hydrogen bonds between the surfaces of cellulose fibres, fibrils and fragments, after they have been brought close enough together by earlier influences. The dramatic effect of this bonding, which firmly joins together the component parts of the sheet, is demonstrated not only by the sudden transformation on the paper machine of a limp, damp extremely weak web into a firm, hard, strong sheet of paper: it is equally demonstrated by counter-transformation of that sheet paper back to a state approaching that of the wet web itself, by simple wetting with water. If the sheet has been wet-strengthened, this reversal will not fully take place, and also it can be delayed by the presence of some sizing material: but the simple soaking of a newspaper will quickly demonstrate what happens in the absence of some protection, with a fairly absorbent sheet of paper.

Robertson [92] has published some impressive data, shown in Fig. 5.17 which typifies how totally paper loses its essential mechanical properties on wetting with water and to a lesser degree with other polar liquids. Its tenacity on water soaking falls to about one twentieth, a level little higher than that of the wet web before drying. Referring back to the discussion at the end of the last section, it is the dramatic difference between the strength of the paper web before and after hydrogen bonding that makes experiments designed to determine the onset of hydrogen bonding so sensitive and correspondingly so difficult to carry out. Preferential bonding through just a min-

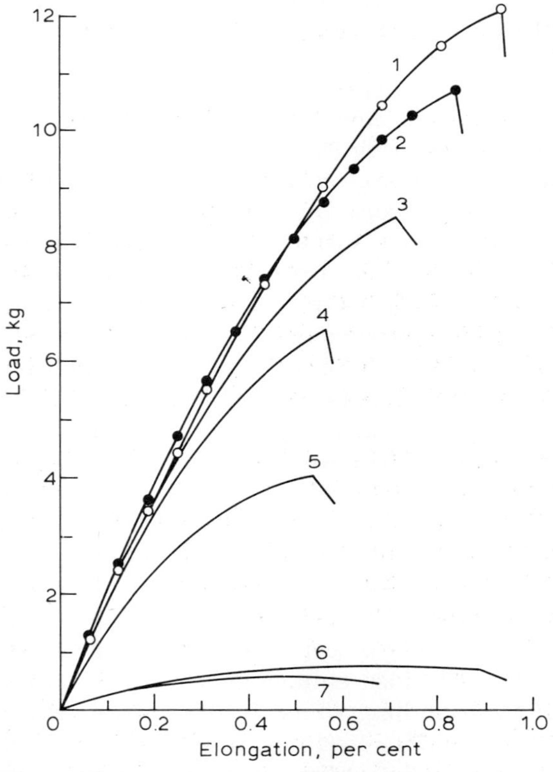


Fig. 5.17. Load-elongation curves obtained for paper samples immersed in different liquids: (1) control; (2) benzene; (3) vinyl acetate; (4) nitromethane; (5) ethanol; (6) formamide; (7) water (from Robertson [92]).

ute fraction of the thickness, at the surface of a drying wet web can immediately show up in a substantial increase in strength properties, simply because of the enormous leverage exerted on the situation by the large change brought about by hydrogen bonding.

Today the fact of hydrogen bonding in the final consolidation of the wet paper web is fully accepted and largely understood, although the complexities of its *modus operandi* are far from easy for the non-specialist to understand.

In this treatise Corte's Chapter 1 on Cellulose-Water-Interactions have fully dealt with this subject, so there is no need for further comment beyond this short note deliberately intended to emphasise the practical importance of the subject. The reader is also recommended to study Nissan's recently published lectures on Fibre Science in Paper [6], for further information and opinion — some of it controversial — on the nature and significance of hydrogen bonding in the consolidation and properties of paper.

## 5.9 ADDITIONAL CONSOLIDATION PROCESSES ON THE PAPER MACHINE: PLANAR COMPACTION AND CALENDERING

These consolidating processes are deemed to fall outside the primary sequence of papermaking through consolidation of the wet web. Even though one of them, planar compaction, is generally applied at a moisture content significantly higher than the normal reel-up moisture content; and even though the other calendering, which is almost universally applied at the end of the machine, can involve application of some water, both of these are optional processes applied to the already formed web of paper.

This is not to minimise their importance. For certain grades of paper, planar compaction is unique in being able to impose required properties on certain grades of paper, notably sack kraft. For most grades of paper, calendering is a *sine qua non*.

The treatment of these subjects here will not be as extensive as that which has been given to most of the events in the papermachine proper. As far as planar compaction is concerned, the process and its effects are fairly readily understood, and the relevant technical literature is scanty. As far as calendering is concerned, the subject is complex, and the relevant literature is substantial. Nevertheless, the calendering process shares to some degree the characteristics that deterred the writer from more than a cursory glance at the engineering technology of the drying section. Much of calendering technology is concerned with proprietory design and operation, and much of it shares common ground with a more general engineering technology spread wider than papermaking. In this respect, calendering falls between wet pressing and thermal drying. It does not merit the detailed attention that was given to wet pressing, whose technology is unique to papermaking: yet it does involve significant elements of science and technology that can only be found within papermaking literature, and therefore ought to be dealt with to some degree in a treatise on Paper Science.

The writer trusts that the relatively brief, but by no means cursory attention he has therefore afforded to the practice and effects of calendering will be acceptable to those who have specialised in this subject. Obviously, this attention cannot include anything like a comprehensive survey of relevant literature, and will largely concentrate upon a few key articles and reviews that can provide references for further study.

### *5.9.1 Planar compaction*

One of the main defects in machine-made paper, for most uses, is its marked degree of planar anisotropy. This is caused by a combination of factors, all arising from the fact that the stock and wet web pass down the machine continuously in one direction.

At the slice and at the point of contact of stock with wire, there are major opportunities for preferential orientation of the fibres in the machine direction, and in practice it is very difficult to get a "square" sheet at the wet end. There is, as noted earlier, even the possibility of some added fibre orientation at the presses (see p. 245), but far more important than this, is the effect of differential tensions on the drying wet web. To a degree, the web is tensioned in the cross direction whenever and wherever it is held on to and around a roll, whether at press roll or drying cylinder. This cross-direction tension is however only really effective for any significant period when wet webs are dried on large M.G. cylinders, where it is in effect stuck onto the roll and restrained from movement in all directions over a critical period of drying. Apart from this, the paper web spends much more time unrestrained in the cross direction, in all the draws between rolls and cylinders, than it spends restrained on rolls and cylinders.

In the machine direction, however, there is positive tension continuously at every stage from wet end to reel up, to such extent that there is usually a positive stretch of the web over the length of the machine. This MD tension effectively inhibits all planar shrinkage factors in the machine direction, whether arising as overall planar shrinkage force from surface tension or from fibre shrinkage effects transmitted to the drying web as already discussed.

The net results of all these factors is that the MD tenacity is normally much higher than the CD tenacity, the tensile stretchability is much less in the MD than in the CD, and so on. These effects with much recorded data, have been discussed at length by Rance [94].

The resultant anisotropy manifests itself undesirably in many paper properties and characteristics, including undue curl tendencies and differential hygro-expansibility, but above all it is the low MD stretchability which is so undesirable because, despite high MD tenacity, it reduces the paper's capability to conform to imposed strains in the machine direction. This shows up especially in paper used for making paper bags and sacks. A filled paper sack dropped from a height is liable to burst open simply because the paper cannot accommodate the imposed strain. It has insufficient capacity to absorb energy in such straining in the machine direction.

One way of overcoming this problem has been the application of crêping, which grossly crimps the paper sheet and thus provides substantial M.D. extensibility. Unfortunately, it also often damages the sheet by delamination and excessive kinking, so that rupture can occur because of these defects despite the high extensibility. Also the paper surface is unsuitable for printing, etc.

In 1953 Sanford Cluett, the inventor of the Sanforising process for compacting textiles to prevent shrinkage after washing and drying, applied a similar system to the planar compaction of paper, the patented Clupak (trademark) process [94]. In principle, this involves passage of the web at a moisture content generally around 30 to 35% through a pressure nip in which a Venturi section is formed between a rubber and a steel roll. The rubber roll rotates more slowly than the steel roll, so that the paper web is compacted in the machine direction. This device is shown diagramatically in Fig. 5.18, from Ihrman and Ohrn [95]. These authors gave their paper at the 1965 Fundamental Research Conference on Consolidation of the Paper Web, and presented the results of experiments involving variations in nip pressure, and roll speed difference. With certain limitations, planar compaction was increased with both increasing nip pressure and increase in differential speed.

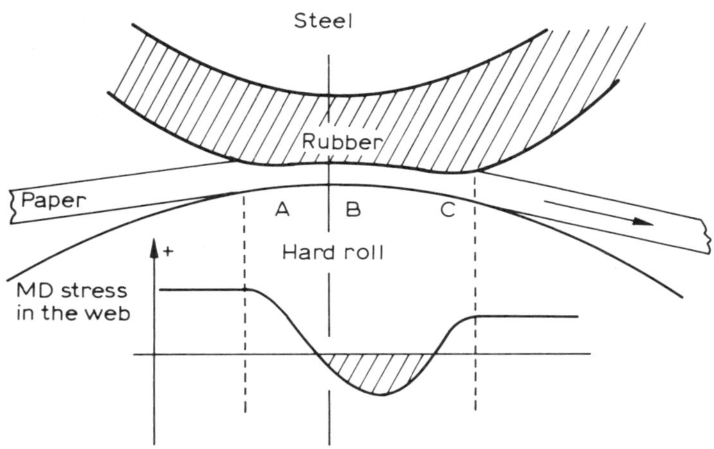


Fig. 5.18. Schematic drawing of pressure nip represented as a Venturi section (from Ihrman and Ohrn [95]).

At a nip pressure of about 18 kg $cm^{-1}$ and a speed differential of about 12%, 15% of MD stretch was imparted to the web.

Compaction always led to a decrease in tenacity, but an increase in rupture energy as well as in extensibility. CD properties were affected to a minor degree. Basically, the same conclusions were reached for rewetted paper as with virgin wet webs. Beating did not significantly influence the relationship.

The surface of compacted kraft paper when photographed in oblique light appeared to indicate an intermediate structure between plain kraft and creped kraft. The wetter the paper entering the nip, the smoother the surface. Cross-sections cut in the thickness direction displayed a structure of curves in this direction. Page [87] states that the fibres in such compacted paper exhibit microcompressions similar to those which he and his co-workers found at the crossing points of fibres in sheets allowed to shrink during drying (see Fig. 5.16). The moisture content for optimum Clupaking has been empirically determined, having regard to: (a) need for a reasonably-consolidated wet web with sufficient integrity to withstand the physical pressures and strains of the nip without disintegrating; (b) need for high extensibility, which falls off if the water content is higher than about 35%; (c) need for the web to be in a condition precedent to any significant degree of hydrogen bonding: if it is too dry, with significant dry strength having already developed, then the compacting nip will disrupt the pre-formed bonded areas, and thus will damage the structure and strength of the sheet. It is significant that the empirically-chosen optimum water content is in fact around 30—35%, a figure of the same order as that considered to be the water content for transition to hydrogen bonding during the drying sequence (see Section 5.7). At either higher or lower water contents, the required pattern of micro-compression of the fibres is not apparently "fixed" within the sheet. Thus, this empirical selection of the optimum water content for Clupak compaction may perhaps be regarded as a crude practical method for determination of the water content for the onset of effective hydrogen bonding on the paper machine.

Ihrman and Ohrn's paper [95] included a mathematical appendix which explains in some detail the mode of operation of the compacting nip.

In practice, the Clupak compacting process has proved highly successful in manufacture in many installations of sack kraft which can withstand rough handling and high drops in loaded condition without rupture. The process is in fact applicable to all types of paper, but the writer is not aware of any commercial applications other than with sack-kraft.

The fact that this compaction process has only been applied with proprietary equipment in a limited field does not reduce its fundamental significance. It provides a mode of consolidation essentially akin to that caused by natural unrestrained cross-web shrinkage of the drying web, but provides it in a way that can be deliberately controlled. It would not be surprising if in due course the principle is more widely applied, when eventually the relevant know-how is fully within the public domain.

### *5.9.2 Calendering*

Calendering has been described as "the papermaker's last chance at altering the properties of paper before it is passed to the consumer" [96].

Machine calendering is carried out on most paper machines between the last dryer and reel up, using calenders comprising a vertical stack of smooth-surfaced iron rolls. The dry paper web, with or without a small amount of water added immediately before entry to the stack, is passed into the first nip and then wraps around the first roll, into the second nip, and so on until it emerges from the last nip in the final calendered state. The stack is driven usually by the bottom (king) roll, and is loaded by the weight of the rolls, plus downward pressure usually applied at the top of the stack. The rolls may be hollow, and may be fitted to acommodate water for cooling, or water or steam for heating.

So-called supercalendering, normally carried out as a separate process off-machine, involves the use of nips between alternate iron rolls and rolls made from softer material, traditionally and generally compressed discs of fibrous material such as cotton. Other variations of calendering such as gloss calendering, matt calendering and friction calendering, are described in the general literature on the technology of papermaking, but cannot be considered here.

The purposes of calendering are several. First of necessity, although not therefore of desirability, is the objective of ensuring uniformity of web thickness so that well-compacted, even reels can be built up at the reel-up station. In achieving this purpose, calendering is simply used as a means of superficially (adjective used literally) overriding imperfections derived from earlier stages of paper formation and consolidation. These include cross-web variations in thickness due to imperfect spread of the stock on the drainage wire, local variations in thickness due to flocculation, and variations due to cross-web variations in consolidation during wet pressing and drying.

Viewed as a corrective operation, calendering is simply a brutal way of literally ironing out these built-in imperfections to ensure even reeling without regard to any subsequent adverse consequences. An imperfection due to cross-web or local variation in weight can never be eliminated. Its adverse consequences can only be postponed. It is well known that compression of local high spots by calendering causes local damage with consequent local weak spots in the sheet, which can cause breaks in the web in subsequent processing (e.g. on the newsprint press). Evening out of cross-web variations due to variations in water removal is less brutal and has a more permanent effect, with less adverse consequence during after-processing.

In addition to the corrective function of calendering, it has deliberately-intended positive functions which add to the qualities of the paper. It is unfortunate that these functions often have to be at least partly subordinated to the corrective function, since the formation of even reels is an overriding practical consideration.

One of the main positive reasons for calendering is to produce an evenness of surface as required by subsequent printing or coating processes. The surface of an uncalendered sheet, even if evenly formed without undue flocculation or cross-web variation, is generally too rough to facilitate high-quality printing. Cellulose fibres are relatively coarse in relation to the dimension of ink deposits on printing, and both the pressing and drying processes can add to and magnify the effects of fibrous coarseness. There is therefore need for surface smoothing by mechanical means to improve this quality.

The efficacy of calendering in smoothing the paper surface for subsequent coating operations depends upon the nature of the coating. Aqueous coating will generally cause the paper surface largely to revert to its uncalendered condition. Therefore, if a smooth surface is specifically required for an aqueous coating, it must be achieved in the pre-calendering stages of papermaking. If, however, the coating process is non-aqueous, then smoothness achieved by calendering can persist through the coating, and can ensure even coating which can be very important when the coating material is expensive. When the coating material is inexpensive, and especially when it is cheaper than cellulose, then imperfections in the surface regularity of the paper substrate are relatively unimportant.

The third purpose of calendering is to confer desired optical properties upon the surface of the paper, especially by increasing its gloss and by ensuring an even gloss. When gloss is an important characteristic, its control is generally achieved through supercalendering, though ordinary machine calendering does significantly affect the optical appearance of the sheet.

It is generally desired to achieve improvements to the surface of the paper in terms of printability and gloss without undue loss of bulk, which renders the sheet less rigid and more difficult to handle. Certain speciality papers however, such as tracing paper, glassine, and condenser tissue paper may specifically require bulk reduction (consolidation) to achieve the bulk properties (e.g. transparency or oil resistance) required for their usage.

Measurement and control of the desired properties of smoothness and gloss offer serious problems. For each of these properties, there is a wide range of test procedures used for empirical control: review of these is beyond the scope of this treatise, and largely inappropriate in view of the empirical nature of most of the tests. Printability as such is generally checked and controlled by direct trials in appropriate printing equipment, so that all questions of correlation between measured properties and performance are by-passed.

The measureable physical effects of calendering are generally quite substantial. The writer is indebted to Dr. John Peel for permission to reproduce the contents of Table 5.4, from an unpublished paper on "Calendering and

Supercalendering — An Appraisal of Current Practice, Understanding and Needs", presented by Peel, Kerekes, and Baumgarten, at a Symposium on Calendering and Supercalendering of Paper held at the University of Manchester Institute of Science and Technology in September, 1975.

Very substantial reductions in bulk and thickness are to be observed, approaching 50% of the uncalendered values in extreme cases. In a real sense, calendering is the last stage of consolidation. A typical thickness reduction on such common grades as newsprint or book paper is 25—30% in one calendering operation. The various roughness and print rating figures shown in the table are arbitrary ratings according to standard tests which are defined in text books on paper and printing technology.

According to Howe and Lambert [97], the mechanism of calendering is considered to combine rolling friction with compressional and shear stresses due to calender roll pressure and rotation over the micro-elastic paper web.

There have been many studies on both the static and dynamic compressibility of paper, e.g. Gavelin [98], Jackson and Ekstrom [99], and Ivarsson [100]. All of these have included the effect of moisture content of the paper sheet, and the general conclusion has been that both elastic compression and plastic deformation increase with increasing pressure, moisture and temperature.

Simulation of the dynamic character of calendering by impulse loading of short duration has been studied by Brecht and Schadler [101], Mardon et al. [102], Peel et al. [103,104], and Wahlstrom et al. [105]. In general, these dynamic simulations showed that less compression is achieved for a given loading than in a corresponding static compression. The work by Peel et al. [103,104] gave results on the influence of pressure, time, moisture and temperature on the compression and recovery of paper sheets containing groundwood, leading to an empirical equation relating the variables and their interactions to the results obtained.

Other studies have been made on full-scale calenders, e.g. Mardon [106], Howe and Lambert [97], and MacKin [107]. Robertson and Haglund [108] combined studies of normal pressing under static and dynamic conditions, with experiments using laboratory and full-sized calenders. They derived a mathematical model based on the work of Mardon et al. [102] and Colley and Peel [104] which largely accounted for the interrelations of the variables found from their studies. This work ignored the shear forces which undoubtedly play a major part in changes in the optical and smoothness properties of paper surface, and concentrated on the normal forces as the prime determinant of compression as such.

A full survey of the results that have been reported on the interactions of pressure, shear, moisture content, temperature and time in calendering is beyond the scope of this work.

The practical effect of these variables is made more complex by the

TABLE 5.4

Changes in the properties of certain grades of paper as a result of calendering under mill conditions

| | Machine calender | | | | | | | | |
|---|---|---|---|---|---|---|---|---|---|
| Grade | Newsprint | | Book [a] | | | Rotary print | | 6-Ply board | |
| Grammage | 54 | | 73 | | | 52 | | 360 | |
| | Before | After | Before | After 1st stack | After 2nd stack | Before | After | Before | After |
| Bulk ($cm^3\ g^{-1}$) | 2.66 | 1.82 | 1.56 | 1.17 | 1.17 | 1.87 | 1.41 | 1.80 | 1.47 |
| Thickness ($\mu m$) | 130 | 89 | 114 | 86 | 81 | 98 | 74 | 650 | 530 |
| Breaking length, MK (km) | 4.8 | 5.0 | | | | 3.8 | 3.3 | 5.6 | 6.0 |
| Tearing resistance, MD (mN) | 23 | 19.5 | | | | 48 | 39 | 100% [c] | 84% [c] |
| Printing opacity (%) | 96.0 | 95.4 | 91.6 | 91.2 | 90.8 | — | (same) | (85 [d]) | (80 [d]) |
| Roughness (TS) | 1250 [e] | 158 [e] | 122 [e] | 65 [f] | 61 [f] | 10 [g] | 56 [g] | | |
| Gloss | | | 18 [h] | 35 [h] | 43 [h] | | | | |
| Print rating (TS) | 33 [j] | 12 [j] | −35 [k] | −10 [k] | −5 [k] | | | | |
| No. of nips | | 7 | | 3 | 7 | | 4 | 3 × 5 | |
| Nip load at bottom $kN\ m^{-1}$ | | 70 | | 32 | 39 | | 80 | 30 | |

TABLE 5.4 (continued)

| | Supercalender | | | | | | | |
|---|---|---|---|---|---|---|---|---|
| Grade | Newsprint | | Machine coated [b] | | Glassine | | Capacitor tissue | |
| Grammage | 54 | | 105 | | 32 | | 11 | |
| | Before | After | Before | After | Before | After | Before | After |
| Bulk ($cm^3\ g^{-1}$) | 2.45 | 1.58 | 1.10 | 0.86 | 1.25 | 0.76 | 1.23 | 0.73 |
| Thickness ($\mu m$) | 133 | 86 | 116 | 90 | 48 | 25 | 13 | 8 |
| Breaking length, MK (km) | 2.2 | 2.2 | 4.6 | 1.1 | 1.1 | 1.3 | — | — |
| Tearing resistance, MD (mN) | 19 | 16 | 55 | 47 | 17 | 15 | | |
| Printing opacity (%) | 93 | 85 | 96.8 | 94.0 | 24.6 | 18.4 | 6.3 | (0) |
| Roughness (TS) | 1030 [e] | 195 [e] | 140 [e] | 30 [e] | 1450 [e] | 27 [e] | | |
| Gloss | 0.3 [i] | 0.7 [i] | 0.6 [i] | 5.0 [i] | | | | |
| Print rating (TS) | | | | | | | | |
| No. of nips | | 8 | | 9 | | | | |
| Nip load at bottom $kN\ m^{-1}$ | | 155 | | 176 | | | | |

[a] Groundwood-free, 60% hardwood, 10% softwood, 30% ash.
[b] Starch/clay coated on 60 $g\ m^{-2}$ base paper.
[c] Stiffness, MD.
[d] Tappi plybond (1bs $in.^{-2}$).
[e] ml $min^{-1}$. Bendsten.
[f] Sheffield units.
[g] s/100 $ml^{-1}$, Bekk.
[h] Ingersoll.
[i] Zeiss GP2; polished black glass (70°) 100.
[j] Subjective rating of wedge proof prints.
[k] Van der Cook (letterpress) proof-presss rating.

incidence of gradients of temperature and moisture through the sheet during the compressive cycle. A recent paper by Lyne [96] reports the advantages of deliberately induced moisture gradients in the calendering of newsprint and multi-ply board. Heat and moisture are thus of major positive importance in calendering, yet both are also sources of unwanted variations as well as deliberately sought effect.

These unwanted variations are compounded by the fact that the initial roughness of the sheet substantially affects the distribution of normal stress through the nip. This leads to local compression and deformation of surface protrusions, resulting in a non-uniform pattern of deformation. In addition, there can be unwanted variations in pressure across the width of the rolls, caused largely by load-induced deflection: however, this source of variation has been largely brought under control in recent times by the introduction of such engineering devices as the Kusters swimming roll.

Calendering is an extremely complicated operation in which the problem of optimising the many variables is compounded with the need for compromise due to incoming web imperfections. The literature on calendering shows a remarkably non-controversial pattern, unlike that on wet pressing. The several teams contributing to studies around the world seem to have recognised the daunting nature of the overall complexity of the process, and the consequent need to collaborate in sorting out the variables in relation to practical needs.

For the present, design and operation in calendering is almost entirely determined by empirical findings relating to particular products from a particular machine, involving acceptable combinations of the corrective and positive functions of the process. More than with any other component of the paper machine, the design and operation of the calender is determined by what has happened earlier in the machine. In such a situation, compromise is bound to outweigh theoretical considerations.

For a more thorough consideration of the subject, the reader is directed to the work of Mardon et al. [102], Kerekes et al. [109], Robertson and Hagglund [108], and Peel et al. [103,104]. When, as is hoped, it becomes available in publication, the earlier mentioned contribution of Peel et al. to the 1975 Manchester Symposium on Calendering with its 130 literature references is obligatory reading for a better understanding of this highly complex end process of consolidation in papermaking.

## 5.10 REFERENCES

1 Consolidation of the Paper Web, in F. Bolam (Ed.), Transactions of the Fundamental Research Symposium held at Cambridge, 1965, Tech. Sect. Brit. Paper Board Maker's Assoc., London, 1966, (2 vols).

2 W.W. Barkas and H.F. Rance, in R. Meredith (Ed.), Mechanical Properties of Wood and Paper, North-Holland, Amsterdam, 1953.

3 Effect of Fiber Characteristics on Paper and Papermaking Properties, papers presented at Fundamental Research Conference, Quebec, Canada, and reprinted from Tappi, 33 (1950) 369—480.
4 Fundamentals of the Paper Machine, papers presented at Fundamental Research Conference, Appleton, WI, and reprinted from Tappi, 37 (1954) 490—685.
5 The Formation and Structure of Paper, in F. Bolam (Ed.), Transactions of the Fundamental Research Symposium held at Oxford, 1961, Tech. Sect. Brit. Paper Board Maker's Assoc., London, 1962 (2 vols).
6 A.H. Nissan, Lectures on Fiber Science in Paper, Joint Textbook Committee of the Paper Industry, Canada, 1974.
7 Fibre-Water Interactions in Papermaking, in Transactions of the 6th Fundamental Research Symposium, held at Oxford, 1977, Fund. Res. Comm., Tech. Div. Brit. Paper Board Ind. Fed., London, 1978 (2 vols).
8 W.B. Campbell, Bull. For. Ser., Dept. Int. Can., 84 (1933).
9 A.J. Stamm, Wood and Cellulose Science, The Ronald Press Co., New York, 1964, Ch. 8.
10 H.F. Lewis, Tappi, 33 (1950) 418.
11 W. Gallay, Tappi, 33 (1950) 425.
12 A.J. Stamm, Tappi, 33 (1950) 435.
13 L.V. Forman, Tappi, 33 (1950) 444.
14 H.F. Rance, Tappi, 37 (1954) 640.
15 H.G. Higgins and J. de Yong, The Beating Process, l.c. [5], p. 651.
16 D. Atack, Advances in Beating and Refining, l.c. [7], p. 261.
17 B. Radvan, C. Dodson and C.G. Skold, Detection and Cause of the Layered Structure of Paper, l.c. [1], p. 189.
18 W.B. Campbell, Pulp Pap. Mag. Can., 48 (1947) no. 3: 103.
19 A.H. Nissan, Tappi, 37 (1954) 597.
20 A. Sedoff, C.V. Holmberg and E.C. Jahn, Pap. Trade J., 109 (1939) T.S. 350.
21 F.C. Stamm, Pap. Trade J., 100 (1943) no. 2: 20.
22 J. Bergstrom, Sven. Paperstidn., 62 (1959) 367.
23 R.E. White and F.W. Berdux, Tappi, 43 (1960) 580.
24 J.S. Sweet, Pulp Pap. Mag. Can., 62 (1961) T 367.
25 P.B. Wahlstrom, Pulp Pap. Mag. Can., 60 (1960) T 379.
26 P.B. Wahlstrom, Pulp Pap. Mag. Can., 60 (1960) T 418.
27 P.E. Wrist, Pulp Pap. Mag. Can., 65 (1964) T 284.
28 S.T. Han, Pulp Pap. Mag. Can., 70 (1969) T 134.
29 W.L. Ingmanson, B. Andrews and R.C. Johnson, Tappi, 42 (1959) 840.
30 W. Ingmanson et al., Tappi, 46 (1963) 150.
31 R.L. Jones, Tappi, 46 (1963) 20.
32 H.D. Turner, J.P. Hohf and S.L. Schwartz, Pap. Trade J., 127 (1948) no. 9:43.
33 H.D. Wilder, Tappi, 43 (1960) 719.
34 F.A. Seborg, F.A. Simmonds and P.R. Baird, Tappi Tech. Assoc. Pap., 22 (1939) 347.
35 C.S. Yih and S.J. McNamara, Tappi, 46 (1963) 204.
36 C.A. Asklof et al., Pulp Pap. Mag. Can., 65 (1964) T 339.
37 J.A. Means, Tappi, 18th Engineering Conference 1963.
38 J.W. Welch, Pap. Mill News, 84 (1961) p. 12 Sept. 11.
39 L.M. Woodside, Pap. Trade J., 145 (1961) no. 45: 48.
40 A.P. Redfern and G. Gavelin, Pap. Technol., 3 (1962) 463.
41 J.D. Lyall, Tappi 18th Engineering Conference, 1963.
42 P.B. Wahlstrom, Pulp Pap. Mag. Can., 70 (1969) T 349.
43 P. Nilsson and K.O. Larsson, Proc. Tech. Sect. C.P.P.A., (1968) T 438.

44 R.E. Morrison, Proc. Tech. Sect. C.P.P.A., (1969) T 35.
45 C. Schiel, Proc. Tech. Sect. C.P.P.A., (1969) T 71.
46 J.E. Wilder, Tappi Engineering Conference, (1967).
47 H.A. Westra, Pap. Tech. Ind., 16 (1975) 165.
48 S. Schmidt, Pap. Tech. Ind., 16 (1975) 227.
49 C.M.W. Wilson and M.J. Birsel, Pap. Tech. Ind., 16 (1975) 239.
50 H.H. Heller, M.A. MacGregor and W.C. Bliesner, Pap. Tech. Ind., 16 (1975) 154.
51 L.H. Busker, Tappi, 54 (1971) 373.
52 J.E. Wilder, Tappi, 51 (1968) 104.
53 N. Wiseman, Trend, Pulp Pap. Res. Inst. Can., 26 (1976) 11.
54 N. Wiseman, Tappi, 60 (1977) 98.
55 R.H. Doughty, Pap. Trade J., 93 (1931) No. 2: 39.
56 R.H. Doughty, Pap. Trade J., 93 (1931) No. 15: 44.
57 R.H, Doughty, Pap. Trade J., 94 (1932) No. 9: 29.
58 R.B. Brown, Pap. Trade J., 94 (1932) No. 13: 35.
59 K.V. Breitveit, Bumazh. Prom. (Paper Industry), 24 (1949) 6.
60 D.J. MacLaurin and J.F. Whalen, Tappi, 37 (1954) 608.
61 I.T. Pye, O.V. Washburn and J.G. Buchanan, Structural Changes in paper on pressing and drying, l.c. [1], p. 353; v. also Proc. Tech. Sect. C.P.P.A., (1964) T 400.
62 E.L. Back and K.G. Norberg, l.c. [1], p. 569.
63 J.F. Oliver and N. Wiseman, l.c. [7], p. 411.
64 L.M. Lyne and W. Gallay, Pulp Pap. Mag. Can., 51 (1950) No. 10: 129.
65 L.M. Lyne and W. Gallay, Pulp Pap. Mag. Can., 55 (1954) No. 11: 128.
66 L.M. Lyne and W. Gallay, Tappi, 37 (1954) 581.
67 H.F. Rance, Proc. Tech. Sect., Br. Pap. Board Maker's Assoc., 33 (1952) 173.
68 E.J. Pritchard, Proc. Tech. Sect., Br. Pap. Board Maker's Assoc., 33 (1954) 31.
69 A.A. Robertson, Measurement and Significance of the water retention of paper-making fibres, l.c. [1], p. 90.
70 Z. Tarnawski, Prezeglad Papierniczy, 28 (1972) 73.
71 Von. L. Byrd, Tappi, 57 (1974) 87.
72 P.E. Wrist, Discussion Contribution, l.c. [1], p. 169.
73 P.A. Tydeman, D.R. Wembridge and D.H. Page, Transverse Shrinkage of Individual Fibres, l.c. [1], p. 119.
74 M.A. Heath and M.M. Johnston, Tappi, 33 (1950) 386.
75 C.O. Seborg, F.A. Simmonds and P.K. Baird, Tappi Tech. Assoc. Pap., 21 (1938) 358.
76 P.K. Christensen and H.W. Giertz. The Cellulose/water relationship, l.c. [1], p. 59.
77 Morrison and Dzieciuch, Can. J. Chem., 37 (1959) 1379.
78 J.E. Stone and A.M. Scallan, J. Polym. Sci. Part C., 11 (1965) 13.
79 J.E. Stone and A.M. Scallan, Pulp Pap. Mag. Can., 66 (1965) No. 8: 407.
80 J.E. Stone and A.M. Scallan, Influence of drying on the pore structures of the cell wall, l.c. [1] p. 145.
81 J.E. Stone and A.M. Scallan, Tappi, 50 (1967) 496.
82 J.E. Stone and A.M. Scallan, Pulp Pap. Mag. Can., 69 (1968) T 288.
83 J.E. Stone, A.M. Scallan and B. Abrahamson, Sven. Paperstidn., 71 (1968) 687.
84 A.M. Scallan and J.E. Carles, Sven. Paperstidn., 75 (1972) 699.
85 F.F. Morehead, Textile Res. J., 17 (1947) 96.
86 D.H. Page and P.A. Tydeman, A new theory of the shrinkage, structure and properties of paper, l.c. [5], p. 397.
87 D.H. Page, Trend, Pulp Pap. Res. Inst. Can., 15 (1969) 7.
88 D.H. Page and P.A. Tydeman, l.c. [1], p. 371.
89 W. Gallay, Discussion Contribution, l.c. [1], p. 393.

90 A.M. Scallan, The Accomodation of Water within Pulp Fibres, l.c. [7], p. 9.
91 D.F. Caulfield, Effect of Cellulose on the structure of water, View II, l.c. [7], p. 63.
92 A.A. Robertson, Pulp Pap. Mag. Can., 65 (1964) T 171.
93 R.J. Kerekes, Trans. Tech. Sect. C.P.P.A., 2 (1976) 88.
94 S. Cluett, U.S. Pat., 2, 624, 245, 1953.
95 C.B. Ihrman and O.E. Ohrn, l.c. [1], p. 410.
96 B. Lyne, The Effect of Moisture and Moisture Gradients on the Calendering of Paper, l.c. [7], p. 641.
97 B.I. Howe and J.E. Lambert, Pulp Pap. Mag. Can., 62 (1961) T 139.
98 G. Gavelin, Sven. Papperstidn., 52 (1949) 413.
99 M. Jackson and L. Ekstrom, Sven. Papperstidn., 67 (1964) 807.
100 B. Ivarsson, Tappi, 39 (1956) 97.
101 W. Brecht and M. Schädler, Das Papier, 17 (1963) 626.
102 J. Mardon et al., Dynamic Consolidation of the Paper Web during Calendering, l.c. [1], p. 576.
103 D. Chapman and J. Peel, Pap. Technol., 10 (1969) 116.
104 J. Colley and J. Peel, Pap. Technol., 13 (1972) 350.
105 P. Wahlstrom et al., Pulp Pap. Mag. Can., 64 (1963) T 205.
106 J. Mardon, Pulp Pap. Mag. Can., 62 (1961) T 139.
107 G. MacKin et al., Tappi Tech. Assoc. Pap., 24 (1941), 169.
108 G. Robertson and L. Hagglund, Svensk. Papperstidn., 77 (1974) 521.
109 R.J. Kerekes and I.T. Pye, Pulp Pap. Mag. Can., 75 (1974) No. 11, 65.

## SUBJECT INDEX